Cultural History and Modern Production Technology of Silage

Yimin Cai • Kazuo Ataku
Editors

Cultural History and Modern Production Technology of Silage

Springer

Editors
Yimin Cai
School of Agricultural Sciences
Zhengzhou University
Zhengzhou, Henan, China

Kazuo Ataku
Rakuno Gakuen University
Ebetsu, Hokkaido, Japan

ISBN 978-981-96-5786-5 ISBN 978-981-96-5787-2 (eBook)
https://doi.org/10.1007/978-981-96-5787-2

This Springer imprint is published by the registered company Springer Nature Singapore Pte Ltd.
The registered company address is: 152 Beach Road, #21-01/04 Gateway East, Singapore 189721, Singapore

Preface

The world is facing a pressing issue of food insecurity, with over 730 million people experiencing hunger globally, as reported by the Food and Agriculture Organization of the United Nations in "State of Food Security and Nutrition in the World 2024".

The increasing demand for food and protein in developing countries highlights the urgent need for innovative food and livestock production technologies. Addressing these challenges is crucial to eradicating poverty and hunger, improving the living standards of people in poor areas, and achieving the Sustainable Development Goals (SDGs).

Silage, a traditional form of fermented feed, with its long history and ancient culture, plays an important role in this pursuit. It is produced through anaerobic fermentation of grass, forage crops, crop residue, and food by-products and is widely used in raising ruminants such as cattle and sheep. Its benefits include nutritional preservation of forage, a stable supply throughout the year, promotion of livestock growth and reproduction, reduction of diseases, and sustainable economic development.

As the global livestock industry continues to develop, the demand for high-quality feed increases. Silage production technology in advanced countries such as Europe and the United States is becoming mature, while tropical regions such as Asia and Africa are also actively promoting the production and processing techniques. Global silage production is rising steadily, driven by its unique advantages, such as long-term storage, high nutritional value, and excellent palatability for livestock. At present, the global market is dominated by corn silage, while other raw materials such as alfalfa, wheat straw, and sorghum are also being used. Additionally, crop residue and food waste are also being utilized to produce low-cost and highly nutritious silage. The availability of diverse and abundant raw materials provides a strong guarantee for the development of the silage industry.

Nevertheless, silage and livestock production are major contributors to greenhouse gas emissions, making decarbonization a mainstream trend in the livestock industry. Many countries around the world are striving to achieve carbon neutrality by 2050 or 2060. To address these global challenges, research on silage fermentation principles and the development and application of emerging agricultural technologies have gained significant attention.

Smart agriculture promotes labor-saving and efficient silage production. By expanding feed production through smart agricultural practices, the production capacity of traditional animal husbandry is enhanced while increasing the added value of agricultural and livestock products. Smart agricultural technologies can rapidly improve production efficiency, reduce the environmental impact and global warming potential, and increase high-quality livestock output when applied in livestock farming. Therefore, the future of silage and livestock production will increasingly rely on smart agricultural technologies and continue to advance in both efficiency and quality.

This comprehensive book brings together world-renowned silage experts to review the culture, history, and development of silage production technology. It summarizes key research on silage fermentation principles, modern production technologies, as well as the utilization of agricultural by-products. Key topics include strategies for reducing greenhouse gas emissions, achieving carbon neutrality, and other cutting-edge research results in green and low-carbon animal husbandry. Furthermore, the book explores the application of smart agricultural technologies in silage production and presents future prospects.

As a valuable resource, this book can be used as a teaching reference book for undergraduates, postgraduates, and teachers in agriculture and animal husbandry, and scientific research institutes. It is also suitable for technicians and experts in animal husbandry-related enterprises.

Zhengzhou, Henan, China — Yimin Cai
Ebetsu, Hokkaido, Japan — Kazuo Ataku

Contents

Editors and Contributors

About the Editors

Yimin Cai holds a PhD in Animal Science from Ehime University, Japan. He was formerly the chief researcher at the National Agricultural Research Organization of Japan. He is currently the senior researcher at the Japan International Research Center for Agricultural Sciences and a distinguished professor at Zhengzhou University. He has won the Japan Grassland Society Award and the Japan Animal Husbandry Award, published more than 200 papers. His current main research directions are agricultural production and food security, development and utilization of agricultural and sideline product resources, carbon neutrality in animal husbandry, and sustainable agricultural development.

Kazuo Ataku is a former president of Rakuno Gakuen University, Japan, and is currently a Professor Emeritus. His specialty is dairy cow feed, particularly research on silage fermentation control. He developed the first lactic acid bacteria product for silage in Japan in 1986. He is a distinguished service member of the Japanese Society of Animal Science.

He has won the Japanese Society of Grassland Science Award, the Livestock Technology Association Award, and the Utsunomiya (Dairy Farming Leadership) Award. He has served as a visiting professor at Inner Mongolia University, Xinjiang Agricultural University, and Inner Mongolia University for Nationalities in China.

Contributors

Yasuhiro Aoki Tokyo University of Agriculture and Technology, Fuchu, Japan

Kazuo Ataku Rakuno Gakuen University, Ebetsu, Hokkaido, Japan

Smerjai Bureenok Rajamangala University of Technology Isan, Nakhon Ratchasima, Thailand

Yimin Cai Zhengzhou University, Zhengzhou, China
Japan International Research Center for Agricultural Sciences, Tsukuba, Japan

Yang Cao Heilongjiang Bayi Agricultural University, Daqing, China

Peter Dele Federal University of Agriculture, Abeokuta, Nigeria

Zhumei Du Yangzhou University, Yangzhou, China

Xusheng Guo Lanzhou University, Lanzhou, China

Delma B. Jethro Centre National de la Recherche Scientifique et Technologique, INERA, Département Production Animale, Ouagadougou, Burkina Faso

Hidenori Kawamoto Tohoku Agricultural Research Center, NARO, Morioka, Japan

Jong Geun Kim Seoul National University, Pyeongchang, South Korea

Hajime Kumagai Kyoto University, Kyoto, Japan

Ping Li Guizhou University, Guiyang, China

Qinhua Liu Yunnan Agricultural University, Kunming, China

Hiroki Matsuyama Yamagata University, Tsuruoka, Japan

Pattaya Napasirth Udon Thani Rajabhat University, Udon Thani, Thailand

Viengsakoun Napasirth National University of Laos, Vientiane, Laos

Damião W. Nguluve Agricultural Research Institute of Mozambique, Matola, Mozambique

Toshiyuki Ohtani Shizuoka Prefectural Research Institute of Animal Industry, Kikugawa, Japan

Huili Pang Zhengzhou University, Zhengzhou, China

Marketta Rinne Natural Resources Institute Finland, Jokioinen, Finland

Lin Sun Inner Mongolia Academy of Agriculture and Animal Husbandry Sciences, Hohhot, China

Zhongfang Tan Zhengzhou University, Zhengzhou, China

Ansah Terry University for Development Studies, Tamale, Ghana

Peng Wang Jilin University, Changchun, China

Yanping Wang Zhengzhou University, Zhengzhou, China

Yantyati Widyastuti National Research and Innovation Agency, Cibinong, Indonesia

Chuncheng Xu China Agricultural University, Beijing, China

Jinsong Yang Hainan University, Haikou, China

Yuli Yanti Sebelas Maret University, Surakarta, Indonesia

Yanhong Yan Sichuan Agricultural University, Chengdu, China

Xianjun Yuan Nanjing Agricultural University, Nanjing, China

Jianguo Zhang South China Agricultural University, Guangzhou, China

Part I
Cultural History and Fermentation Principle of Silage

Chapter 1
Cultural History of Silage

Yimin Cai and Kazuo Ataku

Abstract This chapter defines the concept of silage culture for the first time. Silage culture includes cultural connotations related to silage production and utilization, such as raw materials, preparation methods, storage equipment, fermented feed, livestock raising, and management. Silage has a long history and culture with the development of human beings. The origins of silage can be traced back to 3000 BC, and it has played an extremely important role in domesticating wild animals and raising livestock. Silage farming continues to develop with the advancement of animal husbandry and changes in people's living needs. This chapter focuses on the characteristics of silage culture, as well as the origin and long cultural history of silage.

Keywords Culture · History · Origin · Silage

1 Silage Culture

1.1 What Is "Silage Culture"?

"Silage culture" generally refers to the culture involved in silage production and utilization, including raw materials, preparation methods, storage equipment, and fermented feed, as well as its relation to livestock feeding and management. Therefore, silage culture covers all aspects of silage production, preparation, processing, distribution, circulation, and livestock raising passed down generations by people in an area.

Y. Cai (✉)
Zhengzhou University, Zhengzhou, China

Japan International Research Center for Agricultural Sciences, Tsukuba, Japan
e-mail: caiyimin@zzu.edu.cn

K. Ataku
Rakuno Gakuen University, Ebetsu, Hokkaido, Japan

Y. Cai, K. Ataku (eds.), *Cultural History and Modern Production Technology of Silage*, https://doi.org/10.1007/978-981-96-5787-2_1

There is currently no precise definition and meaning of silage culture, and it is a field that is not yet very mature. The formation of silage culture is closely related to many factors. Natural conditions such as local climate and geography, as well as humanistic conditions such as harvesting, preparation, and preservation techniques of silage raw materials, alongside other social conditions such as livestock raising methods and technological trends, have had a profound impact on the production and utilization of silage, thus forming a unique silage culture in an area. However, as human civilization has evolved and times have changed, some traditional silage cultures have become less prominent, while new silage cultures have emerged. Consequently, silage culture continues to evolve in response to advancements in animal husbandry and the changing needs of people's lives.

1.2 Characteristics of Silage Culture

1.2.1 Digging Pit and Burying Grass

Animal husbandry refers to the act of raising livestock for a certain purpose. For example, in ancient times, people raised livestock to solve the problem of food and clothing. Nowadays, the development of animal husbandry is to obtain economic benefits and meet people's living needs (Woolford 1984; McDonald et al. 1991; Wilkinson et al. 2003).

The original raising of ruminant livestock, such as cattle and sheep, was simple grass grazing. During this period, silage was mainly stored in underground silos, using a method of "digging holes and burying grass." Its main purpose was to preserve fresh forage for winter in cold areas and during dry seasons in tropical areas. It also served as supplementary feed for livestock during times of feed scarcity. Various silage facilities have replaced the method of storing grass by digging pits; however, this storage method is still utilized in some tropical developing countries. In addition, simple silage facilities such as semiunderground silos and stack silos that are widely used in the world may have evolved from underground storage methods (Fig. 1.1).

1.2.2 Development of Silage Facilities

Since the beginning of the twentieth century, silage has rapidly developed. With the expansion of livestock farming, silage production increased significantly alongside hay and cereals, primarily focusing on single-product feeding. During this period, various storage facilities, such as tower silos, trench silos, and bunker silos, were developed to ensure the stability of silage fermentation. The silage production technology in some advanced countries gradually realized mechanization (Fig. 1.2).

Fig. 1.1 Digging pit to bury grass and stack silo. (Photos by Cai)

Fig. 1.2 Tower silo, bunker silo, and forage harvester. (Photos by Cai)

1.2.3 Integrated Preparation and Storage

A mixed feed of roughage and concentrate appeared in the mid-twentieth century and is considered the prototype of TMR. In the late 1970s, wrapped silage machines were developed, and they integrated and mechanized the processes of storage and adjustment, thereby reducing labor intensity. During this period, wrapped silage served as a means of circulating feed (Fig. 1.3).

1.2.4 Total Mixed Ration (TMR) Circulating Feed

The TMR production system was introduced in the 1990s. In modern animal husbandry, the fermentation technology of total mixed rations is widely used. TMR technology meets the nutritional needs of livestock by thoroughly mixing coarse feed, concentrate, minerals, vitamins, and other additives. This technology maximizes the use of agricultural and sideline product resources and reduces feed costs. The TMR production and distribution center was immediately established (Fig. 1.4), playing an active role in promoting ecological agriculture and improving livestock production efficiency.

Fig. 1.3 Preparation and storage of wrapped silage (Kozu farm, Shimota, Japan). (Photos by Cai)

Fig. 1.4 High-density TMR production and distribution center (Nasu TMR center, Nishinasuno, Japan). (Photos by Cai)

1.2.5 Exchange of Regional Culture

To enhance society's understanding and support for science, many scientific research institutions host annual development days, welcoming visitors from all walks of life (Fig. 1.5). Communication between scientific research and the public does not only enhance society's understanding and support for science but also promotes the practical application of research findings and promotes positive interactions between science and society. For example, the silage tower has an attractive design, and the wrapped silage features vibrant colors. Its diverse and colorful shapes have become a striking view on open days. By using clear language and effective metaphors, scientific researchers can communicate complex scientific concepts and research findings about silage to the public. This approach enhances public trust and support for agricultural and animal husbandry research and fosters regional cultural exchanges.

1.2.6 Local Brands of Livestock Products

This aims to create a regional livestock brand leveraging local agricultural resources and unique natural or industrial assets. Generally, livestock production areas have unique environmental resources and a long history of livestock breeding. These regions often have established industrial scales and a high level of organization. They implement information traceability systems and prioritize ecological agriculture and environmental protection. As a result, they have significant development

Fig. 1.5 Silage and regional cultural exchanges. (Photos by Cai)

Fig. 1.6 Creation and marketing of animal product brands such as "Rice Milk" and "Rice Beef". (Photos by Cai)

advantages and strong market potential. For example, Japan is rich in rice, and some regions are referred to as the "hometown of rice." They utilize paddy fields to produce feed rice or fresh rice straw silage and have created animal product brands such as "rice milk" and "rice beef" (Saitama, Japan). Also, the northeastern region of Japan has a long history of producing apples, and apple residues are used to ferment TMR and to create local brands such as "Apple Wagyu" (Hirosaki, Japan). The regional agricultural culture has a rich heritage, reflecting a unique commitment to preserving and promoting exceptional farming practices. Silage embodies a distinct brand core value and story, actively engaging in cultural exchanges and promotions with society and consumers. As a result, regional brands play a crucial role in enhancing farmers' income and fostering economic development (Fig. 1.6).

1.2.7 Smart Agriculture and Smart Silage

Smart agriculture utilizes modern technologies such as the Internet of Things (IoT), artificial intelligence (AI), and big data to enhance agricultural practices. This approach integrates information technology with agriculture, leading to a new method of agricultural production characterized by information perception, precise management, and intelligent control throughout the entire production process. Smart agriculture is developing rapidly, expanding its application areas and significantly transforming silage production and animal husbandry. Figure 1.7 shows that the Ministry of Agriculture, Forestry, and Fisheries of Japan has developed a variety of smart agricultural machinery in recent years that is being applied to feed and livestock production. This includes robot tractors, drones, automatic breast feeders,

Fig. 1.7 Smart agricultural machinery developed in recent years (left 1 and left 2: National Experiment Station for Precision Agriculture, Beijing, China; left 3: Ministry of Agriculture, Forestry and Fisheries of Japan, Tokyo, Japan). (Photos by Cai)

Fig. 1.8 Burning (left 1) and silage preparation (left 2 and left 3) of rice straw. (Photos by Cai)

automatic feeders, cowshed monitoring systems, milking robots, and intelligent deodorization systems. These technologies can be utilized in all aspects of forage and livestock production management. Smart agriculture plays an important role in green and low-carbon silage production, bringing new vitality and opportunities for the development of animal husbandry.

1.2.8 Sustainable Development of Agriculture and Animal Husbandry

Seasonal forage shortages occur in autumn and winter in cold regions, and in the dry season in tropical regions, particularly affecting the availability of fresh green forage. Feeding hay leads to a reduction in the nutritional value and palatability of forage. Through silage production, the silage made is not only fresh and palatable but also alleviates the problem of forage shortage. As shown in Fig. 1.8, after crops are harvested, a large amount of straw is often discarded or burned. This treatment method not only wastes resources but also pollutes the environment. The greenhouse gases emitted by burning the straw contribute negatively to global warming, which affects the sustainable development of the economy and society to a certain extent. Feeding livestock with straw silage fermentation technology greatly influences the sustainable development of both the economy and society. This approach not only reduces feed costs and increases herders' income but also facilitates the return of straw to the field through livestock manure. This process promotes a virtuous cycle in ecological agriculture and supports the sustainable development of both agriculture and animal husbandry.

2 Silage History

Silage has a long history and culture along with the development of people. The origin of silage can be traced back to 3000 B.C., and it has played an extremely important role in the domestication of wild animals and the raising of livestock. This chapter examines the historical data of ancient Egypt and examines ancient Egyptian wall paintings, focusing on the origin and long cultural history of silage.

2.1 Origin of Silage

Cattle are believed to have been domesticated approximately 8000 years ago. Initially, cattle were fed green grass through methods, such as grazing and mowing, and during winter when the grass was scarce, they were fed natural hay. It is thought that the technique of silage production was discovered by chance.

Silage has a long history, with its origins believed to date back more than 3000 years (Woolford 1984; McDonald et al. 1991; Wilkinson et al. 2003).

Egyptian oil paintings from 1000 to 2000 BC prove that the ancient Egyptians were skilled in grain processing. As shown in Fig. 1.9, the murals in the museums of Naples depict the preparation of whole-crop silage, not only for grain but also for millet, as early as 1000–1500 BC. Additionally, silos discovered in the ruins of Carthage suggest that grass silage was being produced around 1200 AD.

2.2 Description of Grass and Silage in the Bible

The Bible has long been the most read book in the world. It is a book of religious, historical, and cultural significance, including descriptions of agricultural practices such as dairy farming and silage. Genesis, the first chapter of the Old Testament, describes the creation of the heavens and the earth. According to the account, God is said to have created the world in 7 days. However, it is not well-known that grass was created as the first living thing and as food for all animals, including humans. Genesis 1:11 states: *Then God said, 'Let the earth produce growing things; let there be on the earth plants that bear seed, and trees bearing fruit each with its own kind of seed.' So it was* (The Revised English Bible 1989).

This event occurred on the third day of creation. Additionally, it is written in Genesis 1:29–31: God also said, 'Throughout the earth I give you all plants that bear seed, and every tree that bears fruit with seed: they shall be yours for food. All green

Fig. 1.9 Murals in the museums of Naples. (Picture by Woolford)

plants I give for food to the wild animals, to all the birds of the air, and to everything that creeps on the earth, every living creature.' So it was, and God saw all that he had made, and it was very good. Evening came, and morning came, the sixth day (The Revised English Bible 1989).

This declaration was made on the sixth day. In the Bible's Creation narrative, grass played an important role as food not only for livestock but also for humans. It was after the events of Noah's Ark that humans began to eat meat, mainly limited to ruminant livestock such as cattle and sheep. Muslims who strictly follow this doctrine still do not eat pork.

Silage is also mentioned in the Old Testament, specifically in Isaiah 30:23–24: *The Lord will give rain for the seed you sow in the ground, and as the produce of your soil he will give you heavy crops. When that day comes your cattle will graze in broad pastures; the oxen and donkeys that plough the land will be fed with well-seasoned fodder, winnowed with shovel and fork* (The Revised English Bible 1989).

The term "well-seasoned fodder" is interpreted as fermented feed with the addition of salt (The Bible, Revised Standard Version, 1980) or silage (The New Interpreter's Bible 2001). The book of Isaiah was written around the eighth century

BC, so silage was already being made and additives were being used at that time. This is the oldest description of silage known to the author.

The origin of silage is unclear, but early methods likely involved burying crops in holes in the earth. It is believed that people by chance discovered that when they dug up the grain that had been buried in the ground, they found a fragrant silage. The Greek word for these pits, Siros, is the origin of the modern term "silo." In the past, grains, which were a valuable foodstuff as well as contraband and other unsuitable items, may have been hidden in these earthen silos. Therefore, they searched for objects hidden inside the silos with a pointed iron spear, known as *Siromastes*. The Old Testament refers to *Siromastes* in 2 Kings 11:10, where the priest gave King David's *Siromastes*, kept in the temple of the Lord, to the captains of the hundreds. *Siromastes* were also used as weapons of war. Isaiah 2:4 admonishes us to beat our swords into plowshares and our spears into sickles and to stop fighting among nations and devote ourselves to agriculture.

The earliest silos were pit and trench silos, which were built underground. Today, the tower silos, as we know, originated from France and the United States in the 1880s and spread to other countries. Currently, bunker silo is the mainstream for large-scale farming, while wrap silo is used for small and medium-sized farms. A key innovation in silage storage emerged in the 1950s with the introduction of plastic technology, which enabled the sealing of various types of silos such as the bunker, trench, and pile silos. By the mid-1970s, round bale wrap silo also emerged, and the beautiful scenery of the ranch which could be seen from the tower silo, changed to round bale silo.

2.3 Development of Silage Technology

Although the concept of silage as a storage technique for forage crops has existed for a long time in various parts of the world, it only became more widely in practice at the end of the nineteenth century.

The first scientific description of silage making was published in 1842 by Grieswald (Woolford 1984; McDonald et al. 1991; Wilkinson et al. 2003). The report states that the grass should be tightly packed into the pit as quickly as possible while pressing down, then it should be immediately covered with a board or similar material and covered with about 45 cm of soil. This technique remains the fundamental principle of silage preparation today.

It is believed that the French farmer Goffart (Woolford 1984; McDonald et al. 1991) made a major contribution to the modernization and spread of silage. Based on his experience, he published a book on the preparation of silage from corn and other crops in 1877, which was translated into English and published in the United States, helping to popularize silage in the United States.

Silage was introduced to Britain after 1882, later than in France and the United States. However, Fry (Woolford 1984; McDonald et al. 1991) wrote a book called "Sweet Ensilage" in 1885, proposing a theory of high-temperature fermentation silage using delayed sealing, which contradicts Goffart et al.'s emphasis of early sealing. It is believed that this delayed the development of silage production in Britain (McDonald et al. 1991).

Research into the mechanisms of silage fermentation and the factors that influence it has progressed rapidly since the beginning of the 1900s. In the 1930s, Virtanen of Finland, in particular, studied in detail the biochemical changes that occur before silage is made and devised a method for preparing it by adding mineral acids (hydrochloric acid or sulfuric acid). This method, known as the AIV method, demonstrated that lowering the pH of the material below 4.0 suppressed plant respiration and the growth of harmful microorganisms, resulting in high-quality silage (Woolford 1984; McDonald et al. 1991; Wilkinson et al. 2003; Virtanen 1933). In 1945, Virtanen further demonstrated that legumes were excellent as raw materials for highly nutritious silage due to their nitrogen fixation and high protein content, and in 1945, he introduced the AIV system (Virtanen 1943) based on livestock feed. He received the Nobel Prize in Chemistry for these achievements. After that, feeding silage treated with hydrochloric and sulfuric acid using the AIV method was discontinued due to health concerns for livestock, and formic acid was subsequently used as a safer alternative.

It has long been known that lactic acid bacteria play the most important role in silage fermentation. In the early 1900s, lactic acid bacteria cultures were first used to prepare silage in France, and then farmers cultivated lactic acid bacteria developed in Germany to prepare silage (McCullough 1978; Watson and Nash 1960). However, many early attempts were unsuccessful due to poor adaptation of bacteria to the silage environment or the lack of viable culture (Spoelstra 1991).

However, technological developments in the 1970s and 1980s led to improvement in commercially available lactic acid bacteria products. In the 1980s and 1990s, lactic acid bacteria products gained significant attention among silage farmers and researchers. Numerous studies have demonstrated the effects of lactic acid bacteria on the fermentation and nutritional value of silage (McCullough 1978; Bolsen 1985).

In Japan, Hokkaido University conducted 10-year-long research, which began in 1960 to find excellent lactic acid bacteria strains for silage, although was not put into practical use. Consequently, Japan began importing lactic acid bacteria products from the United States and Europe for a while in the late 1970s. During this time, Ataku identified two types of lactic acid present in silage L (+) and D (−), and in 1985, a product was developed using lactic acid bacteria (*Lacticaseibacillus casei*) that produced large amounts of L (+) lactic acid, which was easily available to livestock (Ataku et al. 1985). Additionally, a new product was created by combining this LAB with Acremonium-derived cellulase (Ataku et al. 1997). This product has been continuously improved and remains the best-selling silage additive in Japan.

2.4 The Beginning of Silage in Asia

Silage making in Asia has a long history. In China, during the Yuan Dynasty (1271–1368), the use of corn silage is mentioned in Wang Zhen's Agricultural Book: Rongsang Tong Tips and Livestock Cultivation. Similarly, in the Qing Dynasty, alfalfa silage is documented in Volume 3 of Yang Shanshan's Youfeng Guangyi. Both texts describe how to make silage.

In Japan, silage was first made in 1887 at Kozu Farm in Gunma Prefecture, with production beginning around the same time in Hokkaido. In Japan, silage was originally called ensilage or silage, with translations such as silage (fodder), buried grass, salted food, pressed cud, and pressed feed. However, in 1971, the Japanese Society of Grassland Science adopted a name that is globally recognized. Addition, silos were originally referred to as anagura or salt cellar.

Japan's first manual on silage production, "The Theory of Grass (1902)" was by Jiro Ogawa. According to the report, although silage had been widely utilized in Europe and United States for 2–3 decades, its adoption was not yet common in Japan. Readers were therefore urged to experiment with silage production in order to enhance livestock nutrition. The manual also emphasized the importance of practices such as cutting, treading, sealing, and weighting to make silage. Ogawa was a businessman, and although his book was published with the aim of disseminating it to farmers, it was an advanced work that triggered the spread and development of grass cultivation and silage technology.

One of the most influential figures in the practical development of silage production and dairy cow feeding techniques in Japan was Sentaro Utsunomiya. In 1887, Utsunomiya travelled to the United States for training in dairy farming and worked at the farm of the University of Wisconsin. There, he studied under Professor A. H. Henry, the farm manager and Dean of the College of Agriculture. Professor Henry published "Feed and Feeding" (1898), often referred to as the bible of livestock feed science, which advanced the knowledge of livestock feed science globally. Upon returning to Japan, Utsunomiya started his own dairy farm, where he put into practice the techniques he had learned, and played a leading role in disseminating those techniques. At the time, Japanese silos were European-style underground silos, but Utsunomiya built Japan's first American-style aboveground tower silos and made silage by filling them with corn. This innovation was supported by university professors and by the early 2000s, corn fields and tower-shaped silos had become symbols of Hokkaido dairy farming.

Until the end of the war in 1945, Japanese dairy farms were generally small size and relied on grazing in the summer and hay and corn silage in the winter. In response to the food crisis after the war, the Japanese government launched a livestock promotion policy, which led to Japan's first full-scale silage research program at the Hokkaido Agricultural Experiment Station. The research group developed key principles for effective silage production, consisting of (1) cutting at the optimal time, (2) adjusting moisture content, (3) cutting and treading, and (4) using additives. Research on silage fermentation became a global effort in the 1950s, and in

Japan, scientists like Yoshinobu Oyama of the Ministry of Agriculture, Forestry, and Fisheries joined in this effort. By the 1970s, researchers, such as Kazuo Ataku (Rakuno Gakuen University), Takayoshi Masuko (Tokyo University of Agriculture), and others, further made advancements. Today, researchers including Yimin Cai (Japan International Research Center for Agricultural Sciences) and Naoki Nishino (Okayama University) are actively making international contributions, highlighting Japan's continuous influence in this field.

2.5 Conclusion

Humans discovered silage more than 3000 years ago and have long used it as stored feed during the winter. In recent years, it has been used all-year-round by dairy and livestock farmers globally as the main raw material for TMR (total mixed ration). Fermented TMR, which was created in Japan in the late 1900s, is a delicious, highly nutritious silage made from agricultural byproducts and food waste. It is truly a silage that is kind to livestock, people, and nature. A young Chinese researcher is diligently studying its secrets, reflecting on the words of Isaac Newton: "If I have further seen, it is by standing on the shoulders of Giants".

References

Ataku K, Nakamura H, Narasaki N, Kikuchi M. Control of silage fermentation by inoculation with *Lactobacillus casei*. Proc XV Int Grassl Cong. 1985:889–91.

Ataku K, Aniwaru A, Narasaki N, No E, Chase L. Effect of a new cellulase derived from acremonium on silage fermentation. Proc XVIII Int Grassl Cong. 1997;14:23–4.

Bolsen KK. New technology in forage conservation-feeding systems. Proc XV Int Grassl Cong. 1985:82–8.

McCullough ME. Fermentation of silage: a review. West Des Moines: National Feed Ingredients Association; 1978.

McDonald P, Henderson AR, Heron SJE. The biochemistry of silage. Bucks: Chalcombe Publication; 1991.

Spoelstra SF. Chemical and biological additives in forage conservation. In: Forage conservation towards 2000. Sonderheft; 1991. pp. 48–70.

The Bible. Revised standard version. New York: American Bible Society; 1980.

The New Interpreter's Bible. Vol. VI. Nashville: Abingdon Press; 2001.

The Revised English Bible. With apocrypha. Oxford/New York: Oxford University Press/Cambridge University Press; 1989.

Virtanen AI. The empire. J Exp Agric. 1933;1:143–55.

Virtanen AI. AIV system (AIV-JARJESTELMA: Finnish). Helsinki: Pellervo-Seura; 1943.

Watson SJ, Nash MJ. The conservation of grass and forage crops. Edinburgh/London: Oliver and Boyd; 1960.

Wilkinson JM, Bolsen KK, Lin CJ. History of silage. In: Silage science and technology, vol. 42. Madison: The American Society of Agronomy Inc./Crop Science Society of America, Inc./Soil Science Society of America, Inc.; 2003. p. 1–30.

Woolford MK. The silage fermentation. New York: Marcel Dekker Inc.; 1984.

Chapter 2
Development History of Silage Science and Technology

Xusheng Guo

Abstract The recorded history for silage research has lasted for over 180 years. Since the initial studies on ensiling of fresh forages via salting and acidification, significant advances in silage science and technology have been achieved recently. All the achievements not only improved silage production efficiency but also provided animals with high-quality and safe feed resources for the whole year around, which greatly promoted animal husbandry and the livelihood of human beings. This chapter overviews the development of silage technology including ensiling facility, equipment, additives, forage processing and silo management; meanwhile, silage science related to characterization of ensiling biochemistry and microbial process is also discussed.

Keywords Development history · Silage science · Silage technology

1 Brief Introduction of Silage History

Silage is defined as any material stored in a silo, which can take various forms, including pits, bunkers, towers, covered heaps, or wrapped bales. Often referred to as "canned grass," silage serves as a year-round source of green fodder for livestock and plays a crucial role in the preservation of green plants. The study of silage has a historical context spanning nearly 180 years (Wilkinson et al. 2003). The scientific exploration of silage began in 1843 when James Johnston, a chemistry lecturer at the University of Durham, inquired whether it was feasible to conserve forages in a wet status. He interpreted a paper by Grieswald published in the Transactions of the Baltic Association for the Advancement of Agriculture in 1842, which detailed the conservation of grass through salting, a method akin to that used in sauerkraut

X. Guo (✉)
Lanzhou University, Lanzhou, China
e-mail: guoxsh07@lzu.edu.cn

Y. Cai, K. Ataku (eds.), *Cultural History and Modern Production Technology of Silage*, https://doi.org/10.1007/978-981-96-5787-2_2

production. Brassley posited that the practice of ensiling forages likely originated in northern Europe, particularly in the Baltic region or Germany.

Following decades of advancement, significant progress in production technology was achieved after the method gained popularity in various countries by the late nineteenth century. A pivotal moment in this field occurred with the issue of research on the direct acidification of forages during fermentation in Italy (Giglioli 1914) and Germany (by Fingerling), both of whom suggested for the utilization of diluted hydrochloric acid. Another scientist, A.I. Virtanen, also contributed to this area by experimenting with a combination of hydrochloric and sulfuric acids. These acids were manually applied to the crops using sprayers, rendering the process labor-intensive and potentially hazardous. During the 1920s and 1930s, further investigations were conducted into other inorganic and organic acids, as well as acid salts. The first documented add of lactic acid bacterial (LAB) as microbial inoculants occurred in France with fermented sugar beet pulp in the early 1900s (Watson and Nash 1961). In 1926, Kuchler reported a method exploited in Germany for inoculant production that involved cultivating bacteria on farms. Nevertheless, massive early attempts to inoculate forages were unsuccessful due to either the inadequacy of LAB strains for the anaerobic fermentation environment or the lack of viability of the bacterial cultures at the time of application (Spoelstra 1991). Additionally, unsatisfactory outcomes were reported with certain inoculants, specifically in forages with low fermentable substrates (Haigh et al. 1987; Weinberg et al. 1988). In the late twentieth century, Muck (1996) proposed that the incorporation of *Lentilactobacillus buchneri* into silages could enhance aerobic stability by increasing acetic acid levels, which could inhibit yeasts responsible during aerobic exposure This proposition, along with subsequent research, sparked considerable debate, as it challenged the prevailing notion of exclusively utilizing homofermentative LAB as silage inoculants.

The advancement of molecular biotechnology has prompted scientists to investigate the influence of additives on the regulation of fermentation quality, microbes, and aerobic stability. Generally, LAB plays a predominant role in silage fermentation, and changes in microbial communities have been shown to affect the end-products of fermentation, which are closely associated with the quality of silage. Recently, the analysis of multi-omics profiles has emerged as a valuable approach for enhancing the insight into ensiling processes.

2 Silage Technology and Management

2.1 Silage Facility

The advent of tower silos marked a significant milestone in the advancement of silage production (shown in Fig. 2.1). Tower silos gained considerable popularity across various regions in Europe and North America. During the 1920s, numerous

Fig. 2.1 Development of ensiling forms for different forages

small tower silos were constructed using materials such as wood, tile, or concrete, with some located within buildings to safeguard the silage from freezing during winter and to enhance accessibility for ruminants. As agricultural operations expanded and ruminant populations increased, many farmers transitioned to horizontal preservation systems, including bunkers, drive-over piles, and bags, to accommodate their growing feed needs. Nonetheless, at the onset of the twentieth century, tower and pit silos were the predominant storage methods on Brazilian farms, primarily due to the unavailability of plastic sheeting for sealing horizontal silos and the relatively smaller size of livestock herds at that time. Recently, the majority of farmers have adopted bunker silos, which offer the significant advantage of facilitating the rapid handling of large quantities of forage (Savoie and Jofriet 2003). Consequently, bunker silos have become the preferred storage solution in majority of large farms. Another widely utilized method for silage production is the stack silo. In many tropical regions, farmers tend to favor cost-effective strategies, such as stack silos, which require minimal structural investment, primarily involving the use of plastic film.

Recent advancements in forage storage technologies, specifically pressed-bag and wrapped-bale systems, have been adopted in regions with well-established ruminant industries. The wrapped-bale method is increasingly favored due to its facilitation of sales, transportation, and adaptable utilization of silage based on specific requirements. Historically, producers transitioned from utilizing silage bags to employing mechanized wrapping systems, which use several layers of stretch polyethylene film to effectively seal every bales. However, this transition was not without challenges, particularly concerning the use of substandard films that compromised the anaerobic conditions necessary for silage preservation, leading to significant losses in dry matter (DM) and nutritional quality (Savoie and Jofriet

2003). Fortunately, many of these problems have been addressed, resulting in a widespread acceptance of baled silage among dairy and beef producers. At present, there are two main types of bale-wrapping systems that farmers can choose from. The first is the individual wrapping method, where each bale is cocooned in airtight film, creating a little silo. This approach offers great flexibility and makes it easier to market the silage. The second type is the in-line wrapping system, where bales are lined up in a row, and a layer of PE film is wrapped around them like a giant sausage. This method boasts impressive efficiency, uses less film per bale, and can stretch on indefinitely perfect for those with a lot of bales to store. However, in-line systems are best suited for farms where feeding happens on-site, since cutting the film to access the bales introduces oxygen, complicating the marketing of the forage as a cash crop (Coblentz and Akins 2018).

2.2 *Silage Equipment*

2.2.1 Forage Harvester

The predominant practice among forage harvesters involves the use of tractor-pulled models, largely due to the prohibitive costs associated with self-propelled harvesters. This trend contrasts with practices observed in other nations with established traditions in forage conservation. Muck and Shinners (2001) reported a decline in global sales of conventional ensilage machinery, which fell from 7000 units in 1990 to 3500 units in 1995, while sales of self-propelled harvesters rose from 1800 to 2500 units during the same timeframe. The reliance on traditional harvesters often leads to the production of excessively long particles with inconsistent sizes, which can exacerbate the risk of silage spoilage. This is attributed to the rougher particle sizes that impede effective compaction and facilitate increased oxygen infiltration due to the higher porosity of the silage mass. Additionally, excessively long particles may hinder the precision of nutritional management in dairy herds with moderate to high production capabilities. The selection against long particles by livestock can lead to ruminal disturbances (Heinrichs et al. 1999) and subsequent nutrient imbalances. Moreover, inadequate chopping by traditional harvesters results in a lower incidence of grain cracking or damage, as intact kernels generally exhibit reduced starch digestibility (Owens et al. 1986). The mechanical processing of grains by forage harvesters is crucial for breaking down the pericarp and facilitating the exposure of starch granules to digestive processes. To mitigate these issues when employing pull-type harvesters, it is essential to prioritize proper machine setup and maintenance, particularly concerning the sharpening of knives and the clearance between the cutter head and shear bar. A survey conducted in 2013 among 260 Brazilian dairy farmers revealed that only 54.6% of respondents sharpened their harvester knives on a daily basis (Bernardes and Do Rêgo 2014). Approximately half of the farmers indicated they sharpened the knives only at the commencement of the harvest (26.9%), every 2 days (11.2%), or less frequently than every 2 days

(7.3%). As previously noted, the primary maintenance concerns for forage harvesters include maintaining sharp knives and ensuring proper adjustment of the knives and shear bar (O'Dogherty 1982). The use of dull knives and improper equipment adjustments can lead to tearing rather than cutting of the material. In certain instances, machinery may become inoperative due to particles becoming lodged between the knives and shear bar, resulting in wasted fuel and suboptimal cut quality. Such tearing leads to the production of long particles, diminished cut uniformity, and subsequent challenges in the compaction of the silage mass.

2.2.2 Filing, Packing, and Sealing

The timely harvesting and ensiling of crops are crucial to minimize respiration and reduce dry matter loss in silage. Immediate packing is essential when storing crops in bunker silos, with wheeled tractors typically preferred as packing vehicles due to their superior weight distribution compared to tracked vehicles. To enhance packing density, additional weights can be affixed to dual rear-wheeled tractors, or tire pressure can be increased. The presence of moisture on tractor tires serves as an indicator of sufficient moisture for effective packing. Once the silo is filled, it should be sealed with an airtight cover to mitigate spoilage losses caused by air and rainfall infiltration. Bunker silos should be covered with 4- to 6-mil plastic, with edges secured using sandbags or silage, and tires placed edge-to-edge for stability. Utilizing halved tires or those with punched-out sidewalls from larger vehicles can facilitate handling and prevent water accumulation in the tire centers. Research indicates a net return of up to 2:1 for covering corn silage and up to 4:1 for alfalfa silage (Pitt et al. 1991). Unfortunately, no commercially viable alternatives to plastic and tires for covering bunkers or piles have been identified for North American silage producers. In Brazil, double-sided polyethylene film (white-on-black) is predominantly used for sealing horizontal silos, although approximately 20% of producers still utilize black plastic, which is considered suboptimal (Bernardes and Do Rêgo 2014). Snell et al. (2003) highlighted the impact of film color on surface temperatures, revealing that black film can reach temperatures up to 16 °C higher than white film during morning hours. The quality of the film employed in sealing horizontal silos is critical in minimizing silage losses, particularly in the outer regions of the mass. In addition to the plastic sheet, many Brazilian farmers apply a protective layer over the film post-sealing, with soil being the preferred material, followed by tires (Bernardes and Do Rêgo 2014). This practice shields the plastic from solar radiation and adverse weather conditions, thereby reducing degradation, especially in tropical climates. Covering the plastic film diminishes air infiltration, postpones aerobic deterioration in the upper layer of the silo, preserves the nutritional quality of the silage, and can enhance the performance of lactating dairy cows (Amaral et al. 2010, 2014). While effective, the process of protecting the plastic film can be labor-intensive, contingent upon the silo's size and the volume of silage produced. In instances where covering the plastic sheet is unfeasible, the quality of the film becomes paramount.

2.3 *Silage Management*

Silage production is an ancient agricultural practice that dates back over 3000 years (Wilkinson et al. 2003). Its widespread adoption accelerated after the 1940s due to advancements in forage harvesting technology (Wilkinson et al. 2003). While silage making is well established in the temperate regions of North America and Europe, it has only recently become popular and widespread in tropical regions (Wilkins and Wilkinson 2015). The production of high-quality silage depends on both controllable and uncontrollable factors. Controllable silage management factors include the maturity or moisture concentration at harvest, chop length, harvesting and ensiling methods, type of storage structure, use of silage additives, feed-out methods, feed bunk management, and effluent management. Uncontrollable climate-related factors, which may vary between warm and cold regions, can adversely affect silage production and utilization, influencing some of the controllable factors, such as moisture concentration at harvest. In this study, we focused on "harvest maturity and moisture content," "chop length," "silage additives," "feed bunk management," and "effluent management."

2.3.1 Harvest Maturity and Moisture Content

Harvest moisture and maturity recommendations vary among different silage crops. Adequate moisture facilitates silage compaction, which is essential for transitioning the silo to an anaerobic environment. Insufficient moisture can make a crop susceptible to yeast and mold issues, while excessive moisture can result in excessive effluent or prolonged fermentation, leading to elevated acid levels and increased protein degradation. Harvest maturity influences the proportions of fermentable sugars available for silage microbes and nutrient substances for livestock. Maturity significantly influences moisture levels in unwilted forage crops. Another aspect of maturity pertains to ensuring that forage sources contain acceptable levels of fiber and starch. The digestibility of fiber can range from 30% to 50% for traditional forages like alfalfa and corn silage, impacting the energy content of the ration, microbial protein production, and laboratory predictions of energy content. Additionally, fiber digestibility can greatly affect DM intake in certain feeding situations (Allen and Knowlton 1995). Ruminal fermentation of starch varies, reaching up to 90% for grains, depending on factors such as plant source, plant maturity, and processing methods. Variations of rumen-fermentable starch can influence rumen pH, DM intake, and ruminal microbial protein production (Allen and Knowlton 1995). Unlike fiber, starch that escapes ruminal fermentation can be digested in the small intestine. However, there are limits (2 kg d-I) to the amount of ruminal bypass starch that can be utilized in the small intestines of high-producing cows, likely due to constraints in amylase production in the lower gastrointestinal tract (Huntington 1994).

2.3.2 Chop Length

The optimal compromise between silo fermentation requirements and effective rumen function based on different crops suggests a theoretical cutting length of 6–18 mm. Particle sizes within this range will facilitate (i) proper compaction of silage and exclusion of oxygen to help prevent heating, molding, and poor bunk life; (ii) easier unloading from storage structures; and (iii) sufficient effective fiber to reduce the likelihood of digestive disturbances, milk fat depression, and an increased incidence of displaced abomasum. Corn silage should be chopped to a theoretical length of 9–12 mm. Longer chop lengths exceeding 18 mm (unless processed) can lead to several challenges, including (i) difficulties in compaction, (ii) improper functioning of silo unloaders, (iii) an increased number of kernels passing through the manure, and (iv) preferential sorting of stover or larger cob pieces in total mixed rations (TMR). Chopping to a length shorter than 9 mm TLC is conducive to better packing, particularly with late-harvested, dry crops, and can enhance silo capacity. However, this practice requires more power and may slow down the harvesting process. An ecutter screen is generally not recommended unless the silage is excessively dry, as it also increases field power requirements and can hinder harvesting efficiency. Alfalfa silages should be chopped to a TLC of 9 mm, with more than 15–20% of the forage particles exceeding 36 mm in length. Chopping to a TLC of 6 mm is acceptable if long-stem hay constitutes more than 25% of the ration (Shaver 1990). A successful strategy involves increasing the chop length of processed corn silage to between 18 and 25 mm by adjusting shear bar settings or removing knives. The longer material compacts effectively and, perhaps more importantly, enables the nutritionist to extract more effective fiber from the corn silage, rather than depending on a longer cut (9 mm) of the alfalfa silage crop (Bal et al. 1998). The potential for improved fermentation when alfalfa silage is chopped shorter. More coarsely chopped alfalfa silage exhibited an increase of up to 4 percentage points in AOF, indicating a concentration of fiber due to the excessive loss of fermented sugars. Based on this data, selecting a finer chop could yield alfalfa silage with an additional 0.06 Mcal of Net Energy for Lactation (NE-L) per kilogram of organic matter (OM). For a producer ensiling 909 metric tons of alfalfa silage, this would correspond to the energy equivalent of over 11,000 kg of cornmeal (Ruppel et al. 1995). With any silage-based ration, it is essential to ensure that cows receive 15–20% of the total ration (on a weight basis) as particles larger than 36 mm. This is crucial for providing adequate effective fiber, which is necessary for proper saliva production (cud-chewing) and optimal rumen function. While forages may be harvested at the recommended chop lengths, it is crucial that the effective fiber content of the ration is not compromised by reprocessing in silage baggers or by excessive mixing in the TMR mixer wagon.

2.3.3 Silage Additive

Additives have been available for enhancing silage preservation for decades. Producers generally have a wide variety of silage additives at their disposal to assist in forage preservation. According to McDonald and Kung, silage additives can be classified into one or more of four categories based on their effects on silage preservation, as outlined from a European and North American perspective in 1991 and 2003, respectively. These categories include fermentation stimulants, fermentation inhibitors, aerobic deterioration inhibitors, and nutrients and absorbents. At a later stage, an in-depth study of silage additives revealed that these additives can be categorized into six groups from the practical perspective of a producer: (1) homofermentative LAB; (2) obligate heterofermentative LAB; (3) combination inoculants that contain obligate heterofermentative LAB along with homofermentative or facultative heterofermentative LAB; (4) other inoculants (non-LAB species); (5) chemicals; and (6) enzymes.

2.3.3.1 Homofermentative LAB

The oldest and most common bacterial inoculants used for silage production are homofermentative LAB. Today, most bacteria in this group are taxonomically classified as facultative heterofermentative LAB species rather than as obligate homofermentative species (Pahlow et al. 2003). The facultative heterofermentative LAB ferment hexoses, such as glucose, in a manner similar to obligate homofermenters, resulting in the production of exclusive lactic acid. In contrast, obligate heterofermenters generate additional compounds from hexoses alongside lactic acid. A key distinction between facultative heterofermenters and obligate homofermenters is the presence of the enzyme phosphoketolase. This enzyme enables facultative heterofermenters to ferment pentoses, primarily yielding lactic and acetic acids. Common facultative heterofermentative strains include *Lactiplantibacillus plantarum*, *Lacticaseibacillus casei*, *Enterococcus faecium*, and various *Pediococcus* species. Silages treated with one or more of these bacteria typically exhibit lower levels of pH, acetic acid, butyric acid, and ammonia nitrogen (ammonia-N) while showing increased lactic acid content and exhibiting better DM recovery compared to untreated silages. A recent meta-analysis of 130 articles indicated that the effects of these inoculants varied depending on the crop (Oliveira et al. 2017). Inoculation significantly reduced the pH levels of silages derived from temperate and tropical grasses, as well as alfalfa and other legumes; however, it did not affect corn, sorghum, or sugarcane silages. The reduction in acetic acid due to inoculation was notable for all crops, except for alfalfa and other legumes. Dry matter recovery was 2.8 percentage points higher in grass silages compared to untreated corn and sorghum silages, while it decreased by 2.4 percentage points in sugarcane silage. Conversely, the reduction in butyric acid and ammonia nitrogen, along with the increase in lactic acid resulting from inoculation, were not influenced by the type of forage. Animal trials have demonstrated that these bacteria not only enhance silage fermentation but also improve milk production, daily weight gain, and feed efficiency (Weinberg and Muck 1996). A recent meta-analysis of 31 studies involving lactating dairy cattle indicated that inoculation with homofermentative or facultative heterofermentative LAB increased raw milk production by 0.37 kg per day ($P < 0.01$). There was only a trend toward increased DMI and no significant effect on feed efficiency (Oliveira et al. 2017). Additionally, trends were observed for increased concentrations of milk fat and milk protein in cows fed inoculated silage. Muck and Kung (1997) reported that homofermentative inoculants reduced aerobic stability in approximately one-third of the studies they summarized.

2.3.3.2 Obligate Heterofermentative LAB

Heterofermentative (facultative and obligate) LAB belong to the *Lactobacillaceae* family and are classified within the genera *Lactobacillus*, *Oenococcus*, *Leuconostoc*, and *Weissella*. The genus *Lactobacillus* is further divided into four groups: the *Lentilactobacillus buchneri* group, the *Lactiplantibacillus plantarum* group, the *Lb. reuteri* group, and the *Lacticaseibacillus casei* group. Both the *Lentilactobacillus buchneri* and *Lb. reuteri* groups consist of obligate heterofermenters; however, the former has been the exclusive source of obligate heterofermentative strains investigated as potential silage inoculants. The *Lentilactobacillus buchneri* group includes species *Lentilactobacillus buchneri*, *Lb. diolivorans*, *Lb. farraginis*, *Lb. hilgardii*, *Lb. kefiri*, *Lb. kisonensis*, *Lb. otakiensis*, *Lb. parabuchneri*, *Lb. parafarraginis*, *Lb. parakefiri*, *Lb. rapi*, *Lb. sunkii*, *Lb. ferintoschensis* (Hammes and Hertel 2009), *Lb. senioris* (Feichtinger et al. 2016), *Lb. kunkeei*, *Lb. homohiochii*, *Lb. fructivorans*, *Lb. sanfranciscensis*, *Lb. lindneri*, *Lb. spicheri*, *Lb. namurensis*, *Lb. zymae*, *Lb. acidifarinae*, *Lb. hammesii*, and *Lb. brevis* (Pot and Tsakalidou 2009). However, only a limited number of species from the *Lentilactobacillus buchneri* group have been evaluated for their effects on silage fermentation. These primarily include *Lentilactobacillus buchneri*, with less common evaluations of *Lb. brevis*, *Lb. diolivorans*, *Lb. hilgardii*, *Lb. kefiri*, and, more recently, *Lb. parafarraginis*. In the mid-1990s, Muck (1996) suggested that the addition of *Lentilactobacillus buchneri* to silages could enhance aerobic stability by promoting moderate increases in acetic acid, which has the potential to inhibit yeasts responsible for initiating aerobic spoilage. Oude Elferink et al. (2001) demonstrated that the increased acetic acid occurred not through the classical phosphoketolase pathway, but rather through the anaerobic conversion of moderate amounts of lactic acid to acetic acid and 1,2-propanediol. Initially, *Lentilactobacillus buchneri* was thought to be incapable of further metabolizing 1,2-propanediol into propanol and propionic acid. While several organisms, including *Lb. diolivorans* and *Lb. reuteri*, possess this capability, recent research by Zielińska et al. (2017) reported that the novel strain *Lentilactobacillus buchneri* A KKP 2047p can also metabolize 1,2-propanediol to propionic acid, a process facilitated by the presence of glucose and cobalamin. Since its introduction as a microbial inoculant for silages, numerous laboratory and field studies (e.g., Mari et al. 2009; Tabacco et al. 2011) have demonstrated that

Lentilactobacillus buchneri effectively improves the aerobic stability of various silages; however, the effects are strain-specific and dose-dependent.

2.3.3.3 Combination Inoculant

The developing inoculants that combine facultative and obligate heterofermentative LAB aims to harness the benefits of both types of inoculants in a single product. The former would manage the initial active fermentation phase, suppressing enterobacteria, clostridia, and other microorganisms, thereby reducing proteolysis and fermentation DM losses. In this active fermentation phase, the pH is expected to decrease more rapidly and reach a lower value compared to untreated silage. Subsequently, the obligate heterofermentative LAB, most cases *Lentilactobacillus buchneri*, would gradually convert lactic acid to acetic acid after the active silage fermentation period, resulting in an increase in pH and enhancing the aerobic stability of the silage. The most frequently cited pairing in the literature has been with *Pediococcus pentosaceus* (e.g., Schmidt and Kung 2010), which was commercialized. Other species used include *Lactiplantibacillus plantarum*, *P. pentosaceus* and *Lactiplantibacillus plantarum*, *Lactiplantibacillus plantarum* and *Enterococcus faecium*, *Lacticaseibacillus casei*, *P. acidilactici*, *Lactiplantibacillus plantarum* and *Lacticaseibacillus caseii*, *P. acidilactici*, *P. pentosaceus*, *Lactiplantibacillus plantarum* and *E. faecium*, and *Lc. lactis* and *Lactiplantibacillus plantarum*. With a broad spectrum of different species, variations in the relative rates of species inclusion within inoculants, and differences among strains of a given species, it is not surprising that there are variations in the results obtained from combination inoculants. Relatively reports since Driehuis et al. (2001) have focused on the success of the combination inoculant during the first 1 to 2 week of fermentation. Schmidt et al. (2009) investigated the fermentation process in alfalfa silage that was either untreated or treated with *Lentilactobacillus buchneri* or *Lentilactobacillus buchneri* plus *P. pentosaceus*. Over the first 5 days, the pH level was higher in the silage inoculated with *Lentilactobacillus buchneri* compared to the untreated silage. In contrast, the combination inoculant resulted in a more rapid decline in pH than the untreated silage. Addah et al. (2012) compared a combination inoculant containing *Lentilactobacillus buchneri*, *Lactiplantibacillus plantarum*, and *Lacticaseibacillus casei* with an untreated control in whole-crop barley silage; in laboratory silos, the combination inoculant had a more rapid pH decline in the first week of ensiling. Arriola et al. (2015) compared four commercial inoculants versus untreated for making bermudagrass silage in wrapped round bales at approximately 50% DM. One of the inoculants was a combination containing *Lentilactobacillus buchneri* and *P. pentosaceus*. It had the fastest rate of pH decline over both 7 and 30 days. However, another inoculant from the same company containing *P. pentosaceus* and *Propionibacterium freudenreichii* performed similarly. These findings suggest that a combination inoculant can dominate the early stages of fermentation, resulting in a more rapid decline in pH compared to untreated silage. Kang et al. (2009) reported on the effectiveness of a combination inoculant containing *Lb. casei* and

Lentilactobacillus buchneri in corn silage. In one corn cultivar, the inoculant successfully improved aerobic stability. Arriola et al. (2011) compared *Lentilactobacillus buchneri* with or without *P. pentosaceus* in corn silage ensiled for 575 days. The untreated control exhibited aerobic stability for 390 h. *Lentilactobacillus buchneri* alone did not enhance aerobic stability but achieved the highest DM recovery among all treatments. The combination inoculant resulted in reduced aerobic stability, lasting 276 h, and demonstrated an intermediate DM recovery; it was the only treatment with a lactic-to-acetic acid ratio exceeding 1.0. It appeared that the untreated silages underwent a *Lentilactobacillus buchneri*-like fermentation, with the concentration of acetic acid being three times greater than that of lactic acid. Thomas et al. (2013) studied the effects of a combination inoculant (*Lentilactobacillus buchneri, P. acidilactici, P. pentosaceus, Lactiplantibacillus plantarum, E. faecium*) on four sorghum cultivars. The inoculant did not yield silage that was more heterofermentative than the corresponding untreated silage. Overall, the ineffectiveness of combination inoculants may stem from either the natural population displaying significant *Lentilactobacillus buchneri*-like activity or the *Lentilactobacillus buchneri* strain in the inoculant failing to dominate the later stages of fermentation for some reason.

2.3.3.4 Other Inoculants

Several innovative inoculant approaches have been explored in recent years; these include *Streptococcus bovis, Propionibacterium acidipropionici, Bacillus* species, and yeast. The bacterial species have been studied prior to 2000, and commercial inoculants contain *Propionibacterium* or *Bacillus* species. (1) *Streptococcus bovis*, a component of novel inoculants, most closely resembles the LAB strains used in inoculants. It is primarily found in the rumen rather than in the silage. It is recognized as one of the fastest-growing bacterial species in the rumen, with a doubling time as short as 20 min, which is faster than that of LAB used in ensiling. This rapid growth makes *S. bovis* an attractive starter strain in inoculants, as it can quickly lower pH during the early stages of ensiling. However, since *S. bovis* is less tolerant of low pH compared to lactobacilli, its population is expected to decline after the active fermentation of silage, thereby reducing the likelihood of its presence in the rumen following silage consumption. Ferreira et al. (2013) compared two strains of *S. bovis* to untreated and *E. faecium*-treated, negative and positive controls, on unwilted elephant grass in laboratory silos. After 60 days of ensiling, both *S. bovis* treatments significantly reduced pH and ammonia nitrogen levels, resulting in a more homofermentative fermentation compared to the other treatments. The *S. bovis* strains also decreased both gaseous and effluent losses, leading to improved dry matter recovery compared to the untreated or *E. faecium*-treated silages. Therefore, both strains demonstrate potential as inoculant strains for tropical grass. (2) *Propionibacterium*, inoculation of silage with *Propionibacterium* species aims to produce propionic acid during the ensiling process, thereby enhancing aerobic stability. Filya et al. (2004) observed that previous studies involving *Pr. acidipropionici* and *Pr. shermanii* yielded mixed results regarding their effectiveness in improving aerobic stability in corn and small-grain silages. Their research examined the effects of inoculating wheat, corn, and sorghum silages with *Pr. acidipropionici*, *Lactiplantibacillus plantarum*, or a combination of both; the results were consistent across all three crops. The application of *Pr. acidipropionici* alone resulted in increased levels of acetic and propionic acids compared to the untreated control, and it produced silages with the least losses during a 5-day aerobic stability test, as indicated by CO_2 production. In this 5-day aerobic stability test, the combination inoculant exhibited CO_2 losses similar to the control but lower than those observed in *Lb. plantarum*-treated silages. Therefore, *Pr. acidipropionici*, when applied independently, proved to be the most effective treatment for reducing aerobic losses and maintaining aerobic stability by keeping yeast counts low.

2.3.3.5 Chemical

A variety of chemical additives with antifungal properties have been utilized to improve the aerobic stability of silages, with short-chain organic acids being the most prevalent. For instance, buffered propionic acid products are frequently used in the US due to their lower corrosiveness and safer handling compared to straight acid (Kung et al. 2000). The antifungal characteristics of organic acids are attributed to their undissociated (protonated) form, and its prevalence is dependent on pH. Additionally, potassium sorbate and sodium benzoate have been used to enhance the aerobic stability of silages. For instance, applying 0.1% potassium sorbate not only increased dry matter recovery and aerobic stability but also reduced the final ethanol concentration in maize silage (Teller et al. 2012). According to Knicky and Spörndly (2011), an additive with sodium benzoate, potassium sorbate, and sodium nitrate enhanced the aerobic stability of a lot of crops with DM > 35%. Additionally, Stanojevic et al. (2009) noted that sodium nitrate, potassium sorbate, and sodium benzoate exhibited synergistic effects against various fungi and bacteria. At the time of writing this review, the high cost of chemicals designed to enhance aerobic stability is a limiting factor to their broader use. If the production expenses for antifungal chemicals decrease in the future, their usage may rise. Additionally, discovering other antifungal compounds with low application and cost rates could be advantageous, but they would need to pass regulatory evaluations to confirm their safety.

2.3.3.6 Exogenous Enzyme

Fibrolytic enzyme complexes, mainly consisting of cellulase and hemicellulase, were initially added into silage to partially degrade fiber to fermentable water-soluble carbohydrates that LAB can utilize during fermentation. However, an overproduction of soluble carbohydrates is undesirable as it may result in problems such as reduced aerobic stability and increased effluent. Fibrolytic enzymes also have the potential to enhance the digestion of ruminal fiber. The process of fiber digestion by these enzymes is intricate and limited by the interactions between carbohydrates and lignin. Previous research has shown that the application of fibrolytic enzymes may lead to a decrease in fiber digestibility in the silage produced (Nadeau et al. 2000). What actually occurred was digestion of the readily digestible portion of fiber during ensiling, leaving the more recalcitrant portion for feeding. The confirmed cellulolytic activity in LAB suggests significant potential for identifying organisms that can thrive in silage environments and exhibit sufficient cellulolytic activity (Matthews et al. 2006). Several years ago, Nsereko et al. (2008) used a species of *Lentilactobacillus buchneri* producing ferulic acid esterase as an ensilage inoculant to improve fiber digestion. Instead of digest fiber directly, this application aimed to break the connection between hemicellulose and lignin, leading to a fiber fraction that is easier for the rumen to digest. While there have been successes in some cases (Kang et al. 2009), other cases have yielded no results (Lynch et al. 2014, 2015). We hypothesize that the inoculant might struggle to produce adequate levels of ferulic acid esterase under different field conditions, leading to inconsistent results. Currently, directly adding sufficient amounts of ferulic acid esterase is not cost-effective, However, advancements in the production of this enzyme for potential applications in the biofuels industry could reduce costs and alter this scenario in the future. Young et al. (2012) were the pioneers in demonstrating that the inclusion of acidic proteases at the time of harvest could enhance proteolytic activity during storage and improve ruminal starch digestion in less time than usual. Similarly, Windle et al. (2014) found that adding an acidic protease significantly improved ruminal starch digestion in maize silages with 31% dry matter and more mature silages at 40% dry matter. However, advancements in formulation and production costs are necessary before this technology can be implemented in practical applications.

2.3.4 Feed Bunk Management

It's crucial to understand that silage is always in a state of flux. Lowering the pH will preserve silage, but it doesn't guarantee its safety in the feed bunk. Silages that spoil and heat up in the feed bunk can significantly impact daily DM intakes. Cows tend to avoid consuming unstable silages, particularly during the hot summer months when DM intake is already reduced. This is particularly relevant for total mixed rations, where feed is continuously available in the feed bunk and is expected to remain fresh for a long time. The instability of silages is typically caused by microbial activity. Poor compaction and distribution, insufficient moisture, and incorrect chop length lead to excessive air infiltration, resulting in aerobically unstable silage. Effective management is essential for whole-plant corn silage and high-moisture grains, as they are especially prone to aerobic spoilage. To minimize air penetration into the silage, it is important to level off, pack the edges, and seal the top of tower silos, as well as to pack and cover bunkers with plastic and tires. When storing high moisture ear corn in a conventional tower silo or bunker, cob pieces should be no larger than 1.25 cm in diameter to avoid separation during handling. Ensuring an even distribution during the filling process is essential to prevent cobs from gathering in one spot, which can hinder compaction and trap excess oxygen. Additionally, harvested forages that have been exposed to rain prior to chopping and ensiling often face bunklife issues. Rain can cause soil-borne bacteria and fungi (mold) to splash onto the crop, essentially introducing spoilage organisms into the silage that can thrive if conditions allow. Crops that are affected by drought, insect infestations, or hail damage typically have higher levels of fungi, which means that careful management is necessary when ensiling these compromised crops. Corn and cereal silages are more susceptible to mold growth compared to grass or legume silages. In corn, the spoilage process usually begins with aerobic bacteria, like *Bacillus* species, followed by a rise in yeast populations. The opposite is true for legume, cereal, and grass silage. Yeast species such as *Candida* and *Hansenula* pose significant challenges as they are capable of metabolizing lactic acid, leading to an increase in the pH of silage. When the pH rises above 4–5, the environment becomes conducive to the proliferation of various molds and spoilage microorganisms. Yeasts like *Saccharomyces* and *Torulopsis*, which are sediment-type, are less of a concern since they have a limited ability to assimilate lactate and, therefore, do not significantly increase silage pH. Regularly providing forages or total mixed rations and routinely cleaning feed bunk can help mitigate these microbial spoilage issues. Ideally, fermentation should occur quickly enough to minimize "front-end" fermentation losses while also ensuring a satisfactory level of aerobic stability in both the storage and feed bunk. Effective management from harvest to feeding, along with the use of appropriate silage additives, is crucial for balancing these two key aspects of silage quality.

2.3.5 Effluent Management

Silage effluent is characterized by elevated levels of biochemical oxygen demand (BOD), nitrogen (N), and phosphorus (P), along with a low pH. If released without treatment, it poses a significant risk to the quality of both surface and groundwater. A particularly concerning issue is its corrosive nature, which can damage steel and concrete due to the high levels of organic acids, complicating its collection, storage, and disposal. The generation of silage effluent is closely linked to the moisture content of the crops and pressure of silo types. For example, large bale silage typically yields less effluent compared to tower and bunker silos; however, it is advisable for grass bales to maintain a moisture content below 76% to reduce leachate production (Davies and Nicholson 1999). Consequently, to mitigate the generation of silage effluent, various management strategies have been previously implemented, including preventative management, costs of reactive management and reactive effluent management. Here, we will provide a detailed examination of the components of "preventative management."

2.3.6 Silage Moisture Content (MC)

One strategy to mitigate effluent production involves ensiling crops at lower MCs. The recommended MCs are contingent upon the type and dimensions of the silos, as well as the specific crop being used. Generally, it is advised that crop MCs should not exceed 70–75% for bunker silos and 60–65% for tower silos. In instances where crops cannot be harvested at the optimal MC, it is essential to implement practices that reduce the risk of effluent production. For grasses or alfalfa, decreasing crop MC can be accomplished through natural wilting or by employing chemical or thermal treatments. Under sunny and dry conditions, natural wilting can lead to a reduction in swath MC from approximately 80% to 65% within a span of 3 to 6 h (Grant and Stock 1974). In contrast, in moist and temperate climates, the reduction of MC through natural wilting may occur at a slower rate due to lower VPDs (McDonald et al. 1991). Consequently, the application of chemical or mechanical treatments may be necessary to facilitate water loss. Chemical desiccants, including diquat, paraquat, and formic acid, compromise the integrity of the cuticle, thereby enhancing moisture loss. However, formic acid also increases effluent production. Heat treatment, which involves the application of hot air and steam beneath the canopy of a standing crop, serves to rupture the cuticles and promote moisture loss (Calder 1982). It is important to note that treatments that damage the cuticle may have adverse effects if subsequent rainfall occurs, as this can facilitate water penetration into the plant tissue, resulting in increased MC and potential nutrient loss (Tetlow 1983). Therefore, it is essential to carefully consider weather forecasts prior to implementing treatments aimed at reducing MC. An alternative method for decreasing crop MC and effluent production involves the incorporation of absorbent materials during the ensiling process. A comprehensive review of this practice has been previously conducted (Jones and Jones 1996); thus, a succinct summary is presented here. Additives such as ground grains, finely ground corn cobs, sugar beet pulp, mixed grass hay, and oat straw have demonstrated effluent absorption capacities of 600, 1920, 2480, 1950, and 2180 kg per ton, respectively. Notably, barley straw and sugar beet pulp have been recognized as effective absorbents that enhance feed quality, as evidenced by the rapid weight gain observed in livestock (Diana 1988; Jones and Jones 1988).

2.3.7 Silo Construction

Silos must be constructed to be watertight in order to prevent precipitation from infiltrating the structure and compromising ensiled crops. Achieving this objective can be challenging due to the highly corrosive nature of effluent, which poses highly destructive to both metal and concrete materials (O'Donnell et al. 1995). Therefore, it is imperative to use materials that exhibit resistance to corrosion. The mix design of concrete influences the durability of structures that meet silage effluent. Research conducted by Lenehan and O'Kiely (1999) indicated that concrete mixtures containing cement contents of 325 and 375 kg/m^3, along with water-to-cement ratios of

0.55 and 0.50, respectively, demonstrated the least mass loss after a period of 10 years. This finding aligns with the recommendations of O'Donnell et al. (1995), who suggested maintaining a water-to-cement ratio of less than 0.6. The durability of concrete may be further enhanced by incorporating waste materials such as pulverized fuel ash from coal-burning power plants (Lenehan and O'Kiley 1999) or granulated blast-furnace slag (GGBS) from the steel manufacturing sector in conjunction with regular Portland cement. The hydrated components of regular Portland cement contain a greater concentration of $Ca(OH)_2$ and aluminates compared to GGBS, which is significant due to the susceptibility of these compounds to reactions with acids and sulfates. Additionally, the microstructure is enhanced because there are fewer and smaller $Ca(OH)_2$ crystals, leading to a more compact and robust transition zone. Besides their interaction with concrete, the acids found in effluent can also cause corrosion of silo hoops, reinforcing steel, and other equipment.

2.3.8 Silage Biochemistry

The biochemistry involved in the process of ensiling is straightforward; nevertheless, it may become intricate due to the interactions between plant enzymes and the activities of various microbial species. The desired effect is to facilitate the conversion of simple plant sugars, such as glucose and fructose, into lactic acid through anaerobic fermentation conducted by LAB. Once an adequate amount of lactic acid has been generated, microbial activity is effectively inhibited, predominantly due to the presence of undissociated lactic acid. Consequently, the silage can be stored in an anaerobic environment until it is needed for feeding. The silage fermentation process can be categorized into several distinct stages: the aerobic phase, the lag phase, the fermentation or stable phase, and the feed-out phase.

2.3.8.1 Aerobic Phase

The initial phase is marked by the consumption of trapped oxygen throughout the ensiling process. The chopped and stored forage in the feed storage exhibits the presence of entrapped air during ensiling. During this period, forage plants continue to undergo respiration for several hours, and potentially for days, contingent upon the packaging (McDonald 1981). In this phase, plant enzymes remain active until the available trapped oxygen is depleted. Inadequate packing of ensiled forage or the introduction of oxygen can result in unfavorable fermentation processes facilitated by molds and yeasts, which subsequently lead to protein degradation and excessive heat generation. To mitigate these issues, it is essential to chop forage crops to appropriate lengths, ensure tight packing within the storage facility, eliminate all entrapped air from the forage mass, and perform immediate packing, as any delays may adversely affect the quality of the silage (Meeske 2005).

2.3.8.2 Lag Phase

This phase is characterized by the depletion of available oxygen in the ensiled materials through the action of anaerobic bacteria. The conditions within the silage, including the availability of substrates for microbial growth, moisture content, and the existing population of microflora, influence the dominant microorganisms present in the forage. As LAB proliferate and metabolize water-soluble carbohydrates to generate lactic acid, the pH of the silage decreases below a critical threshold, thereby suppressing the growth of other spoilage microorganisms. At this stage, the majority of acid-tolerant Clostridia are inhibited (Jonsson 1991).

2.3.8.3 Fermentation or Stable Phase

A significant reduction in pH levels prevents the proliferation of clostridia and enterobacteria. During this phase, the dry matter is crucial for fermentation processes. An increase in moisture content, exceeding 70%, inhibits the growth of LAB, consequently diminishing both the rate and extent of fermentation. This phase represents the longest duration in the ensiling process and persists until the proliferation of microorganisms is entirely suppressed by elevated acid levels (low pH) (Driehuis et al. 1999). At this stage, the forage achieves a preserved state. It is important to recognize that undesirable bacterial populations may thrive, leading to excessive protein degradation, the production of toxic compounds, and a reduction in dry matter (Ohshima and McDonald 1978). Therefore, wilting forage to a dry matter content of 30–35% prior to ensiling can effectively mitigate the presence of undesirable bacteria.

2.3.8.4 Feed-Out Phase

In this phase, high-quality silage will maintain stability and consistency in its composition. Nevertheless, the pH value alone is insufficient to determine the quality of the silage or the type of the fermentation process. The end products of silage fermentation are typically evaluated through various indicators that assess the quality and composition of the silage. This assessment is conducted by examining physical characteristics such as appearance, texture, color, and aroma, as well as measuring specific compounds including lactic acid, acetic acid, butyric acid, ethanol, acid detergent insoluble nitrogen, and ammonia. In this instance, we will use "anaerobic fermentation" as a case study to thoroughly describe the biochemical process.

2.4 *Anaerobic Fermentation*

Anaerobic fermentation occurs when oxygen is removed, leading to a transition to anaerobic metabolism. This results in a rapid increase in microorganisms that produce both neutral and acidic byproducts. The primary bacteria involved are enterobacteria and LAB. The acidic byproducts lower the pH of the silage, promoting the growth of more acid-resistant LAB. When there is an adequate supply of substrate, LAB dominate the fermentation process, generating lactic acid and further acidifying the silage until the pH reaches a level that inhibits LAB growth, resulting in stable silage. However, if the moisture content exceeds 70%, LAB growth is hindered, which slows down the fermentation process. This phase is the longest in the ensiling process and continues until high acid levels (low pH) completely inhibit microbial growth. It is important to note that undesirable bacteria can flourish during this time, leading to excessive protein breakdown, the production of harmful substances, and a loss of dry matter. To prevent the growth of these unwanted bacteria, it is advisable to wilt forage to a dry matter content of 30–35% before ensiling. During the anaerobic fermentation process, the microorganisms found on the ensiled crops break down sugars to generate energy for their growth, producing acids and neutral byproducts in the process. In most silage fermentations, the primary sugars involved are fructose and glucose. The levels of these sugars at harvest can vary significantly based on environmental factors and the type of forage. However, after cutting, the overall carbohydrate content typically declines. The concentrations of fructose and glucose can only be increased through the hydrolysis of polysaccharides and, to a limited extent, through photosynthesis. There is generally a significant reduction in these sugars due to ongoing plant respiration and microbial fermentation. The silage microflora may also utilize alternative energy sources, such as pentose sugars generated from the hydrolysis of matrix polysaccharides (pectic and hemicellulosic) by either internal or external enzymes, organic acids found in the ensiled material, or amino acids produced from the breakdown of crop proteins. The fermentation of hexoses, pentoses, and organic acids results in various end products, which are produced by different types of bacteria and yeasts.

2.4.1 Utilization of Hexoses by Silage Microorganisms

The production of acid in an anaerobic setting is essential for effective preservation through ensiling. However, regarding the microflora involved in ensiling, acid production serves primarily to maintain a balance of oxidation and reduction (redox) during the anaerobic metabolism of substrates, which generates energy (ATP) for growth and carbon skeletons for biosynthesis. In practice, this redox balance is achieved through the reoxidation of NADH. The genetic composition of bacteria equips them with a diverse range of enzymes and metabolic pathways to maintain this balance. The ability to utilize various substrates and produce different end products is a key factor in classifying bacteria. It is also important to note that adaptive

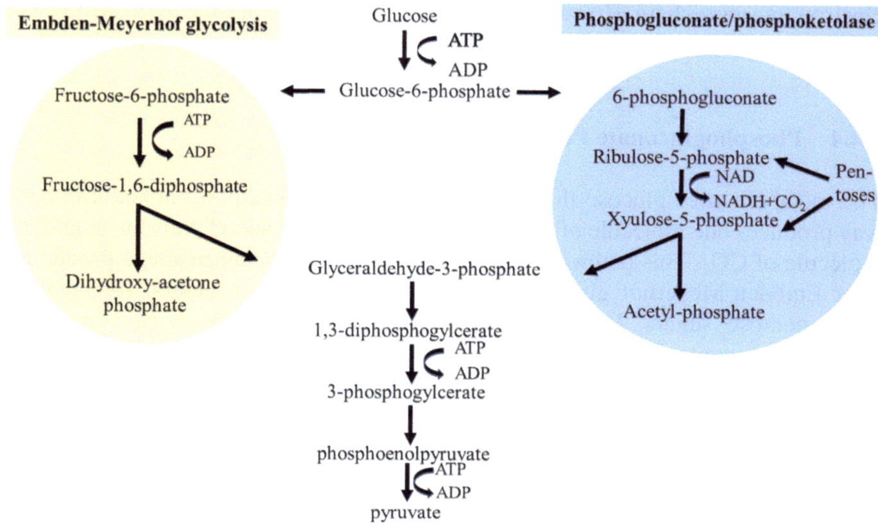

Fig. 2.2 Embden-Meyerhof glycolysis and phosphogluconate/phosphoketolase

mechanisms enable individual bacterial strains to use different substrates and generate various end products based on their environmental conditions, such as pH, moisture levels, redox potential, and oxygen tension.

2.4.2 Glucose Metabolism

There are four metabolic pathways that can break down glucose and other substrates in the absence of oxygen. The primary pathways, known as the Embden-Meyerhof and 6-phosphogluconate/phosphoketolase pathways, are depicted in Fig. 2.2. A brief overview of the pentose phosphate and Entner-Doudoroff pathways will be provided, but for a more detailed explanation, readers are encouraged to refer to standard biochemistry textbooks.

2.4.3 Embden-Meyerhof Glycolytic Pathway

In the classical glycolytic pathway, a single glucose molecule is transformed into two pyruvate molecules, resulting in the net production of two moles of ATP (through substrate-level phosphorylation) and two moles of NADH. The main enzyme involved is fructose bisphosphate aldolase, which facilitates the breakdown of fructose 1, 6-bisphosphate into two triose phosphate molecules. When oxygen is available, the resulting pyruvate enters the tricarboxylic acid cycle, where it is fully oxidized to carbon dioxide and water, producing additional ATP through the electron transport chain. However, under anaerobic conditions, the pyruvate must be reduced to regenerate NADH and preserve redox balance.

2.4.4 Phosphogluconate Phosphoketolase Pathway

The metabolism of glucose through the phosphogluconate/phosphoketolase pathway produces one molecule of pyruvate, one molecule of acetyl phosphate, and one molecule of CO_2. This pathway is the second most significant in silage production after Embden-Meyerhof glycolysis and serves as the main method for breaking down pentose sugars. In this process, glucose 6-phosphate is converted into 6-phosphogluconic acid, which is then decarboxylated to form ribulose 5-phosphate. Ribulose 5-phosphate is subsequently split by phosphoketolase into glyceraldehyde 3-phosphate and acetyl phosphate. The metabolism of glyceraldehyde 3-phosphate continues similarly to glycolysis, producing ATP and pyruvate, which requires the reduction of pyruvate to regenerate NAD. The outcome of acetyl phosphate depends on the cell's redox state. If there is no alternative electron acceptor, NAD is regenerated, resulting in ethanol as the final product. Conversely, if an alternative is present, an additional ATP molecule can be produced by transferring the phosphate group from acetyl phosphate to ADP, leading to acetate as the final product. In bacteria where both glycolysis and the phosphogluconate/phosphoketolase pathway are active, the latter is typically suppressed. Therefore, under anaerobic conditions, pyruvate and acetyl phosphate are primarily used as substrates to regenerate NAD.

2.4.5 Pentose Phosphate Pathway

The main role of the pentose phosphate pathway is to produce NADPH for biosynthetic processes. The crucial step involves glucose 6-phosphate dehydrogenase, which, along with lactonase and 6-phosphogluconate dehydrogenase, leads to the formation of ribulose 5-phosphate, CO_2, and two NADPH molecules. Depending on the cell's metabolic needs, various transformations facilitated by transaldolase and transketolase enzymes can yield different end products, including pyruvate. In terms of anaerobic fermentation, the pentose phosphate pathway results in the production of pyruvate and CO_2.

2.4.6 Entner-Doudoroff Pathway

This pathway resembles the traditional glycolytic pathway as it also converts glucose into two 3-carbon sugars. However, it is distinct in that glucose 6-phosphate is converted into 6-phosphogluconic acid, which is then split by a specific aldolase into glyceraldehyde 3-phosphate and pyruvic acid. Unlike glycolysis, this pathway produces only one ATP molecule through substrate-level phosphorylation and reoxidizes just one NADPH molecule. Initially identified in *Pseudomonas* species, this pathway is likely not significant in bacteria involved in silage production (Conway 1992). Nevertheless, it is found in *Escherichia coli* and may play a role in glucosamine metabolism in bacteria (Iwamoto and Imanaga 1991).

2.5 Silage Microbiome and Metabolomic

2.5.1 Silage Microbiome

The profiles of microbiomes and metabolomes in silage have emerged as valuable tools for examining historical trends that were not thoroughly investigated in conventional studies. The process of ensiling, which has been utilized since the late 1800s to preserve fresh forage, is driven by microorganisms. The microbiome found in freshly harvested forage is essential to this process and consists of a diverse array of bacteria, yeasts, and molds, with colony-forming units ranging from 10^5 to 10^9 cfu/g. When the forage is placed in the silo, the diversity of the microbiome diminishes as obligate aerobes and acid-sensitive microbes are either killed or unable to grow due to the exclusion of oxygen and the resulting decrease in pH from acid production. In these conditions, epiphytic LAB, enterobacteria, yeasts, and molds may enter a viable but unculturable state, making it difficult to identify them using traditional plating techniques. The microbial populations involved in the ensiling process are highly complex, necessitating considerable effort to characterize their members based on morphological, physiological, and biochemical characteristics.

To evaluate epiphytic bacterial communities of plants, a range of methods has been employed over the years, including both culture-based techniques (like most probable number, selective media, and biochemical profiling) and culture-independent methods such as denaturing gradient gel electrophoresis (DGGE), single-strand conformation polymorphisms, terminal RFLP (T-RFLP), real-time PCR, ribosomal intergenic spacer analysis, and terminal restriction fragment length polymorphism (Li and Nishino 2011; Pang et al. 2011; Stevenson et al. 2006). However, these methods only revealed a limited number of microorganisms that researchers intended to identify. Next-generation sequencing (NGS) and PacBio single-molecule real-time sequencing technology (SMRT) provide high-throughput capabilities, allowing for the identification of a much larger portion of microbiota and the relative abundances of various microbes at the genus and species levels. For instance, Guo et al. (2018) were the first to investigate the bacterial community of alfalfa silage at the species level using PacBio SMRT sequencing. This approach was soon adopted in various silage studies to thoroughly uncover the microbial composition of different forages, such as corn, alfalfa, grass, and tropical silages, both before and after the ensiling process (Xu et al. 2020, 2021; Du et al. 2022; Xia et al. 2022).

Among various forages, whole-plant corn silage is the most widely used in ruminant dairy diets globally, thanks to its high water-soluble carbohydrate (WSC) content and adequate levels of epiphytic LAB that facilitate natural fermentation. Typically, the epiphytic bacteria found on fresh whole-plant corn are mostly undesirable species, including *Xanthomonadaceae*, *Sphingobacteriaceae*, *Enterobacteriaceae*, *Moraxellaceae*, and various genera such as *Meyerozyma*, *Acinetobacter*, *Klebsiella*, *Agrobacterium*, *Microbacterium*, *Chryseobacterium*, and *Candida* (da Silva et al. 2020; Guan et al. 2018, 2020; Keshri et al. 2018; Xu et al. 2021). While there is significant variation in the composition of epiphytic microbial communities in whole-plant corn silage, the microbiome that develops during the ensiling process is notably different from the initial epiphytic community (Gharechahi et al. 2017). Consequently, the relative abundance of LAB typically increases after ensiling, and they become the dominant microorganisms during the later stages of fermentation. In contrast, alfalfa silage has a more diverse range of epiphytic microorganisms compared to cereal silages. The undesirable bacteria present on alfalfa before ensiling mainly include *Pantoea*, *Streptococcus*, *Enterobacter*, *Bacillus*, *Pseudomonas*, *Exiguobacterium*, *Massilia*, *Planococcus*, *Sphingobium*, *Acinetobacter*, and *Clostridium* (Bai et al. 2021). As fermentation begins, the dominant microbial community shifts rapidly, transforming from undesirable bacteria to LAB that take over the fermentation process. The primary LAB involved in alfalfa silage fermentation are *Lactiplantibacillus*, *Lentilactobacillus*, *Weissella*, *Lactococcus*, *Enterococcus*, and *Pediococcus* species (Guo et al. 2018; Hu et al. 2020). The addition of LAB significantly influences the composition and succession of microbial communities during the ensiling of alfalfa silage. *Lactiplantibacillus plantarum* is a commonly used LAB additive for alfalfa silage. Furthermore, when alfalfa is inoculated with *Lactiplantibacillus plantarum*, lactobacilli become the predominant bacteria involved in the fermentation process (Yang et al. 2020). Even in high-moisture alfalfa silage, the inoculation of *Lactiplantibacillus plantarum* enhances the abundance of lactobacilli while suppressing *Clostridium* growth (Yang et al. 2020). A specific isolated strain *Lactiplantibacillus plantarum* from rumen fluid and dairy cow feces has been shown to increase the relative abundance of lactobacilli in alfalfa silage (Guo et al. 2020) and grass silage. In addition to whole-plant corn and alfalfa silages, grass silages made from barley, ryegrass, oats, and other grasses are widely utilized as feed for ruminants around the globe. The characteristics of grass silages distinct from those of legumes like alfalfa or whole-plant corn, leading to different microbial compositions and changes in grass silages compared to corn and alfalfa silages during the ensiling process. Furthermore, inoculants significantly influence the microbial community composition in grass silages. For example, LAB are frequently employed to enhance the aerobic stability of barley silage. Liu et al. (2020) found that in barley silage, the use of LAB inoculants (such as *Lactiplantibacillus plantarum*, *Lacticaseibacillus casei*, and *Lentilactobacillus buchneri*) reduced the diversity of bacterial communities during ensiling, while exposure to air increased the diversity of fungal communities. Spontaneously fermented barley silages were primarily composed of *Lactobacilli*, *Weissella*, and *Pediococcus* species. The introduction of LAB inoculants boosted lactobacilli populations but decreased the numbers of *Pediococcus* and *Weissella* during ensiling. Additionally, air exposure led to a reduction in *Acinetobacter* levels. The combination of inoculants and molasses also lowered *Acinetobacter* and *Enterococcus* levels in barley silage after 5 and 7 days of aerobic exposure (Liu et al. 2020). In tropical regions, special care is needed when preparing silage due to specific climatic conditions. *Clostridium* in tropical grass silages thrive in humid environments, which helps prevent pH drops and spoilage. Moreover, warm conditions increase the risk of aerobic spoilage, which is heavily influenced by the epiphytic microbial communities present in tropical forages. Zou et al. (2021) identified *Cronobacter*, *Methylobacterium*, and *Enterococcus* as the dominant genera in naturally fermented stylo silage, while Zi et al. (2022) found *Weissella*, *Enterobacter*, and *Pantoea* to be the main bacterial communities in untreated stylo silage.

Analyzing functional changes, guided by microbiome functional predictions, has allowed for the assessment of metabolic pathways in microbial communities during the ensiling process. Key pathways related to silage fermentation include amino acid, carbohydrate, nucleotide, and energy metabolism. Amino acids are crucial for LAB as they play a role in energy regulation and help the bacteria cope with environmental stress. Recently, metagenomics has been increasingly utilized in silage fermentation systems to explore functional activities (or genes), interactions, and the relationships between microbial communities and environmental factors, underscoring its potential significance for silage development (Wu et al. 2020; Zhang et al. 2023; Wang et al. 2024). Future methodological advancements could enhance our understanding in this field, including the identification of new valuable molecules.

2.5.2 Silage Metabolomic

The end products produced during fermentation can be attributed to either direct microbial activity or the degradation and transformation of substances inherent in the material. During the aerobic phase of silage, plant enzymes (Pahlow et al. 2003) and the variable abundance of LAB strains contribute to the dynamics of the metabolome, resulting in variations in numerous metabolites. Conventional analytical methods typically identify metabolites such as lactic acid, acetic acid, butyric acid, propionic acid, and ethanol, which are employed to evaluate fermentation quality and aerobic stability. LAB strains, which are integral to silage fermentation, are known to produce specific compounds, including 2,3-butadione, reuterin, acetaldehyde, 3-phenyllactic acid, and 3-hydroxydecanoic acid (Schnu 2005; Sjögren et al. 2003). Furthermore, LAB are capable of synthesizing a diverse array of amino acids, fatty acids, vitamins, and oligosaccharides (Sun et al. 2012), indicating that silages encompass a broader spectrum of metabolites. Broberg et al. (2007) also investigated the metabolites generated by LAB in silage and identified p-hydrocoumaric acid, hydroferulic acid, and p-coumaric acid produced by *Lactiplantibacillus plantarum* MiLAB393 and MiLAB14 inoculants in grass.

Metabolomics has recently been utilized in various studies focused on silage to identify novel bioactive metabolites that may serve as indicators of silage quality. These metabolites include amino acids, fatty acids, vitamins, flavor compounds, and other bioactive substances that positively contribute to animal health and welfare. The diverse metabolites present in different silages have the potential to enhance the quality of animal meat or modify the metabolomic profile of milk (Lanza et al. 2021). A multitude of metabolites, primarily comprising organic acids, sugars, amino acids, polyols, and volatile compounds, has been identified in silage (Guan et al. 2020; Guo et al. 2018; Xu et al. 2021). The application of LAB inoculants can influence the silage metabolome by modifying microbial communities (Zhang et al. 2021). Specifically, the inoculants *Lactiplantibacillus plantarum*, *Lentilactobacillus buchneri*, and *Lacticaseibacillus casei* have been shown to enhance the relative concentrations of certain organic acids and polyols in alfalfa silage, with the concentrations of most organic acids exhibiting a positive correlation with the abundance of *Lactobacilli*, *Lactococcus*, *Weissella*, and *Enterococcus* (Guo et al. 2018). In whole-plant silage, the inoculants *Lactiplantibacillus plantarum* or *Lentilactobacillus buchneri* have been found to increase the concentrations of specific amino acids (such as phenylalanine, lysine, tyrosine, and glycine), phenolic acids (including 4-hydroxycinnamic acid and 3,4-dihydroxycinnamic acid), flavor compounds (such as gluconic lactone), and organic acids (including l lauric acid, 3-hydroxypropionic acid, pentadecanoic acid, oxamic acid, and isocitric acid) (Xu et al. 2021). In an investigation of silage odor, Zhang et al. (2021) examined volatile compounds in stylo and rice straw silages, revealing that *Lactiplantibacillus plantarum* reduced the concentrations of certain volatile substances, which were positively correlated with the presence of undesirable bacteria but negatively correlated with the abundance of *Lactobacillus*.

Notably, various bio-functional metabolites, including bacteriostatic, antioxidant, anti-inflammatory compounds, and neurotransmitters, have been identified in silage (Guo et al. 2018; Xia et al. 2022; Xu et al. 2021). These metabolites possess the potential to enhance animal health and welfare, and the application of inoculants may further augment their concentrations. For example, the levels of metabolites with anti- inflammatory (glycitin, lithospermic acid, and psoralidin), antioxidant (ferulic acid, isoferulic acid, sinapinic acid, and moslosooflavone), antimicrobial (phenyllactic acid), and anti-tumor (alnustone) were found to be elevated through the inoculation of LAB in a mixed silage of *Sesbania cannabina* and sweet sorghum (Xia et al. 2022). Furthermore, metabolites exhibiting antimicrobial activity (such as methylbutanoic acid, 1,2-propanediol, and tiglic acid, and tiglic acid) have been identified in Napier grass silage that was treated with LAB (Guan et al. 2020). The application of compounded LAB led to increased relative concentrations of antioxidant compounds, including 7-galloylcatechin, as well as antibacterial substances, including ferulic acid, in the silage derived from the leaves of *Cyperus esculentus* L. leaf. (Sun et al. 2021). Furthermore, our research team conducted a comprehensive analysis of the metabolomic profiles of whole-plant corn silage, both with and without the inoculation of *Lactiplantibacillus plantarum*. This investigation employed an integrated metabolomics approach that utilized microextraction (SPME) coupled with gas chromatography/mass spectrometry (GC/MS), as well as liquid chromatography/mass spectrometry (LC/MS). A total of 2087 metabolites were identified within the corn silage samples. Notably, the inoculation with

Lactiplantibacillus plantarum significantly modified the volatile organic compounds present in the silage and enhanced the concentrations of several essential amino acids, including l-isoleucine, lysine, and threonine, as well as pyridoxine, acetylcholine, and various compounds exhibiting antioxidant properties (such as alpha-linolenic acid and luteolin), anti-inflammatory (notably alpha-linolenic acid), and antimicrobial (including 3-phenyllactic acid and cinnamaldehyde) in the corn silage (Su et al. 2023). However, further investigation is necessary to elucidate the specific mechanisms and effects of LAB on both the microbiome and metabolome in future research endeavors.

References

Addah W, Baah J, Okine EK, McAllister TA. A third-generation esterase inoculant alters fermentation pattern and improves aerobic stability of barley silage and the efficiency of body weight gain of growing feedlot cattle. J Anim Sci. 2012;90(5):1541–52.

Allen MS, Knowlton KF. Non-structural carbohydrates important for ruminants. Feedstuffs. 1995;67(16):13–5.

Amaral RC, Queiroz BC, Garcia EHC, SáNeto A, Bernardes TF, Nussio LG, et al. Aerobic deterioration in maize silages under different covering methods of the plastic film. Grassl Chang World. 2010;15:482.

Amaral RC d, Santos MC, Daniel JLP, Sá Neto A d, Bispo ÁW, Cabezas-Garcia EH, et al. The influence of covering methods on the nutritive value of corn silage for lactating dairy cows. R Bras Zootecn. 2014;43(9):471–8.

Arriola KG, Kim SC, Adesogan AT. Effect of applying inoculants with heterolactic or homolactic and heterolactic bacteria on the fermentation and quality of corn silage. J Dairy Sci. 2011;94(3):1511–6.

Arriola KG, Queiroz OCM, Romero JJ, Casper D, Muniz E, Hamie J, et al. Effect of microbial inoculants on the quality and aerobic stability of bermudagrass round-bale haylage. J Dairy Sci. 2015;98(1):478–85.

Bai J, Ding ZT, Ke WC, Xu DM, Wang MS, Huang WK, et al. Different lactic acid bacteria and their combinations regulated the fermentation process of ensiled alfalfa: ensiling characteristics, dynamics of bacterial community and their functional shifts. Microb Biotechnol. 2021;14:1171–82.

Bal M, Jirovec A, Shaver R, Shinners K, Coors J. What chop for process corn silage? Hoard's Dairyman; 1998.

Bernardes TF, do Rêgo AC. Study on the practices of silage production and utilization on Brazilian dairy farms. J Dairy Sci. 2014;97(3):1852–61.

Broberg A, Jacobsson K, Strom K, Schnurer J. Metabolite profiles of lactic acid bacteria in grass silage. Appl Environ Microbiol. 2007;73(17):5547–52.

Calder FW. Effect of steam-treated grass silage on animal gain. Can J Plant Sci. 1982;62(1):89–94.

Coblentz WK, Akins MS. Silage review: recent advances and future technologies for baled silages. J Dairy Sci. 2018;101(5):4075–92.

Conway T. The Entner-Doudoroff pathway: history, physiology and molecular biology. FEMS Microbiol Rev. 1992;9(1):1–27.

da Silva ÉB, Savage RM, Biddle AS, Polukis SA, Smith ML, Kung L Jr. Effects of a chemical additive on the fermentation, microbial communities, and aerobic stability of corn silage with or without air stress during storage. J Anim Sci. 2020;98(8):skaa246.

Davies OD, Nicholson RJ. Effluent production from grass ensiled in round big bale wrapped with 4 layers of plastic film. In: Pauley T, editor. Proceedings of international silage conference. 12th, Uppsala, Sweden, Swedish University of Agricultural Science; 1999. pp. 5–7.

Diana DL. The effect of absorbent additives on silage quality and on effluent production. In: Stark BA, Wilkinson JM, editors. Silage effluent. Marlow: Chalcombe Publications; 1988. p. 49–50.

Driehuis F, Elferink SJWHO, Spoelstra SF. Anaerobic lactic acid degradation during ensilage of whole crop maize inoculated with *Lactobacillus buchneri* inhibits yeast growth and improves aerobic stability. J Appl Microbiol. 1999;87(4):583–94.

Driehuis F, Oude Elferink SJWH, Van Wikselaar PG. Fermentation characteristics and aerobic stability of grass silage inoculated with *Lactobacillus buchneri* with or without homofermentative lactic acid bacteria. Grass Forage Sci. 2001;56(4):330–43.

Du Z, Sun L, Lin Y, Yang F, Cai Y. Using pacBio SMRT sequencing technology and metabolomics to explore the microbiota-metabolome interaction related to silage fermentation of woody plant. Front Microbiol. 2022;13:857431.

Feichtinger M, Mayrhofer S, Kneifel W, Domig KJ. Tetracycline resistance patterns of *Lactobacillus buchneri* group strains. J Food Prot. 2016;79(10):1741–7.

Ferreira DDJ, Lana RDP, Zanine ADM, Santos EM, Veloso CM, Ribeiro GA. Silage fermentation and chemical composition of elephant grass inoculated with rumen strains of *Streptococcus bovis*. Anim Feed Sci Technol. 2013;183(1–2):22–8.

Filya I, Sucu E, Karabulut A. The effect of *Propionibacterium acidipropionici*, with or without *Lactobacillus plantarum*, on the fermentation and aerobic stability of wheat, sorghum and maize silages. J Appl Microbiol. 2004;97(4):818–26.

Gharechahi J, Kharazian ZA, Sarikhan S, Jouzani GS, Aghdasi M, Hosseini Salekdeh G. The dynamics of the bacterial communities developed in maize silage. Microb Biotechnol. 2017;10(6):1663–76.

Giglioli I. Transactions of 3rd international congress. Trop Agric. 1914;2:660–90.

Grant R, Stock R. G74-142 harvesting and preserving hay crop silage; 1974. p. 1304.

Guan H, Yan YH, Li XL, Li XM, Shuai Y, Feng GY, et al. Microbial communities and natural fermentation of corn silages prepared with farm bunker-silo in Southwest China. Bioresour Technol. 2018;265:282–90.

Guan H, Shuai Y, Yan YH, Ran QF, Wang X, Li DD, et al. Microbial community and fermentation dynamics of corn silage prepared with heat-resistant lactic acid bacteria in a hot environment. Microorganisms. 2020;8(5):719.

Guo XS, Ke WC, Ding WR, Ding LM, Xu DM, Wang WW, et al. Profiling of metabolome and bacterial community dynamics in ensiled Medicago sativa inoculated without or with *Lactobacillus plantarum* or *Lactobacillus buchneri*. Sci Rep. 2018;8(1):357.

Guo L, Yao DD, Li DX, Lin YL, Bureenok S, Ni KK, et al. Effects of lactic acid bacteria isolated from rumen fluid and faces of dairy cows on fermentation quality, microbial community, and in vitro digestibility of alfalfa silage. Front Microbiol. 2020;10:2998.

Haigh PM, Appletont M, Clench SF. Effect of commercial inoculant and formic acid±formalin silage additives on silage fermentation and intake and on liveweight change of young cattle. Grass Forage Sci. 1987;42(4):405–10.

Hammes W P, Hertel C. Genus I. Lactobacillus Beijerinck 1901. 212AL. In: De Vos P, Garrity GM, Jones D, Krieg NR, Ludwig W, Rainey FA, Schleifer KH, Whitman WB (Eds.), Bergey's Manual of Systematic Bacteriology, vol. 3: The Firmicutes (2nd edition), Springer, New York. 2009; pp. 465-511.

Heinrichs AJ, Buckmaster DR, Lammers BP. Processing, mixing, and particle size reduction of forages for dairy cattle. J Anim Sci. 1999;77(1):180–6.

Hu ZF, Niu HX, Tong Q, Chang J, Yu JH, Li SG, et al. The microbiota dynamics of alfalfa silage during ensiling and after air exposure, and the metabolomics after air exposure are affected by *Lactobacillus casei* and cellulase addition. Front Microbiol. 2020;11:519121.

Huntington GB. Ruminant starch utilization progress has been extensive. Feedstuffs. 1994;66:16–45.

Iwamoto R, Imanaga Y. Direct evidence of the Entner-Doudoroff pathway operating in the metabolism of D-glucosamine in bacteria. J Biochem. 1991;109(1):66–9.
Jones R, Jones DIH. Effect of absorbent on effluent production and silage quality. In: Stark BA, Wilkinson JM, editors. Silage effluent. Marlow: Chalcombe Publications; 1988. p. 47–8.
Jones DIH, Jones R. The effect of in-silo effluent absorbents on effluent production and silage quality. J. Agric. Eng. Res. 1996; 4:173–186.
Jonsson A. Growth of *Clostridium tyrobutyricum* during fermentation and aerobic deterioration of grass silage. J Sci Food Agric. 1991;54(4):557–68.
Kang TW, Adesogan AT, Kim SC, Lee SS. Effects of an esterase-producing inoculant on fermentation, aerobic stability, and neutral detergent fiber digestibility of corn silage. J. Dairy Sci. 2009; 92:732-738.
Keshri J, Chen Y, Pinto R, Kroupitski Y, Weinberg ZG, Sela Saldinger S. Microbiome dynamics during ensiling of corn with and without *Lactobacillus plantarum* inoculant. Appl Microbiol Biotechnol. 2018;102:4025–37.
Knicky M, Spörndly R. The ensiling capability of a mixture of sodium benzoate, potassium sorbate, and sodium nitrite. J Dairy Sci. 2011;94(2):824–31.
Kung L, Robinson JR, Ranjit NK, Chen JH, Golt CM, Pesek JD. Microbial populations, fermentation end-products, and aerobic stability of corn silage treated with ammonia or a propionic acid-based preservative. J Dairy Sci. 2000;83:1479–86.
Lanza I, Lolli V, Segato S, Caligiani A, Contiero B, Lotto A, et al. Use of GC-MS and 1H NMR low-level data fusion as an advanced and comprehensive metabolomic approach to discriminate milk from dairy chains based on different types of forage. Int Dairy J. 2021;123:105174.
Lenehan J, O'Kiely P. Evaluation of mix specification and PFA as a cement replacer in concretes used in silage storage structures. End of project report; 1999.
Li Y, Nishino N. Monitoring the bacterial community of maize silage stored in a bunker silo inoculated with *Enterococcus faecium*, *Lactobacillus plantarum* and *Lactobacillus buchneri*. J Appl Microbiol. 2011;110(6):1561–70.
Liu BY, Yang ZQ, Huan HL, Gu HR, Xu NX, Ding CL. Impact of molasses and microbial inoculants on fermentation quality, aerobic stability, and bacterial and fungal microbiomes of barley silage. Sci Rep. 2020;10(1):1–10.
Lynch JP, Jin L, Lara EC, Baah J, Beauchemin KA. The effect of exogenous fibrolytic enzymes and a ferulic acid esterase-producing inoculant on the fibre degradability, chemical composition and conservation characteristics of alfalfa silage. Anim Feed Sci Technol. 2014;193:21–31.
Lynch JP, Baah J, Beauchemin KA. Conservation, fiber digestibility, and nutritive value of corn harvested at 2 cutting heights and ensiled with fibrolytic enzymes, either alone or with a ferulic acid esterase-producing inoculant. J Dairy Sci. 2015;98(2):1214–24.
Mari LJ, Schmidt RJ, Nussio LG, Hallada CM, Kung L. Short communication: an evaluation of the effectiveness of *Lactobacillus buchneri* 40788 to alter fermentation and improve the aerobic stability of corn silage in farm silos. J Dairy Sci. 2009;92(3):1174–6.
Matthews A, Grbin PR, Jiranek V. A survey of lactic acid bacteria for enzymes of interest to oenology. Aust J Grape Wine Res. 2006;12(3):235–44.
McDonald P, Henderson AR, Heron S. The Biochemistry of Silage. Chalcombe Publications, Marlow, UK, 1981.
McDonald P, Henderson AR, Heron S. The Biochemistry of Silage (Second Edition). Chalcombe Publications, Marlow, Bucks, UK. 1991.
Meeske R. Silage additives: do they make a difference? S Afr J Anim Sci. 2005;6:49–55.
Muck RE. A lactic acid bacteria strain to improve aerobic stability of silages. In: U.S. Dairy Forage Research Center 1996 research summaries. Madison: U.S. Dairy Forage Research Center; 1996. p. 42–3.
Muck RE, Kung Jr L. Effects of silage additives ensiling. In: Proceedings of silage: field to feedbunk. NRAES-99. Ithaca: Natural Resource, Agriculture, and Engineering Service; 1997. pp. 187–99.

Muck RE, Shinners KJ. Conserved forage (silage and hay): progress and priorities. In: Proceedings of the XIX international Grassland Congress: grassland ecosystems: an outlook into the 21st century, 11–21 February 2001, Sao Pedro, Sao Paulo, Brazil; 2001. pp. 753–62.

Nadeau EMG, Buxton DR, Russell JR, Allison MJ, Young JW. Enzyme, bacterial inoculant, and formic acid effects on silage composition of orchardgrass and alfalfa. J Dairy Sci. 2000;83(7):1487–502.

Nsereko VL, Smiley BK, Rutherford WM, Spielbauer A, Forrester KJ, Hettinger GH, et al. Influence of inoculating forage with lactic acid bacterial strains that produce ferulate esterase on ensilage and ruminal degradation of fiber. Anim Feed Sci Technol. 2008;145(1–4):122–35.

O'Dogherty MJ. A review of research on forage chopping. J Agric Eng Res. 1982;27:267–89.

O'Donnell C, Dodd VA, O'Kiely P, Richardson M. A study of the effects of silage effluent on concrete: part 1, significance of concrete characteristics. J Agric Eng Res. 1995;60:83–92.

Ohshima M, McDonald P. A review of the changes in nitrogenous compounds of herbage during ensilage. J Sci Food Agric. 1978;29:497–505.

Oliveira AS, Weinberg ZG, Ogunade IM, Cervantes AAP, Arriola KG, Jiang Y, et al. Meta-analysis of effects of inoculation with homofermentative and facultative heterofermentative lactic acid bacteria on silage fermentation, aerobic stability, and the performance of dairy cows. J Dairy Sci. 2017;100:4587–603.

Oude Elferink SJWH, Krooneman J, Gottschal JC, Spoelstra SF, Faber F, Driehuis F. Anaerobic conversion of lactic acid to acetic acid and 1,2-propanediol by *Lactobacillus buchneri*. Appl Environ Microbiol. 2001;67:125–32.

Owens FN, Zinn RA, Kim YK. Limits to starch digestion in the ruminant small intestine. J Anim Sci. 1986;63:1634–48.

Pahlow G, Muck RE, Driehuis F, Elferink SO, Spoelstra SF. Microbiology of ensiling. Agron Monogr. 2003;42:31–94.

Pang H, Zhang M, Qin G, Tan Z, Li Z, Wang Y, et al. Identification of lactic acid bacteria isolated from corn stovers. Anim Sci J. 2011;82(5):642–53.

Pitt RE, Muck RE, Pickering NB. A model of aerobic fungal growth in silage: 2. aerobic stability. Grass Forage Sci. 1991;46:301–12.

Pot B and Tsakalidou E. Taxonomy and metabolism of Lactobacillus. In: Ljungh A and Wadstrom T (ed), Lactobacillus molecular biology: from genomics to probiotics. Caister Academic Press, Norfolk, United Kingdom. 2009; pp3-58.

Ruppel KA, Pitt RE, Chase LE, Galton DM. Bunker Silo Management and Its Relationship to Forage Preservation on Dairy Farms. J Dairy Sci. 1995; 78(1):141-153.

Savoie P, Jofriet JC. Silage storage. In: Silage science technology. Madison, Wisconsin, USA. 2003; pp. 405–68.

Schmidt RJ, Hu W, Mills JA, L. Kung L. The development of lactic acid bacteria and *Lactobacillus buchneri* and their effects on the fermentation of alfalfa silage. J. Dairy Sci. 2009; 92:5005-5010.

Schmidt RJ, Kung L. The effects of *Lactobacillus buchneri* with or without a homolactic bacterium on the fermentation and aerobic stability of corn silages made at different locations. J Dairy Sci. 2010;93(4):1616–24.

Schnu J. Antifungal lactic acid bacteria as biopreservatives. Trends Food Sci Technol. 2005;16(1–3):70–8.

Shaver RD. Forage Particle Length in dairy rations. In: Proceedings of NRAES dairy feeding systems symposium. Ithaca: NRAES; 1990. p. 58–64.

Sjögren J, Magnusson J, Broberg A, Schnürer J, Kenne L. Antifungal 3-hydroxy fatty acids from *Lactobacillus plantarum* MiLAB 14. Appl Environ Microbiol. 2003;69:7554–7.

Snell HGJ, Oberndorfer C, Lucke W, Van den Weghe HFA. Effects of polyethylene color and thickness on grass silage quality. Grass Forage Sci. 2003;58:239–48.

Spoelstra SF. Chemical and biological additives in forage conservation. In: Landbauforsch. Voelkenrode Sonderheft (Germany, FR); 1991. p. 123.

Stanojevic D, Comic L, Stefanovic O, Solujic S. Antimicrobial effects of sodium benzoate, sodium nitrite and potassium sorbate and their synergistic action in vitro. Bulg J Agric Sci. 2009;15:308–12.

Stevenson DM, Muck RE, Shinners KJ, Weimer PJ. Use of real time PCR to determine population profiles of individual species of lactic acid bacteria in alfalfa silage and stored corn stover. Appl Microbiol Biotechnol. 2006;71(3):329–38.

Su R, Ke WC, Bai J, Wang MS, Usman S, Xie DM, et al. Comprehensive profiling of the metabolome in corn silage inoculated with or without *Lactiplantibacillus plantarum* using different untargeted metabolomics analyses. Arch Anim Nutr. 2023;77:323–41.

Sun MC, Li AL, Huo GC, Meng XC. Progress on the metabolomics of lactic acid bacteria. Microbiol Tongbao. 2012;39(10):1499–505.

Sun J, Wang T, Huang F, Liu Y, Shi W, Ma C. Silage fermentation: a potential microbial approach for the forage utilization of *Cyperus esculentus* L. by-product. Fermentation. 2021;7(4):273.

Tabacco E, Piano S, Revello-Chion A, Borreani G. Effect of *Lactobacillus buchneri* LN4637 and *Lactobacillus buchneri* LN40177 on the aerobic stability, fermentation products, and microbial populations of corn silage under farm conditions. J Dairy Sci. 2011;94(11):5589–98.

Teller RS, Schmidt RJ, Whitlow LW, Kung L. Effect of physical damage to ears of corn before harvest and treatment with various additives on the concentration of mycotoxins, silage fermentation, and aerobic stability of corn silage. J Dairy Sci. 2012;95:1428–36.

Tetlow RM. Pre-harvest desiccation of crops for conservation: 2. Effect of level of formic acid applied and the volume of solution used on the moisture concentration and chemical composition of a perennial ryegrass sward. Grass Forage Sci. 1983;38:27–32.

Thomas ME, Foster JL, McCuistion KC, Redmon LA, Jessup RW. Nutritive value, fermentation characteristics, and in situ disappearance kinetics of sorghum silage treated with inoculants. J Dairy Sci. 2013;96:7120–31.

Wang Y, Gao Y, Wang X, Lin Y, Xu G, Yang F, et al. Insights into the phage community structure and potential function in silage fermentation. J Environ Manag. 2024;358:120837.

Watson SJ, Nash MJ. The conservation of grass and forage crops. Soil Sci. 1961;92:149.

Weinberg ZG, Muck RE. New trends and opportunities in the development and use of inoculants for silage. FEMS Microbiol Rev. 1996;19:53–68.

Weinberg ZG, Ashbell G, Azrieli A. The effect of applying lactic acid bacteria at ensilage on the chemical and microbiological composition of vetch, wheat and alfalfa silages. J Appl Bacteriol. 1988;64:1–7.

Wilkins R, Wilkinson M. Major contributions in 45 years of international silage conferences. In: Proceedings of XVII international silage conference, Piracicaba, Brazil. Piracicaba: University of São Paulo; 2015. p. 26–51.

Wilkinson JM, Bolsen KK, Lin CJ. History of silage. In: Silage science and technology. Madison: American Society of Agronomy; 2003. p. 1–30.

Windle MC, Walker N, Kung L. Effects of an exogenous protease on the fermentation and nutritive value of corn silage harvested at different dry matter contents and ensiled for various lengths of time. J Dairy Sci. 2014;97:3053–60.

Wu Z, Luo Y, Bao J, Luo Y, Yu Z. Additives affect the distribution of metabolic profile, microbial communities and antibiotic resistance genes in high-moisture sweet corn kernel silage. Bioresour Technol. 2020;315:123821.

Xia T, Wang T, Sun J, Shi W, Liu Y, Huang F, et al. Modulation of fermentation quality and metabolome in co-ensiling of sesbania cannabina and sweet sorghum by lactic acid bacterial inoculants. Front Microbiol. 2022;13:851271.

Xu D, Ding Z, Wang M, Bai J, Ke W, Zhang Y, et al. Characterization of the microbial community, metabolome and biotransformation of phenolic compounds of sainfoin (*Onobrychis viciifolia*) silage ensiled with or without inoculation of *Lactobacillus plantarum*. Bioresour Technol. 2020;316:123910.

Xu D, Wang N, Rinne M, Ke W, Weinberg ZG, Da M, et al. The bacterial community and metabolome dynamics and their interactions modulate fermentation process of whole crop corn silage prepared with or without inoculants. Microb Biotechnol. 2021;14:561–76.

Yang F, Wang Y, Zhao S, Wang Y. *Lactobacillus plantarum* inoculants delay spoilage of high moisture alfalfa silages by regulating bacterial community composition. Front Microbiol. 2020;11:1989.

Young KM, Lim JM, Der Bedrosian MC, Kung L. Effect of exogenous protease enzymes on the fermentation and nutritive value of corn silage. J Dairy Sci. 2012;95:6687–94.

Zhang Q, Guo X, Zheng MY, Chen DK, Chen XY. Altering microbial communities: a possible way of lactic acid bacteria inoculants changing smell of silage. Anim Feed Sci Technol. 2021;279:114998.

Zhang X, Ding Z, Usman S, Zhang J, Chen M, Guo X. Metagenomics insights into the effects of lactic acid bacteria inoculation on the biological reduction of antibiotic resistance genes in alfalfa silage. J Hazard Mater. 2023;443:130329.

Zi XJ, Liu Y, Chen T, Li M, Zhou HL, Tang J. Effects of sucrose, glucose and molasses on fermentation quality and bacterial community of stylo silage. Fermentation. 2022;8(5):191.

Zielińska K, Fabiszewska A, Świątek M, Szymanowska-Powałowska D. Evaluation of the ability to metabolize 1,2-propanediol by heterofermentative bacteria of the genus Lactobacillus. Electron J Biotechn. 2017; 26:60-63.

Zou X, Chen D, Lv H, Zhang Q, Zheng P. Effect of ellagic acid on fermentation quality and bacterial community of stylo silage. Fermentation. 2021;7(4):1–12.

Chapter 3
Principle of Silage Fermentation

Zhongfang Tan, Huili Pang, and Yimin Cai

Abstract Silage is a kind of livestock storage feed, and its fermentation principle is through the fermentation of lactic acid bacteria, which produces lactic acid, lowers the pH value, and inhibits spoilage bacteria. Silage can be stored for a long time, reducing the loss of nutrients during the process and enhancing the stability of feed quality. By controlling the fermentation process, the growth of miscellaneous bacteria can be inhibited, and mildew and rot can be prevented. This chapter focuses on the principles of fermentation and storage related to changes in microbial population structure and metabolites during silage fermentation.

Keywords Fermentation principle · Fermentation stage · Organic acid · Silage

Silage is a stored feed that is prepared and processed using fresh forage crop, grass, and agricultural by-product resources as material. Silage fermentation is a dynamic process accompanied by microbiological activity and biochemical reaction. At the beginning of silage fermentation, organic matter is oxidized and decomposed through plant cell respiration, producing carbon dioxide, water, and heat (Raza et al. 2023; Paul 2016). After ensiling, the continuous decrease of oxygen and increase of carbon dioxide in the silo environment leads to the rapid demise of aerobic microorganisms that are harmful to silage fermentation. Lactic acid bacteria (LAB) can transfer the water-soluble carbohydrate (WSC) in the silage material into lactic acid, forming an acidic anaerobic environment, inhibiting the growth of harmful microorganisms, and achieving the purpose of preserving feed (Ávila and Carvalho 2020; Dunière et al. 2013; Gouhis 2017; Kung et al. 2018; Du et al. 2023). At present, silage fermentation technology has been widely promoted and applied in livestock production in the world, and silage can play an increasingly important role in the effective use of agricultural by-product resources, alleviate the shortage of feed and promote the sustainable production of livestock.

Z. Tan (✉) · H. Pang · Y. Cai
Zhengzhou University, Zhengzhou, China
e-mail: tzhongfang@zzu.edu.cn; pang@zzu.edu.cn; caiy0234@jircas.go.jp

Y. Cai, K. Ataku (eds.), *Cultural History and Modern Production Technology of Silage*, https://doi.org/10.1007/978-981-96-5787-2_3

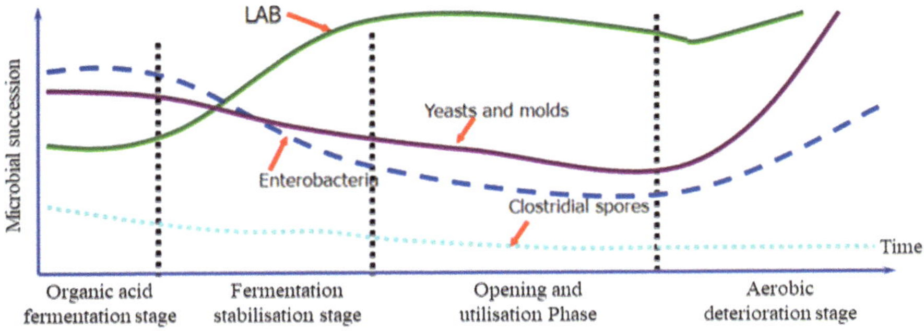

Fig. 3.1 Changes in microbial community during ensiling

The silage fermentation process can be generally divided into the following stages, i.e., the early stage of fermentation, the lactic or butyric acid fermentation stage, the fermentation stabilization stage, and the feeding stage (Fig. 3.1) (Okoye et al. 2023; McDonald 1982).

Early stage of fermentation: after the silage material is chopped, filled and sealed, the plant cells of the material make use of the residual oxygen in the silo to carry out respiratory metabolism, oxidizing and decomposing WSC and producing carbon dioxide, water, and heat. This process continues until the silo is depleted of oxygen and anaerobic conditions develop.

Excessive residual oxygen in the silo prolongs the respiration period of the plant cells and consumes a large amount of glycogen, which not only affects lactic acid fermentation but also leads to higher temperatures in the silo and increased losses of various nutrients. For this reason, compaction and air removal from the material during ensiling are very important to promote lactic acid fermentation and reduce nutrient losses.

At the early stage of fermentation, undesirable microorganisms such as aerobic bacteria, coliform bacteria, yeast, and mold presented to the material multiply rapidly using the juice discharged from the material subjected to mechanical pressing (Caplice and Fitzgerald 1999). These aerobic bacteria decompose the sugars and proteins in the silage, forming indoles, gases, and acetic acid, etc. Microbial activity, plant cell respiration, and enzymatic degradation consume oxygen in the interstitial spaces of the material and release a certain amount of heat, thus forming an anaerobic and warm environment, which creates favorable conditions for the proliferation of LAB.

The first stage ends when the anaerobic environment is reached. If the silage material is favorable and well-managed, the process can be completed in as little as 24 h. If the sugar and moisture of the material are not favorable and not well-managed, the process can last for several days or weeks. The principle of silage fermentation is, therefore, to minimize the duration of the first stage of fermentation and to move as quickly as possible to the lactic acid fermentation stage.

1 Organic Acid Fermentation Stage

This stage can be divided into the lactic acid fermentation stage and the butyric acid fermentation stage.

Lactic acid fermentation: after the silo forms an anaerobic state, the material begins to undergo anaerobic respiration and lactic acid fermentation. Plant intermolecular respiration mainly consumes oxygen under the action of intracellular enzymes, produces carbon dioxide, water, and organic acids and releases heat at the same time. During this phase, LAB become the dominant population and grow rapidly, utilizing the WSC in the material to produce large amounts of lactic acid, which rapidly lowers the pH, leading to the inhibition or death of spoilage bacteria, clostridia, and mold. As shown in the molecular formula below, the LAB related to silage fermentation can be divided into two major types: homofermentation and heterofermentation (Du et al. 2023).

Homofermentation by LAB breaks down one molecule of the glucose into one molecule of lactic acid. Heterofermentation is to decompose glucose into lactic acid, ethanol and carbon dioxide, fructose into lactic acid, mannitol, acetic acid and carbon dioxide, and five-carbon sugar into lactic acid and acetic acid (Fidan et al. 2022). The LAB associated with silage fermentation are cocci of the genera *Streptococcus*, *Pediococcus*, *Enterococcus*, and *Lactococcus* and rod of the genus *Lactobacillus.* These cocci producing lactic acid can grow vigorously in the early stage of fermentation and form the dominant population, and with the decrease of pH value, its ability to grow is gradually weakened. Thereafter, acid-resistant *lactobacilli* multiplies and dominates, further producing lactic acid and enabling long-term preservation of the silage.

2 Butyric Acid Fermentation

If the forage material has high moisture and insufficient WSC, LAB cannot produce enough lactic acid to lower the pH value and inhibit the harmful bacteria at this stage, and *clostridia* will *proliferative* and lead to butyric acid fermentation (Bangar et al. 2022; Coelho et al. 2022; Cotter and Hill 2003; Okoye et al. 2023). *Clostridia* can decompose sugar and lactic acid to produce butyric acid and also decompose protein into ammonia and biogenic amine, thereby forming a bad odor, resulting in the loss of nutrients in the feed and the decline of palatability of livestock, and ultimately reducing the quality of silage fermentation. In this process, propionic acid bacteria will produce part of the propionic acid. Therefore, this poor-quality fermentation will contain some propionic acid and butyric acid in addition to high levels of ammonia nitrogen.

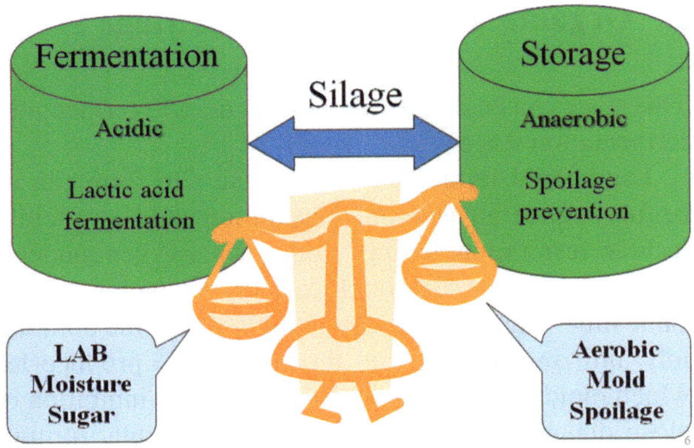

Fig. 3.2 Basic principle of silage fermentation

3 Fermentation Stabilization Stage

Silage through organic acid fermentation, the formation of an acidic and anaerobic double duress silage environment, the propagation of harmful microorganisms is inhibited, silage fermentation into a relatively stable state (Fig. 3.2). Under ideal fermentation conditions, lactic acid generation can generally reach more than 1.0% of the material, and if the moisture is 60%, i.e., the lactic acid content is equivalent to more than 6.0% of the dry matter (DM), the pH value can be reduced to less than 4.2, and the silage can be preserved for a long time (Kung et al. 2018; Okoye et al. 2023). This stage requires that air cannot enter the silo, maintaining a strict anaerobic state, and silage production can generally be opened for use after the stabilization stage, theoretically the silage maturity period takes about 1 month, if the external conditions have not changed, it can be preserved for a long and stable period. At this stage, the microbial diversity decreases sharply and LAB evolve into the dominant microbial population, and with the accumulation of lactic acid to a certain content, the diversity of LAB also decreases gradually. The dominant populations in mature silage are acid-tolerant LAB and yeast, with small amounts of *Bacillus* and *Clostridium* frequently detected.

4 Opening and Utilization Phase

Once the silage has matured and is ready to be fed to livestock, it is ready for the opening and utilization phase. When opening the silo, the condition of the silage fermentation must be carefully checked, usually the surface of the silage is prone to mold, and the moldy part needs to be removed in time before utilization. After taking the required silage amount from one side of the silo according to the daily livestock feeding quantity, promptly cover it with plastic film to maintain good sealing condition.

Fig. 3.3 Aerobic deterioration and storage losses of silage due to damage to the wrapping film in the farm. (Photos by Cai)

5 Aerobic Deterioration Stage

After silage is opened, if it is not managed properly, it will cause aerobic spoilage. This stage is due to the fact that when the silage is opened and exposed to the air, aerobic bacteria and yeasts begin to multiply, degrading organic acid and increasing the pH value, leading to a proliferation of mold. The DM loss of silage due to aerobic deterioration can be more than 5% (Borreani et al. 2018; Liu et al. 2013). Aerobic deterioration will not only reduce the nutritional value of feed and palatability of livestock but some pathogenic spoilage bacteria will also affect feed safety and livestock health (Fig. 3.3).

The aerobic spoilage phase of silage fermentation refers to a series of processes of silage spoilage and deterioration caused by oxygen entering the cellar during silage fermentation. Silage can be stabilized for a long time by anaerobic fermentation under good sealing conditions (Gouhis 2017). However, when silage is exposed to air, especially after poor sealing or opening, oxygen enters the cellar and causes aerobic microorganisms to multiply, leading to aerobic deterioration. The stages of aerobic deterioration are as follows.

6 Proliferation of Aerobic Microorganisms

The first step of aerobic deterioration is after oxygen enters the silage, aerobic microorganisms (yeasts, molds, and aerobic bacteria, etc.) grow rapidly in an aerobic environment and metabolize using the nutrients in the silage.

6.1 Yeast

Yeasts are abundant in high-quality silage and, in the presence of oxygen, utilize the soluble sugars in the silage to carry out respiration, producing carbon dioxide and water, thereby leading to a loss of DM and energy from the feed (Borreani et al.

2018). Yeast reproduction also consumes lactic acid in silage, destroying the acidic environment in silage and gradually increasing the pH value, which in turn creates conditions for the growth of other spoilage microorganisms.

6.2 Mold

Molds, which mainly include species such as *Aspergillus* and *Penicillium*, begin to grow when silage is exposed to air (Chen et al. 2015). Molds are capable of breaking down large organic molecules, including cellulose and proteins, to produce mycotoxins, which pose a risk to feed safety and animal health. In addition, the growth of molds and yeasts releases heat, causing an increase in silage temperature and accelerating the aerobic spoilage process.

6.3 Aerobic Bacteria

Aerobic bacteria such as *Acetobacter* can oxidize the ethanol in the feed to produce acetic acid, which further decomposes and destroys the nutrients of silage (Gomes et al. 2018; He et al. 2022). This decomposition not only reduces the nutritional value of silage but also produces a rancid odor and reduces the palatability of feed.

7 Decomposition of Nutrient

While aerobic microorganisms grow in silage, they also begin to decompose nutrients such as carbohydrates, proteins, and fats in the feed (Dunière et al. 2013).

7.1 Decomposition of Carbohydrates

Carbohydrates in the material are the fermentation substrate for LAB but under aerobic conditions. Yeast and other aerobic microorganisms will use these carbohydrates to carry out the respiratory metabolism and produce carbon dioxide and water (Abedi and Hashemi 2020). In particular, water-soluble sugars (e.g., glucose, fructose) are rapidly decomposed in the presence of oxygen, resulting in energy loss from the feed.

7.2 Protein Breakdown

Aerobic microorganisms, especially molds, use proteins as a source of nitrogen, breaking them down into amino acids and small peptides, which are further converted into ammonia nitrogen and other nitrogenous wastes (Adeniyi et al. 2023; Jain et al. 2022). The production of ammonia nitrogen during this process can make silage smell bad and, at the same time, tend to raise the silage pH, making it suitable for the growth of spoilage bacteria (Kung et al. 2018; Ward 2000; Kung and Shaver 2001). In addition, the breakdown of proteins leads to a decrease in the nitrogen content of the silage, thus reducing its feeding value.

7.3 Decomposition of Fat

Fat is easily oxidized under aerobic conditions, especially polyunsaturated fatty acids, which will produce peroxides and free radicals during oxidation and eventually generate volatile substances such as aldehydes and ketones. These oxidation products have a strong odor, which makes silage palatability and feed intake decrease, and even adversely affects the health of animals.

8 Oxidation of Organic Acids and pH Rise

Once the anaerobic environment of silage fermentation is damaged, yeasts and molds are able to break down lactic acid for respiratory metabolism under aerobic conditions, leading to a gradual decrease in the acidity of silage and a rise in pH. Once the pH value rises to a certain critical point, other aerobic spoilage microorganisms, such as *coliform* bacteria, will start to grow, which will further aggravate the spoilage of the feed.

9 Secondary Heat Phenomenon

During aerobic spoilage of silage, the metabolic activity of aerobic microorganisms releases a large amount of heat, thereby leading to an increase in silage temperature. This phenomenon is called "secondary fermentation or secondary heating" and is a distinctive feature of aerobic spoilage. Elevated silage temperatures not only accelerate the metabolic rate of aerobic microorganisms but also cause damage to some heat-sensitive nutrients. The high-temperature environment also accelerates enzymatic reactions, increasing the rate of decomposition of carbohydrates, proteins, and fats in the silage, thereby leading to further nutrient losses. In addition, elevated temperatures may inactivate sensitive components such as vitamins and minerals in silage, reducing the nutritional value of the feed (Vu et al. 2019).

10 Mycotoxin Production

While molds multiply under aerobic conditions, they produce various *mycotoxins*, such as *aflatoxins* and *ochratoxins*. These *mycotoxins* are highly toxic and can be harmful to livestock even at low concentrations, e.g., triggering problems such as liver damage and immunosuppression. The accumulation of *mycotoxins* did not only affect the palatability of the feed but also has a negative impact on the growth and health of the animals and may even endanger human health through the food chain (Wang et al. 2024).

11 Decrease in the Organoleptic Quality of the Feed

Various metabolites, such as ammonia, aldehydes, ketones, and *mycotoxins* produced during the aerobic spoilage process of silage fermentation, can cause the feed to produce a pungent odor, and the color of the feed gradually changes from lime green to brown or even black (Kung et al. 2018). In addition, the secondary heat will also cause the texture of the feed to become soft and loose in structure, which will further affect its palatability and feeding value. Livestock will refuse to eat or eat less of deteriorated silage, seriously affecting the efficiency of feed utilization and the productivity of livestock.

12 Aerobic Spoilage Prevention and Control Measures

12.1 Sealing Management

There is a need to ensure that silage is well sealed during fermentation and storage so as to effectively prevent the propagation of aerobic microorganisms.

12.2 Feed Compaction

During the process of silage preparation, the raw materials should be compacted as much as possible to improve the density in the silo and reduce the air content in the raw materials.

12.3 Add Organic Acid Product

Some organic acids can inhibit the growth of aerobic microorganisms and slow down the aerobic spoilage process. For example, an appropriate amount of acetic acid, propionic acid, etc. can prolong the storage time of silage (Du et al. 2023).

12.4 Regular Checking and Timely Sealing

In the storage process of silage, the sealing should be checked regularly to avoid the entry of outside air. At the same time, the silage should be removed as soon as possible each time and then promptly sealed to avoid prolonged exposure to air.

Aerobic spoilage of silage is a complex biochemical process involving the propagation and metabolic activities of a variety of aerobic microorganisms (Shuo et al. 2024). In the presence of oxygen, these microorganisms consume the carbohydrates, proteins, and fats in the silage, thereby producing spoilage products such as carbon dioxide, water, heat, and *mycotoxins*.

Therefore, all aspects of silage fermentation and storage are closely related, and understanding the principles of silage fermentation and adopting a scientific approach to production are the keys to reducing silage fermentation losses and improving livestock productivity.

References

Abedi E, Hashemi SMB. Lactic acid production–producing microorganisms and substrates sources-state of art. Heliyon. 2020;6(10):e04974.

Adeniyi A, Bello I, Mukaila T, Sarker NC, Hammed A. Trends in biological ammonia production. Biotech. 2023;12(2):41.

Ávila CLS, Carvalho BF. Silage fermentation—updates focusing on the performance of microorganisms. J Appl Microbiol. 2020;128(4):966–84.

Bangar SP, Suri S, Trif M, Ozogul F. Organic acids production from lactic acid bacteria: a preservation approach. Food Biosci. 2022;46:101615.

Borreani G, Tabacco E, Schmidt RJ, Holmes BJ, Muck RE. Silage review: factors affecting dry matter and quality losses in silages. J Dairy Sci. 2018;101(5):3952–79.

Caplice E, Fitzgerald GF. Food fermentations: role of microorganisms in food production and preservation. Int J Food Microbiol. 1999;50(1–2):131–49.

Chen SCA, Sorrell TC, Meyer W. *Aspergillus* and *Penicillium*. In: Manual of clinical microbiology. Washington, DC: ASM Press; 2015. p. 2030–56.

Coelho MC, Malcata FX, Silva CCG. Lactic acid bacteria in raw-milk cheeses: from starter cultures to probiotic functions. Food Secur. 2022;11(15):2276.

Cotter PD, Hill C. Surviving the acid test: responses of gram-positive bacteria to low pH. Microbiol Mol Biol Rev. 2003;67(3):429–53.

Du Z, Yamasaki S, Oya T, Cai Y. Cellulase–lactic acid bacteria synergy action regulates silage fermentation of woody plant. Biotechnol Biofuels Bioprod. 2023;16(1):125.

Dunière L, Sindou J, Chaucheyras-Durand F, Chevallier I, Thévenot-Sergentet D. Silage processing and strategies to prevent persistence of undesirable microorganisms. Anim Feed Sci Technol. 2013;182(1–4):1–15.
Fidan H, Esatbeyoglu T, Simat V, Trif M, Tabanelli G, Kostka T, et al. Recent developments of lactic acid bacteria and their metabolites on foodborne pathogens and spoilage bacteria: facts and gaps. Food Biosci. 2022;47:101741.
Gomes RJ, de Fatima Borges M, de Freitas Rosa M, Castro-Gómez RJH, Spinosa WA. Acetic acid bacteria in the food industry: systematics, characteristics and applications. Food Technol Biotechnol. 2018;56(2):139.
Gouhis F, Moujahed N, Nefzaoui A, Louhaichi M. Successful silage: keeping forages for a long time while maintaining good quality (French). 2017, 10: 1–2.
He Y, Xie Z, Zhang H, Liebl W, Toyama H, Chen F. Oxidative fermentation of acetic acid bacteria and its products. Front Microbiol. 2022;13:879246.
Jain P, Dhanker R, Bhardwaj A, Singhla G, Saxena K, Hussain T. Aerobic and anaerobic ammonia-oxidizing bacteria: a resilient challenger or innate collaborator. In: Development in wastewater treatment research and processes. Amsterdam: Elsevier; 2022. p. 195–222.
Kung L Jr, Shaver RD, Grant RJ, Schmidt RJ. Silage review: interpretation of chemical, microbial, and organoleptic components of silages. J Dairy Sci. 2018;101(5):4020–33.
Kung L, Shaver R. Interpretation and use of silage fermentation analysis reports. Focus Forage. 2001;3(13):1–5.
Liu QH, Shao T, Zhang JG. Determination of aerobic deterioration of corn stalk silage caused by aerobic bacteria. Anim Feed Sci Technol. 2013;183(3–4):124–31.
McDonald P. Silage fermentation. Trends Biochem Sci. 1982;7(5):164–6.
Okoye CO, Wang Y, Gao L, Wu Y, Li X, Sun J, et al. The performance of lactic acid bacteria in silage production: a review of modern biotechnology for silage improvement. Microbiol Res. 2023;266:127212.
Paul EA. The nature and dynamics of soil organic matter: plant inputs, microbial transformations, and organic matter stabilization. Soil Biol Biochem. 2016;98:109–26.
Raza T, Qadir MF, Khan KS, Eash NS, Yousuf M, Manzoor R, et al. Unrevealing the potential of microbes in decomposition of organic matter and release of carbon in the ecosystem. J Environ Manag. 2023;344:118529.
Shuo W, Peishan H, Chao Z, Wei Z, Xiaoyang C, Qing Z. Novel strategy to understand the aerobic deterioration of corn silage and the influence of *Neolamarckia cadamba* essential oil by multi-omics analysis. Chem Eng J. 2024;482:148715.
Vu VH, Li X, Wang M, Liu R, Zhang G, Liu W, et al. Dynamics of fungal community during silage fermentation of elephant grass (*Pennisetum purpureum*) produced in northern Vietnam. Anim Biosci. 2019;32(7):996.
Wang X, He L, Ullah H, Shi X, Hou J, Liu Y, et al. Combined effects of mixing ratios and tree size: how do mixed forests respond to climate and drought events? Front Plant Sci. 2024;15:1477640.
Ward RT. Fermentation analysis: use and interpretation. Tri-state dairy nutrition conference, Fort Wayne, Indiana, USA; 2000. pp. 117–35.

Chapter 4
Microbial Network and Fermentation Information of Silage

Zhumei Du and Yimin Cai

Abstract The ensiling microbial network plays a central role in the fermentation process of silage, and the dynamic changes and interactions between microorganisms directly determine the fermentation quality and feed stability. During ensiling, the fermentation network led by lactic acid bacteria rapidly reduces pH through metabolic coordination and inhibits the growth of undesirable microorganisms. However, potentially competitive microorganisms such as yeasts, molds, and clostridia may disturb the balance of microbial networks and lead to abnormal fermentation. The structure of microbial community network is affected by the regulation of material characteristics, environmental conditions, and exogenous additives such as lactic acid bacteria inoculants, and the functional correlation and metabolic pathways among different microorganisms constitute a complex ecological network. The relationship between microbial network and fermentation quality was analyzed by multi-omics technique, which provided a new perspective for precise regulation of microbial community. This not only helps to optimize the silage process and improve feed quality but also lays a theoretical foundation for the in-depth study of fermentation micro ecology.

Keywords Ensiling process · Fermentation information · Microbial community · Regulatory mechanism · Symbiotic network

The silage symbiotic microbial network is a complex network system formed by the interactions between microorganisms presented to forage crops and grasses, which is of great significance for silage fermentation regulation and storage stability (Du et al. 2024a). During ensiling, various types of microorganisms form a dynamic and

Z. Du (✉)
Yangzhou University, Yangzhou, China

Y. Cai
Zhengzhou University, Zhengzhou, China

Japan International Research Center for Agricultural Sciences, Tsukuba, Japan
e-mail: caiy0234@jircas.go.jp

Y. Cai, K. Ataku (eds.), *Cultural History and Modern Production Technology of Silage*, https://doi.org/10.1007/978-981-96-5787-2_4

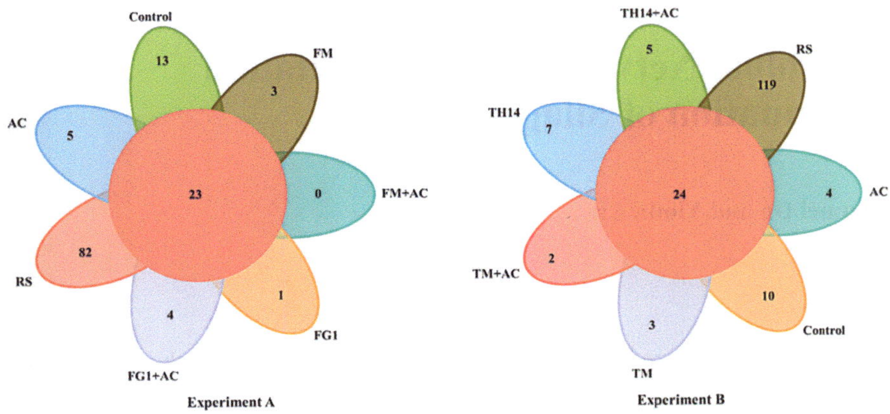

Fig. 4.1 Abundance and diversity of microorganisms in rice straw silage. (Data and Fig. from Du)

self-regulating symbiotic network through their interaction, information exchange, and metabolic activities (Liu et al. 2019). The silage microbial network system is an ecological network of interactions between different microbial community (e.g., lactic acid bacteria (LAB), *clostridia*, yeasts, molds, etc.) during silage fermentation (Li et al. 2024). In this network system, all kinds of microorganisms act together on silage material through symbiosis, competition, synergy, and other relationships to complete the fermentation process. The interrelationship of microorganisms determines the stability and fermentation efficiency of the system and affects the quality of silage (Borreani et al. 2018). As shown in Fig. 4.1, the microorganisms in this network system are diverse, and the microorganisms in this network system communicate through metabolites and information molecules to establish a dynamic ecological balance to adapt and maintain the stability of the silage environment.

1 The Main Microbial Community of the Symbiotic Network System of Silage Fermentation

1.1 Lactobacilli

LAB, such as *Lactobacillus acidophilus* and *Lactiplantibacillus plantarum*, form dominant populations in the co-culture network system of biological preservation. As shown in Fig. 4.2, before silage, the main bacterial species in the rice straw raw material is aerobic bacteria. After silage, LAB become the dominant population and dominate the silage fermentation.

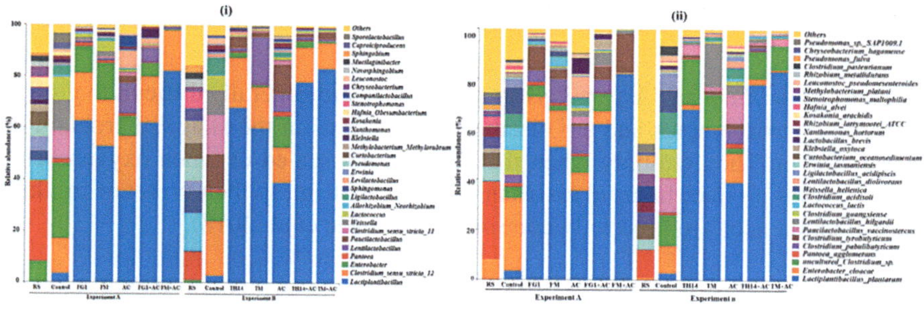

Fig. 4.2 Microbial community structure of rice straw silage. (Data and Fig. from Du)

1.2 *Enterobacter*

The *Enterobater* species are harmful to silage fermentation and can break down carbohydrates into carbon dioxide and water through the carbohydrate metabolic pathway, increasing dry matter (DM) and energy losses (Okoye et al. 2023; Kung et al. 2018; Du et al. 2021). These bacteria compete with LAB for nutrients and act antagonistically with them in the biosymbiotic network system of silage.

1.3 *Clostridia*

Clostridium species responds positively to the high moisture and weakly acidic environment of silage, proliferating and forming a network of dominant populations (Wells and Wilkins 1996). It breaks down proteins while competing with LAB for fermentation substrates, thus degrading amino acids to ammonia nitrogen, leading to nutrient losses. *Clostridia* can ferment into poor-quality silage, negatively affecting palatability and feed safety for livestock.

1.4 *Yeast*

They produce ethanol and carbon dioxide by fermenting sugars during silage fermentation and sometimes create a symbiotic relationship with LAB, creating an anaerobic environment suitable for their growth (Yang et al. 2007).

1.5 Mold

Such as *Penicillium* and *Aspergillus* species, which are generally difficult to grow in anaerobic environments and are disadvantaged strains in the silage microbial network system. They multiply when the silage is not sealed or when oxygen penetrates. Molds produce harmful toxins in their metabolism, so their presence in silage fermentation needs to be strictly controlled (Ogunade et al. 2018).

1.6 Other Microorganisms

These include propionic acid bacteria, acetic acid bacteria, etc., which play different roles at different stages of the microbial network system, and some microbial communities play the role of inhibitors when conditions are unfavorable in order to maintain the balance of the overall network (Ushijima et al. 1984).

2 Microbial Interrelationships

Microbial interrelationships in the silage microbial symbiotic network are manifested in the following ways.

2.1 Synergistic

LAB can form synergistic relationships with, for example, yeasts. The metabolism of LAB lowers the pH, making the environment more suitable for the survival of certain acid-tolerant yeasts (Badotti et al. 2012). At the same time, the carbon dioxide produced by yeast in their metabolism helps maintain an anaerobic environment, preventing the growth of unfavorable microorganisms.

2.2 Competition

Competition exists between different LAB species, between *lactobacilli* and *clostridia*, and between *lactobacilli* and mold. *Lactobacilli* inhibit the growth of *clostridia* and mold by rapidly metabolizing sugars and producing acid to lower the environmental pH.

2.3 Parasitism and Inhibition

Under certain aerobic conditions, molds and bacteria may parasitize or inhibit other microorganisms. For example, following oxygen penetration, certain molds can invade silage and affect the ability of LAB to reproduce; certain aerobic bacteria parasitize silage, multiply vigorously, and inhibit the growth of LAB (Badotti et al. 2012).

3 Microbial Metabolism and Symbiotic Network Regulation

In the symbiotic microbial network system of silage fermentation, microbial metabolism is the core mechanism to achieve the dynamic regulation of the network. Through their own metabolic activities, microorganisms continuously produce metabolites, thus changing the physicochemical properties of the silage environment, providing or restricting the growth conditions for different microbial populations, and ultimately realizing dynamic equilibrium. Figure 4.3 shows the symbiotic network of microorganisms in rice straw silage, such as *enterococci, lactobacocci, weissella*, and other LAB, can rapidly multiply under the weak oxygen conditions at the early stage of silage fermentation to produce lactic acid and antimicrobial substances and inhibit aerobic microorganisms, which creates a favorable

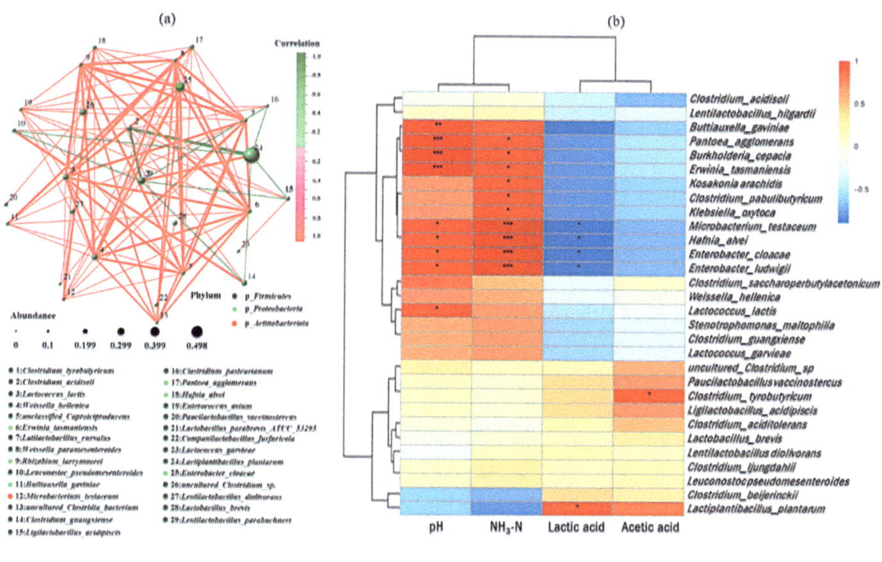

Fig. 4.3 The microbial co-occurrence network (**a**) and correlation analyses between bacterial community and terminal fermentation products (**b**) at species level in paper mulberry silage. (Data and Fig. from Du)

environment for lactobacilli to grow. These lactic acid-producing cocci gradually died out with the change of anaerobic and acidic environmental conditions, and then *Lactobacillus* species responded to the change of silage environmental conditions and formed a symbiotic network with LAB as the dominant community (Fig. 4.3a) (Du et al. 2024a, b). In this network, *lactobacilli* produce lactic acid and reduce pH and ammonia nitrogen content (Fig. 4.3b).

3.1 Regulation of Silage Environment by Metabolites

During ensiling, microorganisms change the conditions of the environment mainly through acid production, oxygen consumption, and decomposition of macromolecules, so as to adapt to or limit the growth of specific microorganisms.

3.1.1 Regulation of Lactic Acid Production

LAB are the main microbial population in silage fermentation and are able to rapidly metabolize sugars and produce lactic acid in the early stages of silage fermentation (Du et al. 2023). This lactic acid significantly lowers the pH of the environment, making it more acidic and thus inhibiting the growth of most anaerobic bacteria and molds, creating an environment suitable for the growth of the lactic acid bacteria themselves (Badotti et al. 2012).

3.1.2 Regulation of the Oxygen Consumption Process

In the early stages of fermentation, the presence of trace amounts of oxygen promotes the growth of yeast, which rapidly consumes oxygen, creating a completely anaerobic environment (Badotti et al. 2012). This process effectively prevents the growth of molds and other aerobic bacteria that require oxygen (Du et al. 2024a, b).

3.1.3 Acetic Acid and Ethanol Production

As the silage fermentation progresses, the yeasts gradually enter a metabolic phase, producing ethanol and carbon dioxide through ethanol fermentation (Ávila and Carvalho 2020). Ethanol and acetic acid have a bacteriostatic effect to a certain extent, which helps to stabilize the microbial community of the silage and prevent the proliferation of spoilage bacteria and other harmful bacteria (Hua et al. 2024).

3.2 *Information Exchange Between Microbial Populations*

The microbial network system during silage fermentation is not a simple coexistence relationship but is regulated and balanced through the transfer of information molecules (Du et al. 2022).

3.2.1 Information Transfer of Acidic Environment

Acid production by LAB causes a rapid decrease in environmental pH, and this change triggers other acid-tolerant microorganisms (e.g., yeast) to enter a stable growth state, while pH-sensitive microorganisms are inhibited (Adamberg et al. 2003).

3.2.2 Signaling by Metabolites

Ethanol, a metabolite of yeast, acts as an information molecule that influences the growth rate of surrounding LAB, causing their metabolic activity to diminish and maintaining the equilibrium of the microbial community.

3.2.3 Oxygen Concentration Signaling

A decrease in oxygen levels leads to a transition from aerobic to anaerobic silage environments, a message that encourages the rapid multiplication of anaerobic LAB while inhibiting the growth of aerobic bacteria and molds (Du et al. 2023).

3.3 *Regulatory Mechanisms of Microbial Metabolism and Symbiotic Networks*

During ensiling, microorganisms have a "dialogue" with the environment through metabolic activities so that the silage ecosystem can adapt to environmental changes and self-regulation.

3.3.1 Rapid Response Mechanism

When the silage material is sealed, LAB in the early stage of fermentation will rapidly start the metabolism of sugar, produce lactic acid, and reduce the pH value (Badotti et al. 2012). This rapid response mechanism allows the environment to rapidly evolve toward conditions suitable for anaerobic growth, avoiding the growth of spoilage microorganisms.

3.3.2 Self-Sustaining Mechanism

With the accumulation of lactic acid, the growth of LAB slows down when the pH drops to a certain level, and other acid-tolerant microorganisms, such as yeasts and acetic acid bacteria, begin to occupy a certain ecological niche (Badotti et al. 2012). The interdependence of different microbial communities forms a stable network that prevents the overpopulation of a single microorganism.

3.3.3 Feedback Mechanism

During the fermentation process, microbial populations automatically adjust their population density and metabolic activities in response to oxygen infiltration or fluctuations in environmental conditions (Giraffa 2004; Guan et al. 2017). For example, the brief infiltration of oxygen will promote the activity of yeast, which will rapidly consume oxygen and restore the anaerobic environment, maintaining the dynamic balance of the silage.

4 Application Value of Microbial Symbiotic Network

By studying the microbial metabolism of silage fermentation and the dynamic regulation mechanism of the symbiotic network, the silage fermentation process can be optimized, and the feed quality can be improved.

4.1 Optimization of Strain Combination

The introduction of specific strain combinations in silage fermentation can enhance the self-regulation ability of the system. For example, the addition of high-quality LAB and yeast combinations can shorten the fermentation time and improve the fermentation efficiency.

4.2 Controlling the Fermentation Environment

The metabolic regulatory mechanisms of microorganisms can be enhanced by temperature, humidity, and oxygen control in the fermentation environment. For example, the optimal metabolic activity of LAB is maintained, and the establishment of an acidic environment is promoted by controlling temperature changes (Badotti et al. 2012).

5 Ecological Functions of Microbial Symbiotic Network Systems

5.1 Environmental Regulation

The key to silage fermentation is to create a stable anaerobic environment for the feed so that harmful microorganisms cannot grow, thus preventing spoilage. The microbial symbiotic network system plays an important role in processes such as pH regulation and oxygen control.

5.2 Nutrient Conversion

The silage microbial network not only protects the stability of the environment but also converts the macromolecules in the silage material into more digestible nutrients through metabolism, increasing the nutrient and palatability of the feed.

5.2.1 Cellulose and Starch Decomposition

Some microbial communities, such as LAB and specific fungi, are able to degrade cellulose and starch, converting them into short-chain organic acids, simple sugars, and other nutrients that are easily digested and absorbed by livestock.

5.2.2 Partial Degradation of Protein

During the fermentation process, certain strains of microbial communities promote the hydrolysis of proteins into amino acids or small peptides, enhancing the protein digestibility of the feed.

5.2.3 Increase in Metabolite

Metabolites, such as lactic acid, acetic acid, and ethanol, not only inhibit harmful microorganisms but also enhance the flavor of silage and increase the feed intake of livestock (Du et al. 2023).

5.3 *Increased Resistance Among Microorganisms*

The microbial symbiotic network system of silage fermentation can enhance the overall resilience in the face of environmental changes through mutual synergy and competition among microbial communities.

5.3.1 Dynamic Equilibrium and Self-Repair

The microbial symbiotic system is able to rapidly adjust the population structure of microorganisms in the face of changes in the fermentation environment (e.g., oxygen infiltration or temperature fluctuations). For example, when oxygen infiltration occurs, yeast will rapidly multiply to consume oxygen and reestablish anaerobic conditions, ensuring the growth advantage of lactic acid bacteria.

5.3.2 Protective Role of Resistant Microorganisms

In silage systems, some acid-resistant microorganisms can survive extreme conditions and provide a protective barrier for the whole microbial community. For example, when pH is low, acid-tolerant LAB continue to grow, while acid-intolerant stray bacteria are suppressed.

5.4 *Bacteriostatic Functions During Fermentation*

Silage microbial network system exerts bacteriostatic function through the metabolites of specific microorganisms to prevent the breeding of harmful microorganisms.

5.4.1 Inhibition of Spoilage Bacteria by an Acidic Environment

Lactic acid produced by LAB gives the environment a low pH value, which effectively inhibits the reproduction of spoilage bacteria, pathogenic bacteria, and other undesirable microorganisms and ensures that the feed does not spoil during the fermentation process (Vinderola et al. 2019).

5.4.2 Bacteriostatic Properties of Metabolites

Metabolites such as acetic acid and ethanol have a natural bacteriostatic effect. These substances form a chemical barrier against harmful microorganisms in silage fermentation, effectively reducing the risk of contamination by stray bacteria and ensuring the quality of silage.

5.5 Self-Regulation Function of Ecosystem

The symbiotic microbial system of silage fermentation shows a strong self-regulation function, which makes the silage system still able to maintain the fermentation effect during long-time preservation.

5.5.1 Population Regulation

At different stages of fermentation, microorganisms such as LAB, yeasts and acetic acid bacteria take turns to dominate the metabolic activity under different conditions. This dynamic regulation of microbial populations effectively reduces the overpopulation of single microorganisms and avoids resource depletion and fermentation anomalies.

5.5.2 Metabolic Feedback

Microorganisms in silage fermentation will regulate their own activities through metabolic feedback; for example, LAB will stop overpopulation after the pH value reaches a specific range, in order to avoid the unfavorable conditions triggered by too low pH. At the same time, the production of ethanol and carbon dioxide by the yeast metabolism helps to maintain the balance (Badotti et al. 2012).

5.6 Forage Shelf Life and Quality Stability

The microbial symbiotic network system enables silage to maintain its quality and nutrient content over a long period of time.

5.6.1 Inhibition of Spoilage and Prevention of Nutrient Loss

The symbiotic microbial network system effectively prolongs the preservation period of the silage and prevents spoilage through the use of a low pH environment, anaerobic conditions, and bacteriostatic metabolites.

5.6.2 Quality Stability

During silage fermentation, the microbial metabolites and the symbiotic network system can be regulated to enhance the stability of silage and ensure that the feed maintains its nutritional value during preservation.

An in-depth understanding of the symbiotic microbial network system of silage fermentation can not only help to optimize the silage process but can also be applied to other areas (Saha et al. 2020). (1) Improving feed nutrition: by artificially optimizing the composition of the silage microbial network, higher quality and more nutritious silage can be produced. (2) Reducing feed losses: the symbiotic network of silage microorganisms can minimize feed spoilage losses and save costs for the farming industry. (3) Ecological agricultural development: the natural fermentation method based on microbial network system can reduce the use of chemical additives in the silage process and promote the development of green ecological agriculture.

The microbial symbiotic network system of silage fermentation provides systematic support for the stability of the fermentation environment, the nutritional enhancement of the feed, and the long-term preservation of the silage through its multiple ecological functions. Understanding and utilizing the functions of this ecosystem provides a scientific basis for further optimizing silage technology and improving feed quality.

References

Adamberg K, Kask S, Laht TM, Paalme T. The effect of temperature and pH on the growth of lactic acid bacteria: a pH-auxostat study. Int J Food Microbiol. 2003;85(1–2):171–83.

Ávila CLS, Carvalho BF. Silage fermentation—updates focusing on the performance of microorganisms. J Appl Microbiol. 2020;128(4):966–84.

Badotti F, Gomes FCO, Rosa CA, Hui Y. Handbook of plant-based fermented food and beverage technology. Boca Raton: CRC Press; 2012. p. 639–48.

Borreani G, Tabacco E, Schmidt RJ, Holmes BJ, Muck RE. Silage review: factors affecting dry matter and quality losses in silages. J Dairy Sci. 2018;101(5):3952–79.

Du Z, Sun L, Chen C, Lin J, Yang F, Cai Y. Exploring microbial community structure and metabolic gene clusters during silage fermentation of paper mulberry, a high-protein woody plant. Anim Feed Sci Technol. 2021;275:114766.

Du Z, Yamasaki S, Oya T, Nguluve D, Euridse D, Tinga B, et al. Microbial co-occurrence network and fermentation information of natural woody-plant silage prepared with grass and crop by-product in Southern Africa. Front Microbiol. 2022;13:756209.

Du Z, Yang F, Fang J, Yamasaki S, Oya T, Nguluve D, et al. Silage preparation and sustainable livestock production of natural woody plant. Front Plant Sci. 2023;14:1253178.

Du Z, Nakagawa A, Fang J, Ridwan R, Astuti WD, Sarwono KA, et al. Cleaner anaerobic fermentation and greenhouse gas reduction of crop straw. Microbiol Spectr. 2024a:e00520–4.

Du Z, Yamasaki S, Oya T, Nguluve D, Euridse D, Tinga B, et al. Microbial network and fermentation modulation of Napier grass and sugarcane top silage in Southern Africa. Microbiol Spectr. 2024b;12(1):e03032–23.

Giraffa G. Studying the dynamics of microbial populations during food fermentation. FEMS Microbiol Rev. 2004;28(2):251–60.

Guan N, Li J, Shin HD, Du G, Chen J, Liu L. Microbial response to environmental stresses: from fundamental mechanisms to practical applications. Appl Microbiol Biotechnol. 2017;101:3991–4008.

Hua S, Wang Y, Wang L, Zhou Q, Li Z, Liu P, et al. Regulatory mechanisms of acetic acid, ethanol and high temperature tolerances of acetic acid bacteria during vinegar production. Microb Cell Factories. 2024;23(1):1–17.

Kung L Jr, Shaver RD, Grant RJ, Schmidt RJ. Silage review: interpretation of chemical, microbial, and organoleptic components of silages. J Dairy Sci. 2018;101(5):4020–33.

Li M, Zi X, Sun R, Ou W, Chen S, Hou G, et al. Co-ensiling whole-plant cassava with corn stalk for excellent silage production: fermentation characteristics, bacterial community, function profile, and microbial ecological network features. Agronomy. 2024;14(3):501.

Liu B, Huan H, Gu H, Xu N, Shen Q, Ding C. Dynamics of a microbial community during ensiling and upon aerobic exposure in lactic acid bacteria inoculation-treated and untreated barley silages. Bioresour Technol. 2019;273:212–9.

Ogunade IM, Martinez-Tuppia C, Queiroz OCM, Jiang Y, Drouin P, Wu F, et al. Silage review: mycotoxins in silage: occurrence, effects, prevention, and mitigation. J Dairy Sci. 2018;101(5):4034–59.

Okoye CO, Wang Y, Gao L, Wu Y, Li X, Sun J, et al. The performance of lactic acid bacteria in silage production: a review of modern biotechnology for silage improvement. Microbiol Res. 2023;266:127212.

Saha S, Basak B, Hwang JH, Salama ES, Chatterjee PK, Jeon BH. Microbial symbiosis: a network towards biomethanation. Trends Microbiol. 2020;28(12):968–84.

Ushijima T, Takahashi M, Ozaki Y. Acetic, propionic, and oleic acid as the possible factors influencing the predominant residence of some species of Propionibacterium and coagulase-negative Staphylococcus on normal human skin. Can J Microbiol. 1984;30(5):647–52.

Vinderola G, Ouwehand A, Salminen S, von Wright A, editors. Lactic acid bacteria: microbiological and functional aspects. Boca Raton: CRC Press; 2019.

Wells CL, Wilkins TD. Clostridia: sporeforming anaerobic bacilli. 4th ed. J Med Microbiol; 1996.

Yang ST, Liu X, Zhang Y. Metabolic engineering–applications, methods, and challenges. In: Bioprocessing for value-added products from renewable resources. Amsterdam: Elsevier; 2007. p. 73–118.

Chapter 5
The Evolution of Silage Facilities and Agricultural Machinery

Hidenori Kawamoto

Abstract This chapter covers the main types and characteristics of silage facilities including silo and the evolution of silage agricultural machinery. Round balers are now indispensable as ensiling machinery. Among the evolutions in silage facilities and machines, the most notable is the round baler. This chapter introduces the two types of round balers that have evolved uniquely in Japan. The first is the round baler for maize, which can bale even fine-chopped materials in response to requests from small and mid-sized dairy farms. The second is a direct-cutting round baler allowing to be harvested in muddy conditions of paddy fields, which has been developed specifically for forage rice. This chapter also introduces a silage unloader that has excellent durability and reliability due to its simple structure, it has been on the market for over 30 years since its development in Japan.

Keywords Forage rice · Maize · Round baler · Silage facility · Silage unloader

The origins of silage date back to ancient Egyptian times, with recorded ensiling for crop preservation in Japan dating to 1887 (Nakui 2009). As silage gained recognition as a crop-conservation technique, tower silos emerged, evolving from wood to brick, concrete blocks, and eventually steel constructions. Presently, the image of a brick tower silo alongside a cowshed persists on milk containers in the market, symbolizing the dairy farm aesthetic (Fig. 5.1). However, tower silos are rarely used in Japan, and bale silages are the mainstream except for bunker silos on large farms.

In the 1970s, following the importation of round balers to Japan, research began on ensiling by putting round bales into bag silos (Manda 1994). The introduction of bale wrappers in 1987 further popularized bale silage on farms. Round balers and bale wrappers are now indispensable as ensiling machinery on small and mid-sized farms. The most notable evolution in silage facilities and machines in Japan is the round baler.

H. Kawamoto (✉)
Tohoku Agricultural Research Center, NARO, Morioka, Japan
e-mail: hkawamo@affrc.go.jp

Y. Cai, K. Ataku (eds.), *Cultural History and Modern Production Technology of Silage*, https://doi.org/10.1007/978-981-96-5787-2_5

Fig. 5.1 Japan's oldest extant silo was built in 1907 and is designated as an important national cultural property (Koiwai Farm, Iwate, Japan). (Photo by Kawamoto)

1 The Evolution of Round Baler

1.1 Round Baler for Chopped Material

Initially designed for grass baling, round balers were adapted in 1999 to meet the demands of small and mid-sized dairy farms for maize baling, even accommodating fine-chopped material (Shito and Yamana 2000, 2002). This baler can bale the maize a cutting length of 12–15 mm into the size of 85 cm in diameter, 90 cm in width, and 300 kg in weight. Baling loss is minimized to less than 1.2% by effectively arranging the rubber belt inside the bale-forming chamber and binding the bale with a net. The resulting dry matter (DM) density of the bale ranges from 177–236 kg/m^3, comparable to the density at the bottom of a tower silo standing 5–6 m tall. Moreover, its front hopper allows for continuous harvesting without interruption, even during bale release (Fig. 5.2).

Comparing the bunker silo system and the roll baler using the same equipment showed similar field processing rates. However, the bunker silo required six workers, while the roll baler only needed three, reducing total working hours to 58% (Shito et al. 2005). The roll baler not only saves labor but also consistently produces high-quality maize silage regardless of operator skill. Furthermore, when applied to other crops for chopped material, the round baler increases bale density, promotes lactic acid (LAC) fermentation, and suppresses ethanol fermentation, improving fermentation quality compared to conventional round baler (Kawamoto et al. 2007).

Fig. 5.2 Round baler for chopped material. It was launched in the market in 2004 (MR-810, Takakita Co. Ltd., Nabari, Japan). (Photo by Kawamoto)

LT-Master (Goweil, Austria) Multi Compactor (Orkel, Norway)

Fig. 5.3 Round balers for chopped material in the global market. (Photos by Kawamoto)

Currently, some global manufacturers produce large round balers for chopped material, which are also used in Japan (Fig. 5.3).

Recently, the number of "Feed production contractors" undertaking ensiling operations for dairy farms in Japan has increased. Some have evolved into total mixed ration (TMR) supply centers. Harvesting crops from farm fields, ensiling, and delivering TMR-bale silage. Industrial balers are commonly used, along with large round balers for chopped material (Fig. 5.4).

In contrast to their overseas counterparts, Japanese manufacturers specialize in compact-sized round balers designed for chopped material. They offer self-propelled harvesting models with exchangeable heads suitable for various crops such as maize, ear corn, forage, etc. (Shito et al. 2010) (Fig. 5.5).

Fig. 5.4 Large round baler for chopped material operating at TMR supply center (Multi Compactor, Orkel, Fannrem, Norway). (Photo by Kawamoto)

Fig. 5.5 Harvesting round baler for chopped material with snapper head for ear-corn bale silage (SMR-1021, Takakita Co., Ltd., Nabari, Japan). (Photo by Kawamoto)

1.2 Round Baler for Forage Rice

In Japan, a 400,000 km network of waterways has long maintained paddy fields (MAFF 2012). As rice consumption declined, feed production from surplus paddy fields was promoted as a national project. Forage rice, particularly suited to poorly drained paddy fields and rainy weather at nearly twice the global average. Initially, forage rice was harvested and ensiled by a conventional round baler system. However, operational stability on muddy paddy fields proved challenging, inevitably leading to mud contamination when picking forage rice from the ground. Therefore, a round baler specifically for forage rice was developed (Urakawa and Yoshimura 2003). A combine harvester with a cutting head specialized for rice harvesting is generally used in Japan. The round baler for forage rice is based on this combine harvester; it was equipped with a disc cutter and a bale-forming chamber instead of the threshing section and grain tank (Fig. 5.6).

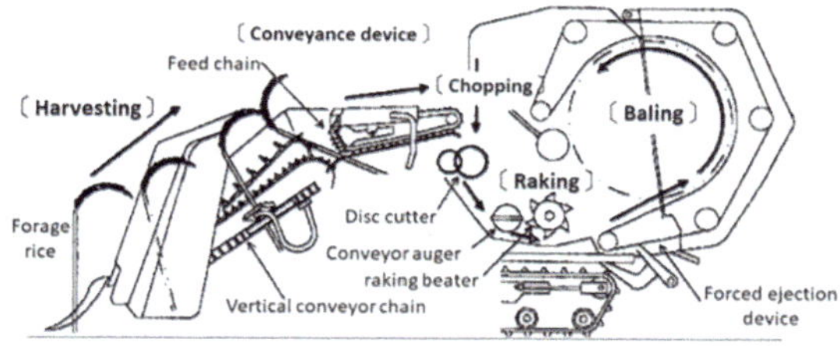

Fig. 5.6 Harvesting round baler for forage rice silage making. (Left: photo by Kawamoto; right: design drawing from Urakawa and Yoshimura 2003)

Fig. 5.7 Loading into a square underground silo. A typical square underground silo has a length of 3 m and a depth of 5 m. (Photo by Kawamoto)

This baler is designed for utilization by rice farmers. Its operating speed, at 1.1 m/s, aligns closely with that of a rice combine harvester, ensuring a seamless transition for rice farmers accustomed to such a combine harvester. Since its commercial introduction in 2000, it has become the predominant choice for harvesting forage rice in Japan. The latest model has improvements to the disk cutter and baling chamber, enabling a chopping length of 30 mm.

2 Silage Unloader "Silo-Crane"

Currently, bale silage is the mainstream in small and mid-sized dairy farms in Japan, but square underground (terrain-based) silos are also used (Fig. 5.7).

In 1987, a revolutionary silage unloader "silo-crane" was developed and made commercially available (Segawa 1999). The use of underground silos in Japan is due to the silo-crane. This device has a simple structure, with a pantograph attached

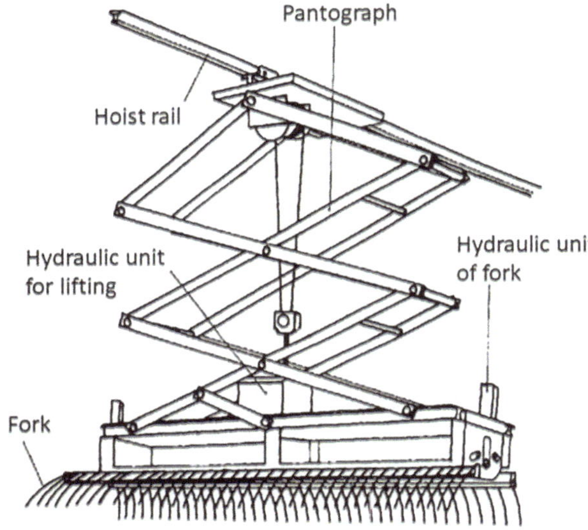

Fig. 5.8 Structure diagram of silo-crane. (Design drawing from Ichito 2024). Two-stage pantograph model is shown. There are up to five stages depending on the lifting height

Attached to the roof rail of silo Stretching towards the silo Grabbing of silage

Fig. 5.9 Silo-crane system. (Photos by Kawamoto)

to a hoist installed on the silo roof that keeps the body horizontal, and suspends a fork-shaped grab that hydraulically opens and closes (Figs. 5.8 and 5.9).

The hoist moves freely along a rail connecting the silo roof to the cowshed. Silo-crane operations involve simply moving the hoist up and down and opening and closing the grab. Each action is controlled by turning on and off the electromagnetic switch, enabling simple sequence control for automating silage transportation from the silo to the feed bunk. Fork-shaped grab, with the same length as one side of the silo, can handle anything from finely chopped maize to long-cut grass. One grip is approximately 70 cm wide and 10 cm deep, minimizing disturbance to the silage surface. Even in the dense layer at the silo bottom, the amount grasped can remain nearly constant by repeatedly opening, closing, raising, and lowering the grab two or three times automatically. In 1992, the silo-crane was connected to a feed mixer and automatic feeding device, developing a fully automated TMR feeding system (Fujita 1999). Silo-crane has excellent durability and reliability due to its simple structure. Therefore, it has been on the market for over 30 years since its development (Okamoto SCL 2024).

References

Fujita N. Evaluation and process to introduce of dairy farming under the mixing-automatic feeding system with silo-crane. HJAE. 1999;8(1):1–20.

Ichito K. Silo-crane. J Jpn Soc Grassl Sci. 2024;51(1):114–20.

Kawamoto H, Otani R, Oshibe A, Yamaguchi H, Deguchi S, Tanaka O, et al. Ensilage of wilted whole crop rice (*Oryza sativa L.*) using a roll baler for chopped material: silage quality in long-term storage. Grassl Sci. 2007;53(2):85–90.

Manda T. Basics and practice of round-bale silage systems. Dairy Total Research Institute, Sapporo, Japan; Sanyo Printing Co., Ltd.; 1994. pp. 28–87.

Ministry of Agriculture, Forestry and Fisheries (MAFF). History and current status of agricultural irrigation. 2012. Accessed 13 May 2024.

Nakui T. Silo and silage story. Rakuno Gakuen University Extension Center, Hokkaido, Japan; Hokkaido Rehabilly Social Welfare Co.; 2009. pp. 6–17.

Okamoto Seisakusho Co., Ltd. SP silo-crane. 2024. Accessed 13 May 2024.

Segawa K. Silage unloader. In: DAIRYMAN Co., Sapporo, Japan, editor. Systemization and machinery for roughage production. Kokoku Printing Co., Ltd.; 1999. pp. 118–21.

Shito H, Yamana N. Development of cutting roll baler for maize harvesting. J Jpn Soc Agric Mach. 2000;62(3):157–9.

Shito H, Yamana N. Development of labor saving technique of maize harvesting and ensiling (part 1). Development of the roll baler for chopped material. J Jpn Soc Agric Mach. 2002;64(4):96–101.

Shito H, Takahashi K, Yamana N, Shibuya Y, Okumura M, Syouda M, et al. Development of labor saving technique of maize harvesting and ensiling (part 3)-development of the secondary prototype and the practical test. J Jpn Soc Agric Mach. 2005;67(3):106–13.

Shito H, Tachibana Y, Kawaide T, Takahashi K, Okajima H, Kitanaka T, et al. Development of self-propelled harvesting roll baler and construction of maize harvesting system. J Jpn Soc Grassl Sci. 2010;56(3):187–93.

Urakawa S, Yoshimura Y. Development of the cutting roll baler for the rice whole crop silage. J Jpn Soc Grassl Sci. 2003;49(1):43–8.

Part II
Fermentation Regulation of Silage

Chapter 6
Dominant Factors Affecting Silage Fermentation

Huili Pang, Zhongfang Tan, and Yanping Wang

Abstract Silage, as a traditional method for preserving animal feed, can be a major source of energy for animals throughout the year. It is widely used in commercial livestock production in most developed countries. High-quality silage is an important foundation for the development of animal husbandry. The main fermentation mechanism of silage is that fresh forage crops and grasses are cut, chopped, and sealed in silos, forming a relatively anaerobic condition for effective preservation. The lactic acid bacteria use the water-soluble carbohydrates in the raw materials for fermentation, producing lactic acid, causing a decrease in pH value and inhibiting the growth of aerobic bacteria, effectively preserving high nutrient content and improving feed palatability. Moisture, water-soluble carbohydrates, lactic acid buffering capacity, and epiphytic microorganisms in forage crops and grasses are the dominant factors affecting silage fermentation, and this chapter focuses on the principle of interaction of these factors.

Keywords Epiphytic microorganisms · Moisture · Lactic acid buffering capacity · Silage fermentation · Water-soluble carbohydrate

Ensiling is a complex biochemical process, and many factors can affect the quality of fermentation. The physicochemical properties of silage raw materials are crucial factors affecting the fermentation quality of silage, and under the premise of ensuring scientific processing and blending, the influence of raw material characteristics on silage quality is particularly important. Silage can improve fermentation quality while still retaining the nutritional components of natural grass, thereby increasing feed value. In addition, the lower pH value of silage can inhibit the survival of pathogenic bacteria and parasites and reduce the pathogenicity of ruminant livestock.

H. Pang (✉) · Z. Tan · Y. Wang
Zhengzhou University, Zhengzhou, China
e-mail: pang@zzu.edu.cn; tzhongfang@zzu.edu.cn; wyp@zzu.edu.cn

Y. Cai, K. Ataku (eds.), *Cultural History and Modern Production Technology of Silage*, https://doi.org/10.1007/978-981-96-5787-2_6

Moisture, water-soluble carbohydrates (WSC), lactate buffering capacity (LBC), and epiphytic microorganisms in forage crops and grasses are the dominant factors affecting the fermentation quality of silage.

1 Moisture

Moisture is an important factor required to ensure optimal condition for the normal activity of LAB. Excessive or insufficient moisture content can affect the fermentation process of silage and the quality of silage; the moisture content in forage plays an important role in the fermentation process of silage, as it affects the type of fermentation products, juice production, and degree of aerobic decay.

The most suitable moisture for the reproductive activity of LAB is 65–75%; suitable moisture content for silage raw materials is 65–70%; for poaceae forage, it is 60–70%; and for leguminous forage, such as alfalfa, it is 55–65%. If the moisture content of the harvested raw materials is high, reaching over 75%, even if the WSC content of the raw materials is high, the WSC content in the cell fluid is too diluted, which cannot meet the sugar concentration required for LAB fermentation and guarantee the preparation of high-quality silage. If the moisture content is too high during ensiling, it is easy to compact and clump and produce clostridia fermentation and breaks down proteins through dehydrogenation, decarboxylation, and redox processes, leading to abnormal fermentation of silage, causing the silage to become smelly and the quality to deteriorate; at the same time, the juice of plant cells is squeezed and lost, resulting in nutrient loss (Yan et al. 2019). When the moisture content in silage raw materials is over 80%, the dry matter (DM) lost in leakage can be about 6.7% of the DM in silage raw materials. With a moisture content of about 70%, there is very minimal liquid juice discharged with less amount of DM lost. Low moisture content makes it difficult to compact silage materials, resulting in large gaps and excessive residual air and moisture; this can lead to the proliferation of aerobic bacteria, affecting the quality of silage and causing loss of nutrients.

The suitable moisture content of silage raw materials varies depending on their texture. Coarse and hard raw materials can have a moisture content of up to 80%, while early harvested, tender, and juicy raw materials are more suitable at 60%. A simple way to determine the moisture content of silage raw materials is to grip the chopped raw materials tightly in your hand and then release them naturally. If they still remain spherical and have wet marks on your hand, their moisture content should be between 68% and 75%; if the grass ball slowly expands and there are no wet marks on the hand, its moisture content is between 60% and 67%. If the hand is released and the grass ball immediately expands, then the moisture content is below 60%, which is only suitable for low-moisture silage. Manually evaluate the moisture content of silage raw materials (Table 6.1).

Silage materials with high or low water content should be treated or adjusted during ensiling. Feed with excessive moisture should be slightly dried and withered before ensiling to achieve the required moisture content. If the appropriate moisture

Table 6.1 Manually evaluate the moisture content of silage raw materials

Squeeze silage raw materials by hand	Moisture (%)
Water is easily squeezed out, and the grass takes shape	≥80
Water can just be squeezed out, grass takes shape	75–80
Can only squeeze out a small amount of water (or unable to squeeze out), but the grass has formed	68–75
Unable to squeeze out water, grass slowly separates	60–67
Unable to squeeze out water, grass quickly separates	≤60

Fig. 6.1 Grass and wilting during harvest. (Photos by Pang)

content cannot be reached after withering, dry materials should be added to mix the silage. It is also possible to mix raw materials with high and low moisture content in appropriate proportions for ensiling. Crop by-products are an important biomass resource crucial in green, sustainable, and intensive agricultural production. They are also promising adsorbent for reducing silage moisture. Combining high-moisture materials with crop by-products can lower water content and reduce competition for food between humans and livestock. Adding wheat bran to mulberry paper for mixed ensiling accelerated the response of microorganisms to anaerobic stress, and LAB became the dominant microbial community, thereby enhancing lactic acid fermentation, affecting the metabolic pathways of microorganisms, and improving the quality of silage (Du et al. 2021). Wang et al. (2023) achieved the goal of obtaining high-quality silage feed by adding different proportions of high moisture content *Artemisia argyi* to wheat straw with extremely low water content (Fig. 6.1).

2 Soluble Carbohydrate

Many factors affect the quality of silage, but the soluble carbohydrate content in the raw materials is an important factor affecting silage quality. Carbohydrates are the main products of photosynthesis and can usually be classified as nonstructural carbohydrates (involved in substance metabolism and energy metabolism) and structural carbohydrates (involved in morphological composition). The nonstructural carbohydrates in the raw materials mainly include WSC and starch, which provide substrates for LAB fermentation during the ensiling process and ensure the normal progress of fermentation. It is generally believed that successful ensiling can only be achieved when the WSC content of silage raw materials is not less than 2% DM. In addition, the content of nonstructural carbohydrates also varies, depending on the variety of raw materials, which affects the quality of silage fermentation. In a study by Dong (2011) on the relationship between the nonstructural carbohydrate content of different varieties of rice straw and silage quality, the nonstructural carbohydrate content was affected by varietal differences. The content of nonstructural carbohydrates in raw materials showed a significant negative correlation with silage pH value and the ammonia nitrogen to total nitrogen ratio, while it exhibited a significant positive correlation with lactate content.

The structural carbohydrates in raw materials mainly include cellulose, hemicellulose, lignin, etc. There are differences in the physical structure between cellulose and hemicellulose. Cellulose is a tightly packed structure formed by the connection of cellulose disaccharides and is enveloped by a composite structure formed by lignin and hemicellulose, which is not easily hydrolyzed by acid. On the other hand, hemicellulose is composed of a main chain of xylan and side chains of arabinose and is more susceptible to acid hydrolysis. Silage fermentation can soften the cellulose in raw materials, and under the action of microorganisms and enzymes, some fibers are degraded into simple sugars that LAB can utilize, improving the digestibility of DM and organic matter. Xue et al. (2018) found that the hemicellulose content and neutral detergent fiber content of poaceae grass significantly decreased after ensiling fermentation. This indicates that hemicellulose is degraded into WSC that can be directly utilized by microorganisms during silage fermentation, promoting silage fermentation.

For successful ensiling, it is essential to have an adequate content of WSC, such as glucose, fructose, fructan, sucrose, trisaccharides, and stachyose. LAB require a sufficient amount of WSC, at least 2% of fresh forage or 8% of DM to produce enough lactic acid. If there is very little WSC in the raw material, even if all other conditions are met, it cannot be made into high-quality silage. The protein and alkaline substances in silage raw materials will neutralize some lactic acid, and only when the pH value in silage raw materials is below 4.2, can it inhibit the activity of harmful microorganisms. Therefore, the sugar content required for LAB to form lactic acid and achieve a pH of 4.2 is a crucial condition, commonly referred to as the minimum required sugar content.

Soluble carbohydrate is the key to the success or failure of ensiling, and this indicator is mainly affected by plant moisture content, crop variety, growth stage, climate, terrain, fertilization, etc. Generally, poaceae forage grasses have a high content of WSC, while legumes have a low content. Therefore, legume forage is difficult to store compared to poaceae. When the actual content of soluble carbohydrates in raw materials exceeds the minimum required level, it is considered a positive soluble carbohydrates difference in silage. In contrast, if the actual sugar content is below the minimum required level, this is referred to as a negative soluble carbohydrate difference in silage. Feed grasses that show a positive difference in silage soluble carbohydrates are generally easier to ensile, and the greater the positive difference, the more favorable the conditions for ensiling. On the other hand, raw materials with a negative silage sugar difference may pose more challenges for ensiling. By focusing on selecting grasses with positive soluble carbohydrate differences, the ensiling process would be improved and the overall quality of the feed be enhanced. The stem of general plants has a positive soluble carbohydrates difference, while the leaves have a negative one, and the whole plant of corn has a positive.

Ensiling high-quality raw materials is possible as long as the DM content is high. However, if the DM content is low and the WSC content is also low, the quality of the ensiling process will be significantly reduced. Many experiments have shown that the quality of silage is directly related to its soluble carbohydrates. Generally, good silage requires raw materials with a WSC content of at least 30 g/kg^{-1} FM, and 50 g/kg^{-1} DM to have good fermentation quality (Ni et al. 2018). When the WSC content is greater than 2%, the probability of poor-quality silage is 5%, while when the sugar content is less than 2%, the probability of poor-quality silage is 44%. Therefore, when the WSC content of silage is less than 3%, other substances with high sugar content should be added. Wang et al. (2023) increased the content of WSC in wheat straw silage by adding LAB and different proportions of *Artemisia argyi* to wheat straw. Du et al. (2023) combined cellulase with LAB in mulberry ensiling, which exhibited a synergistic effect, through which cellulases such as glycanase, pectinase, and carboxymethyl cellulase decomposed cellulose and hemicellulose into sugars. LAB converted these sugars into lactic acid through the glycolytic pathway, thereby improving mulberry silage's microbial community structure, metabolism, and fermentation quality.

3 LBC

The buffering capacity of raw materials refers to their ability to resist changes in pH value, which is an important factor affecting the success of ensiling, usually expressed in milligrams of alkali equivalent required to change the pH value of 1 kg of DM. The buffering capacity of raw materials can affect silage's lactic acid fermentation process. McDonald et al. (1992) found that the higher the buffering energy of raw materials, the slower the pH value of silage decreases, resulting in low acidity and easy butyric acid fermentation, affecting silage fermentation quality. Under the same conditions, silage is more likely to succeed when the buffering energy is low. In addition, the buffering energy is affected by the crude protein content of the raw material, and the higher the crude protein content, the higher the buffering energy. Research by Xue et al. (2018) revealed that the crude protein content of typical natural grasses, poaceae, and leguminous grasses are significantly positively correlated with the buffering energy value.

The buffering performance of forage crops is achieved by anions (organic acids, orthophosphate, sulfate, nitrate, chloride) and plant proteins; thus, feed grass have buffering energy, mainly due to the balance between acidic and alkaline substances contained in them. Various components, such as dietary fiber in silage raw materials, will generate a large amount of organic acid, such as lactic acid, acetic acid, etc. These organic acids can combine with calcium, magnesium, and aluminum ions to form stable buffer salts and stabilize pH values in acidic environments. In rice straw silage, the pH value of silage is significantly positively correlated with the buffering energy value and DM content of raw materials; the lactate content and V-score are significantly negatively correlated with the buffering energy value of the raw materials. Therefore, forages should have low buffering energy so that the pH can rapidly decrease with the production of lactic acid in the early stage of fermentation, inhibiting the activity of harmful microorganisms earlier and reducing the loss of nutrients. A high buffering energy is not conducive to reducing the pH of the entire silage system in the early stage of ensiling, making it difficult to inhibit the growth of harmful microorganisms and making it difficult for LAB to dominate the fermentation system.

Due to the low WSC content, high protein content, and buffering energy of leguminous forage, it is difficult to produce high-quality silage through conventional ensiling. However, due to its high nutritional value, it can provide sufficient protein for animals, can promote their growth and development, and is often referred to as high-quality "protein feed." Domestic and foreign scholars often improve the fermentation quality of legume forage silage by adding chemical and biological additives. The addition of molasses or LAB plus molasses has the most significant effect on improving the fermentation quality of alfalfa silage with high buffering energy.

Compared with leguminous forage, grasses of the poaceae family have higher crude fiber digestibility, have lower cell wall content and lignification degree, and are rich in minerals and microorganisms. They also have higher WSC content and lower buffering energy and usually do not require any additives to obtain high-quality silage. Poaceae forage can be used as roughage to replace some grains and provide animals with certain energy and protein.

4 Epiphytic Microorganisms in Forage Crops and Grasses

The silage fermentation system is a microbial ecological environment that involves complex biochemical and microbial interactions (Pahlow et al. 2003). There are significant differences in microbial communities between fresh grass and silage. There are a large number of anaerobic and aerobic microorganisms attached to grass, which are classified into aerobic bacteria, LAB, *intestinal* bacteria, yeast, mold, *clostridia*, *bacilli*, *acetobacter*, and fungi using classical taxonomic methods (Table 6.2).

Among them, LAB rapidly reproduces under anaerobic conditions using WSC in silage raw materials, producing a large amount of lactic acid and reducing the pH

Table 6.2 Population of epiphytic microorganisms in forage crops and grasses

Microorganism	Number (cfu/g of fresh matter)
Aerobic bacteria	>10,000,000
Lactic acid bacteria	10–1,000,000
Enterobacteria	1000–1,000,000
Yeast	1000–100,000
Mold	1000–10,000
Clostridia	100–1000
Coliform bacteria	100–1000
Acetic acid bacteria	100–1000
Propionic bacteria	10–1,000,000

value of feed. With the formation of anaerobic and acidic environments, the activity of harmful microorganisms such as *coliform* bacteria and *clostridia* is gradually inhibited, and aerobic microorganisms such as yeast and mold are also in a dormant state of inactivity, thus avoiding the loss of nutrients in silage caused by harmful microorganisms.

LAB associated with silage include species of *Lactobacillus, Lactococcus, Enterococcus, Pediococcus, Leuconostoc, Streptococcus*, and *Weissella*. In addition to producing a large amount of lactic acid, e.g., homofermentation LAB use 1 molecule of glucose to produce two molecules of lactic acid; some LAB also produce certain amounts of ethanol and carbon dioxide during the fermentation process: heterofermentation LAB use one molecule of glucose to produce one molecule of lactic acid, one molecule of carbon dioxide, and one molecule of acetic acid or ethanol. The number of LAB attached to the surface of forage such as alfalfa and mulberry was very low, about 10–10^4 cfu/g, while the number of LAB attached to the surface of whole plant cereal crops, such as corn and sweet sorghum, was relatively high, up to 10^7 cfu/g. The number of LAB attached to the surface of silage raw materials significantly impacts the quality of silage fermentation in the early stage. Cai and Kumai (1994) found that for high-quality lactic acid fermentation, the number of LAB in silage raw materials must reach 5.0 lg cfu g^{-1} FW (fresh weight). Forages with low numbers of attached LAB, WSC, and moisture content make natural ensiling unfavorable for the proliferation of LAB and lack sufficient substrates. Its microecological environment is not conducive to ensiling fermentation, making direct ensiling difficult to succeed. in order to ensure that silage can enter the anaerobic fermentation stage dominated by LAB as early as possible, the addition of exogenous LAB is often used to improve the micro ecosystem of silage. The presence of LAB had an important effect on the microbial community and metabolic pathways during silage fermentation, and a "co-occurrence microbial network" formed with LAB, effectively inhibiting the growth of harmful microorganisms and dominating the anaerobic fermentation process (Du et al. 2023).

Yeast is the most important aerobic microorganism in the ensiling process, which can utilize WSC and lactic acid. Although many types of yeast are attached to the surface of forage, *Candida* and *Pichia* are the main types after silage fermentation. In general, yeast only reproduces in the early stages of fermentation, and its growth is quickly inhibited as oxygen is consumed and organic acids such as lactic acid accumulate. After aerobic exposure, yeast recovers its activity and proliferates extensively using sugars and lactic acid produced by silage fermentation, producing carbon dioxide and heat. Due to the consumption of lactic acid by yeast, the pH of the feed increases, and other aerobic microorganisms gradually recover their activity (pH > 4.5). Some yeasts are also active in anaerobic environments and can ferment sugar substances such as ethanol. When the activity of LAB is inhibited in a low pH environment, yeast can use the remaining sugar substances to produce ethanol, forming high ethanol silage (>20 g/kg DM, usually sugar cane silage).

Mold is an aerobic filamentous fungus that mostly exists on the silage surface and in the parts that come into contact with air. It is the main harmful microorganism that causes aerobic spoilage of silage. When subjected to environmental stress, fungi can produce large amounts of *mycotoxins* such as *aflatoxin, zearalenone, vomitoxin*, and T2 toxin. Consuming feed containing *mycotoxins* can cause acute or chronic damage to animals, damaging their liver, kidneys, and reproductive function. Good management can effectively suppress the production of fungal toxins during the ensiling process. Mold growth mostly prefers humid environments. When the moisture content in silage raw materials is too high, it will accelerate mold growth, reproduction, and toxin production. Therefore, when making silage, avoid excessive moisture in the raw materials, which should be controlled between 55% and 65%. This will effectively reduce the number of molds and control the production of *mycotoxins*.

Acetobacter is an anaerobic microorganism that can produce lactic acid from ethanol under low pH conditions. When ethanol is depleted, *acetobacter* consumes acetic acid to produce carbon dioxide and water. This will increase feed pH, activate other aerobic microorganisms, and cause a significant loss of nutrients. At present, no acetobacter has been found in C3 plants and leguminous grasses; it has only been reported in C4 plants.

Bacteroides can be divided into aerobic and anaerobic types. Although these bacteria can also utilize carbohydrates and organic acids, they have no significant effect on silage fermentation. When silage is exposed to air, yeast and *acetobacter* consume lactic acid silage, causing the pH of the feed to rise above 4.5. The heat generated by the bacteria accelerates the decay of the feed, making it sticky.

Enterococcus uses nitrate as an electron acceptor to convert nitrate into nitrite and nitrogen oxides, mainly acting under anaerobic conditions. *Enterococcus* can also use carbohydrates to produce acetic acid, succinic acid, and 2,3-butanediol. It is generally believed that high acetic acid content will reduce the feed intake of silage. However, previous studies have shown that the acetic acid produced by *L. buchneri* does not reduce the feed intake of silage, and *enterococci* may have produced other chemicals that affect the palatability of silage (Kleinschmit and Kung 2006).

Clostridium is an obligate anaerobic bacterium commonly found in high moisture silage, mainly divided into three types: protein hydrolyzing *clostridia* (mainly utilizing amino acids), butyric acid producing *clostridia* (utilizing WSC and acetic acid). Among them, protein-hydrolyzing bacteria decompose amino acids to produce a large amount of ammonia, amines, and carbon dioxide. The main products of *Clostridium butyricum* are butyric acid, acetic acid, hydrogen, and carbon dioxide. *Clostridia* fermentation can cause significant DM loss and reduce the palatability of feed, and long-term ingestion of *clostridia-fermented* feed can also lead to animal ketosis. pH and moisture are two important factors that affect the growth and reproduction of *clostridia*. The optimal pH for the growth of *clostridia* is 7.0–7.4. When the pH value is below 4.5, the growth of *clostridia* will be inhibited; when the silage material wilts to a moisture content of about 55–70%, it can effectively reduce the number of *clostridia*. Therefore, in production practice, measures such as wilting of raw materials can be taken to reduce the number of *clostridia*.

Silage fermentation is a process in which microorganisms cooperate and compete with each other and is a dynamic silage microbial ecosystem. The types and quantities of microorganisms in silage are dynamically changing and have certain rules. Generally speaking, after the silage raw materials are sealed, the changes in microbial communities during the fermentation process are generally similar. Based on the changes in microbial species, the fermentation stage of silage is divided into aerobic stage, fermentation stage, stable stage, and feeding stage, as shown in Table 6.3.

At the beginning of sealing, due to the presence of oxygen, aerobic bacteria dominate the silage. As oxygen is depleted, silage enters the anaerobic fermentation stage. In the early anaerobic stage, facultative or obligate anaerobic microorganisms (such as coliform bacteria, yeast, clostridia, etc.) compete fiercely with LAB for plant nutrients. The emergence of anaerobic and acidic silage environments facilitates the reproduction of acid tolerant LAB (such as *Lactiplantibacillus* (*L.*) *plantarum*, *L. brevis*, and *Lentilactobacillus* (*Le.*) *buchneri*) using soluble carbohydrates as substrates to produce metabolites primarily composed of lactic acid. This process may last for several days or weeks. In the process of producing high standard silage, LAB quickly dominate fermentation to produce a large amount of lactic acid, rapidly reducing the pH value of silage and forming a microecological environment conducive to the storage of silage feed. In the case of soil contamination caused by low stubble height during harvesting, prolonged aerobic phase, or slow acidification, the microbial community in silage will mainly consist of *clostridia*, yeast, and mold. These bacteria can use soluble sugars, organic acids, and proteins to produce harmful substances, such as butyric acid and ammonia nitrogen, which can cause

Table 6.3 Changes of microbial communities during ensiling

Stage	Dominant microorganism
Aerobic stage	Aerobic bacteria, yeast, mold, and coliform bacteria
Fermentation stage	Lactic acid bacteria mainly composed of cocci, including *Leuconostoc*, *Streptococcus*, and so on
Stable stage	Lactic acid bacteria mainly composed of rods, including *Lactiplantibacillus* (*L.*) *plantarum* and *Lentilactobacillus buchneri*
Feeding stage	Various lactic acid bacteria, yeast, and mold

the pH value of silage to rise. At the same time, pathogenic microorganisms such as *Listeria* may also be accidentally added, increasing the feeding risk of silage feed.

After oxygen consumption, facultative or obligate anaerobic microorganisms, such as yeast and *coliform* bacteria, compete with LAB for substrates. As the acidic environment gradually forms, LAB such as *Leuconostoc* and *Streptococcus* begin to move, and as the pH value decreases, they are replaced by more acid-resistant LAB such as *L. plantarum* and *Le. buchneri*. When feeding and utilizing, air enters, the infiltration of air provides favorable conditions for aerobic microorganisms to utilize WSC, lactic acid, etc., leading to decrease of acidity and increase in pH, and the silage environment is suitable for the proliferation of miscellaneous bacteria, ultimately resulting in a decline in quality.

When using forage with high WSC content for silage fermentation, the survival of a large number of acid-resistant yeast leads to high ethanol content in the fermentation product, resulting in significant DM and energy loss. The changes in microbial community structure in silage are important reasons for the changes in fermentation products. Monitoring the changes in fermentation products and microbial community structure during the silage process can provide a theoretical basis for avoiding the formation of adverse fermentation products.

Microbial quantity measurement using the three-part methods according Cai and Kumai (1994). Take 10 g of fresh natural grass or silage sample and mix it with 90 mL of sterile water, and then mix thoroughly. After that, use a tapping sterile *homogenizer* to tap for 2 min to obtain a 10^{-1} dilution solution; dilute continuously with sterilized distilled water at a ratio of 1:9 to make 10^{-2}, 10^{-3}, 10^{-4} and 10^{-5} gradient dilutions were taken, and 20 μL of 10^{-1}, 10^{-3}, and 10^{-5} gradient dilutions were coated onto three different culture media. MRS (de Man, Rogosa, Sharpe), medium was anaerobically incubated at 30 °C for 48 hours (h) to determine the number of LAB, while PDA (potato dextrose agar), BLB (Blue light broth agar) or EMB (eosin-methylene blue agar) medium were incubated at 30 °C for 48 h to determine yeast and mold (with 10% tartaric acid), *coliform* bacteria, respectively; NA (nutrient agar) medium was incubated at 30 °C for 24 h to determine general aerobic bacteria. Counting was performed using the plate counting method based on different culture media and their formation. Identification of different bacterial species based on colony morphology characteristics, and determination of microbial quantity through conversion based on dilution ratio of agar solution.

In conclusion, to obtain high-quality silage, the original material must meet the following conditions: the core of silage fermentation is lactic acid fermentation by LAB to produce lactic acid, so sufficient and highly active LAB and abundant fermentable substrates are essential; the DM content of silage raw materials should be maintained at a high level (greater than 200 g kg^{-1} FM); silage raw materials should have low buffering energy, so that the pH can rapidly decrease with the production of lactic acid; a good physical structure is also crucial for the production of high-quality silage. A suitable physical structure can make the material easier to compact in the silage cellar, better exhaust the air in the entire fermentation system, and enable LAB to start anaerobic respiration and produce lactic acid earlier. However, in actual production and life, not all silage materials can meet the above conditions at the same time, so the vast majority of silage will improve the fermentation quality of silage by adding silage additives during preparation. In addition, mixed silage, fermentation of two or more complementary ingredients in an anaerobic environment in a certain proportion, can not only effectively solve the problem of high moisture content, high buffering energy, and low WSC content in leguminous forage that makes it difficult to silage successfully but also increase the utilization rate of agricultural and light industry by-products, expand feed sources, reduce environmental pressure, and ensure the development of animal husbandry (Ni et al. 2018; Wang et al. 2023). The mixed silage raw materials mainly include feed crops, crop straw, crop by-products, and light industry by-products. The regulation of moisture content and soluble sugar content in silage raw materials, as well as the complementarity of nutritional components, are important factors for successfully regulating mixed silage.

References

Cai Y, Kumai S. The proportion of lactate isomers in farm silage and the influence of inoculation with lactic acid bacteria on the proportion of L-lactate in silage. Anim Sci Technol. 1994;65(9):788–95.

Dong C. Dynamics of feeding value related traits of rice (*Oryza sativa* L.) straw during late growth period and the effects of gibberellic acid and cutting heights on the feeding value of rice straw. Doctoral dissertation. 2011; Nanjing Agricultural University.

Du Z, Lin Y, Sun L, Yang F, Cai Y. Microbial community structure, co-occurrence network and fermentation characteristics of woody plant silage. J Sci Food Agric. 2021;102(3):1193–204.

Du Z, Yamasaki S, Oya T, Cai Y. Cellulase-lactic acid bacteria synergy action regulates silage fermentation of woody plant. BFB. 2023;16(1):125.

Kleinschmit DH, Kung L Jr. A meta-analysis of the effects of *Lactobacillus buchneri* on the fermentation and aerobic stability of corn and grass and small-grain silages. J Dairy Sci. 2006;89(10):4005–13.

McDonald PA, Henderson AR, Heron SJE. The biochemistry of silage, vol. 28. 2nd ed. Marlow: Chalcombe Publication; 1992. p. 125.

Ni K, Zhao J, Zhu B, Su R, Pan Y, Ma J, et al. Assessing the fermentation quality and microbial community of the mixed silage of forage soybean with crop corn or sorghum. Bioresour Technol. 2018;265:563–7.

Pahlow G, Muck RE, Driehuis F, Elferink ST, Spoelstra SF. Microbiology of ensiling. SST; 2003. pp. 31–93.

Wang Z, Tan Z, Wu G, Wang L, Qin G, Wang Y, et al. Investigation on fermentation characteristics and microbial communities of wheat straw silage with different proportion *Artemisia argyi*. Toxins. 2023;15(5):330.

Xue Y, Bai C, Sun J, Sun L, Chang S, Sun Q, et al. Effects of locations and growth stages on nutritive value and silage fermentation quality of *Leymus chinensis* in Eurasian steppe of northern China. Grassl Sci. 2018;64:40–50.

Yan Y, Li X, Guan H, Huang L, Ma X, Peng Y, et al. Microbial community and fermentation characteristic of Italian ryegrass silage prepared with corn stover and lactic acid bacteria. Bioresour Technol. 2019;279:166–73.

Chapter 7
Silage Preparation Technology

Zhumei Du and Yimin Cai

Abstract Silage can effectively preserve the nutrients in forages, improve the palatability of animals, and eliminate germs and pests. Silage provides green and juicy feed for livestock in the winter and dry seasons in the cold regions and the tropics. Good preparation technology can be made into high-quality, shelf-stable, and nutritious feed, which is of great significance for the development of animal husbandry. This chapter introduces the basic silage preparation methods, including forage crops and grasses, timely harvesting of forage, moisture adjustment, chopping, air removal, sealing, and other general preparation methods.

Keywords Animal husbandry · Forage · Preparation technology · Silage fermentation

Silage is a livestock storage feed prepared by microbial fermentation under anaerobic conditions using fresh grasses, forage crops, crop residues, or food by-products (Du et al. 2023). Silage is widely used in many countries worldwide to make up for the lack of feed in livestock production during the winter or dry season (Bilal et al. 2021). Silage is susceptible to changes in fermentation quality due to various conditions, such as material, climate, and storage facilities (Okoye et al. 2023). Correct silage preparation is essential to improve fermentation quality, minimize forage nutrient losses, and improve palatability and intake by livestock.

The technical steps of silage preparation mainly include pre-storage preparation, mowing, moisture adjustment, chopping, packing, compaction to exclude air, sealing and preservation, and opening (Fig. 7.1) (Grant and Ferraretto 2018).

Z. Du (✉)
Yangzhou University, Yangzhou, China

Y. Cai
Zhengzhou University, Zhengzhou, China

Japan International Research Center for Agricultural Sciences, Tsukuba, Japan
e-mail: caiy0234@jircas.go.jp

Y. Cai, K. Ataku (eds.), *Cultural History and Modern Production Technology of Silage*, https://doi.org/10.1007/978-981-96-5787-2_7

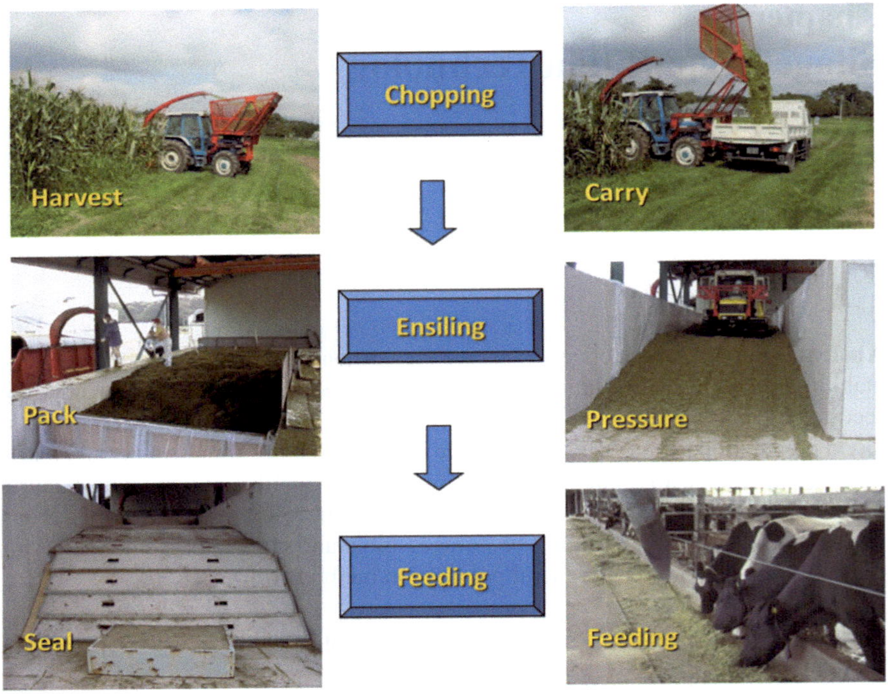

Fig. 7.1 Silage preparation process and livestock feeding. (Photos by Cai)

1 Pre-Silage Preparation

Before ensiling, depending on the type of material and the amount of silage produced, silage facilities need to be prepared in advance (Franco et al. 2016). Design a reasonable type of silo according to the size of the farm to ensure stability and safety and, at the same time, do a good job of waterproofing to avoid surface runoff water flowing into the silo. If old silos are used, they will need to be cleaned, sterilized, or restored prior to storage to ensure their storage effectiveness. When using silage facilities, attention should be paid to regular inspection and maintenance to ensure that the environment in the silo is dry, clean, and free of odor. To prevent waste and deterioration, the silage should be used judiciously in following the livestock's actual demands. When taking feed, care should be taken to keep the facilities intact and sealed to avoid mold growth and feed deterioration from exposure to air. To eradicate infections like dangerous microbes and parasite eggs, silage facilities should also be cleansed and disinfected often.

1.1 Tower Silo

Tower silo is permanent cylindrical structure, usually constructed from strong materials such as brick, cement, and steel. Their height and diameter are designed according to the size and needs of the feedlot, with towers generally ranging from a few meters to a dozen meters in height and several meters in diameter. This design is not only robust but also effective in withstanding the weight and pressure of the silage (Kolli et al. 2023).

Its advantages are (1) good silage effect and low loss rate. As the tower is closed and tight, the internal anaerobic environment is stable, which is conducive to the fermentation of lactic acid bacteria (LAB), thus inhibiting the growth of other stray bacteria, so the silage can be preserved for a longer time (Kim et al. 2021). At the same time, the silage tower is convenient for mechanized operation, whether it is loading or picking up, it can be completed by mechanical equipment, which greatly improves working efficiency. (2) Tower silo also has a certain adaptability and flexibility. Depending on the requirements, feedlots of varying sizes and locations can be planned and constructed. The silage tower is the best option for silage facilities in low-lying, high-water-table regions.

The disadvantages are (1) the construction of tower silos requires large investment due to its relatively high cost of construction. However, this investment is usually considered worthwhile in some countries and regions, taking into account the long-term benefits of its use and its importance for livestock production. (2) Tower silos are susceptible to natural disasters such as tornadoes, typhoons, and earthquakes due to their height.

However, tower silo is suitable for all sizes and types of farms. Through rational design and construction, tower silo can provide livestock with high-quality silage, guaranteeing their healthy growth and performance. In recent years, due to the popularity of bunker and roll bale silage modification techniques, the utilization of tower silos has shown a decreasing trend year by year (Savoie and Jofriet 2003).

1.2 Bunker Silo

Banker silo usually consists of three walls and a bottom, forming a semi-enclosed space. The walls and bottom are usually constructed using strong materials such as masonry and cement to ensure they are strong, durable, and impervious to water. The width, depth, and length of the bunker silo are designed according to the size and needs of the feedlot and usually 4–6 m wide, 2–4 m high, and up to 20–40 m or longer (Huhnke 2004).

Its advantages are as follows: (1) The bunker silo has a large storage capacity, which can meet the large demand for silages in large-scale feedlots. (2) The design of the silo facilitates mechanized loading and feeding, which can greatly improve work efficiency, reduce labor intensity and save manpower costs. (3) The bottom is usually designed to be inclined, which is conducive to drainage and prevents the accumulation of water, leading to mold and feed deterioration. (4) The bunker silos are suitable for feedlots in different climatic conditions, especially in arid and cold areas, and can effectively alleviate the problems of feed shortage and so on (Hofmann 2011).

The disadvantages are (1) the surface of the bunker silo is usually covered with plastic film, which is weakly closed, and the silage surface is prone to aerobic spoilage and leads to losses in silage fermentation. (2) The air at the opening is exposed to form an aerobic environment, which is prone to aerobic spoilage under high-temperature conditions (Hofmann 2011).

1.3 Trench Silo

The trench silo is an underground or semisubterranean facility specializing in the storage of silages in a variety of shapes, but circular and rectangular are the most common (Dvorachek 1914). A round silage cellar is generally a bucket structure with a large top and a small bottom, while a rectangular silage cellar has a wider bottom with an inclination to facilitate drainage and pick-up. The walls and bottom of the trench silo are usually made of masonry, cement, and other materials and are required to be smooth, solid, and impermeable to prevent mold and deterioration of the feed.

The advantages are as follows: (1) Trench silo effectively preserves the nutrients in the forage and reduce losses caused by oxidation and decomposition. Through sealed fermentation, the protein, vitamins, and other nutrients in silage are retained and improve the utilization rate and nutritional value of the feed. (2) Trench silo makes full use of underground or semi-underground space and reduces the occupied area of the ground, which is conducive to the reasonable planning and management of the feedlot. (3) The construction and use of trench silo are relatively simple, and it is easy for feeders to carry out daily management and maintenance. At the same time, the trench silo is also relatively convenient and can be flexibly adjusted according to the actual needs of livestock. (4) The geographical location and scale of the trench silo can be adjusted according to the actual situation of livestock feeding on pasture, and the best construction plan can be found.

The disadvantages are as follows: (1) The trench silo is not conducive to mechanized operation, and the silage material is not easy to be compacted. (2) The trench silo usually needs to be loaded manually, which increases labor intensity (Hill and Spielman 1943).

1.4 Silage Bag

Silage bags are plastic bags used for ensiling, mainly for the storage of forages and crop residues, and other feedstuffs, to maintain the freshness and nutrition of the feedstuffs through fermentation. Silage bags are usually made of polyethylene tube film, with a thickness between 0.8 and 1 mm (Savoie and Jofriet 2003).

The advantages are the following: (1) Silage bags are simple to make, low cost, a fast return on investment, and have higher comprehensive benefits. (2) Silage bags can effectively reduce mold, liquid flow, and feed loss during feeding. (3) Silage bags are not affected by season, sunshine and rainfall, and can be stacked in the open air, saving the cost of building silos and land. (4) Silage bags are convenient for centralized harvesting, drying and storage, convenient feeding and labor saving. (5) Silage bags can be recycled, do not pollute the environment, and are easy to transport and commercialize.

The disadvantages are as follows: (1) The plastic bags are not suitable for mass production of silages and mechanized operations. (2) Plastic bags are less durable and increases silage deterioration, hence not suitable for long-term silage preservation (Kung and Muck 2017).

1.5 Roll Bale Silage

Roll baler was created in Europe in the late 1970s. With the development of technology, roll balers and wrappers were applied to make roll bale silage. The wrapping was used with multilayer polyethylene (PE) films. Nowadays, wrapped silage has been widely accepted by dairy and beef cattle farms and has become an important modification technology in livestock production. Roll bale silage is random and easy to feed and can also be used as a distributed feed to commercialize the product.

Roll bale silage means that the harvested material is compacted and baled at high density by a baling machine and then wrapped in special silage plastic stretch film to form a sealed state and create an anaerobic fermentation environment. Roll bale silage production is recognized as a technology that compares favorably with traditional hay or silage (Coblentz and Akins 2018).

Its advantages are (1) flexible operation, easy to use, and not limited by weather factors: (2) randomness, the same equipment can be used for hay or roll bale silage making; (3) convenient and cost-saving transport, with no restriction on storage locations; (4) reduced field loss rate (legume leaves); (5) Long-term storage with good sealing; (6) less loss of juice from silage; (and 7) reduces labor intensity and saves time (Coblentz and Akins 2018).

The disadvantages are as follows: (1) Roll bale silage depends on the sealing of polythene film, and the quality of the film seal can easily affect the fermentation quality of silage. (2) The price of polythene film is high, which can easily increase the cost of feed production. (3) The used polyethylene film needs to be recycled, and improper recycling procedures can cause environmental pollution (Coblentz and Akins 2018).

1.6 Sausage Silage

Sausage silage is also a type of wrapped silage. A filling machine continuously, quickly, and consistently fills bags with chopped green fodder and seals under high pressure to create sausage silage.

The sausage silage is stored in the open air, making use of the air-blocking and light-shading functions of the stretch film bags to provide a better fermentation environment for LAB.

The advantages are as follows: (1) Mechanized operation with high efficiency and unlimited quantity of silage material. (2) Compared with roll bale silage, the amount of wrapping film is less, which can save the cost of feed production (Grant and Ferraretto 2018).

The disadvantages are (1) the setup place needs to be fixed and the storage place is limited. (2) Bird and rodent infestation cause breaches in the wrap which leads to silage deterioration (Grant and Ferraretto 2018).

2 Raw Material Mowing

The fermentation quality and nutritional value of silage are closely related to the stage of raw material mowing. An excessively early mowing time will result in a high moisture content, low dry matter (DM) production, and low nutrient content, while an excessively late mowing time will result in high lignin and cellulose content, which affect palatability and digestibility in livestock.

Silage preparation should be finished as quickly as possible if the forage is mowed during the proper harvesting season. This is to prevent the uncontrollably high growth of aerobic bacteria, which could compromise the silage preparation process.

The best mowing stage of different forages should be determined according to the dry matter yield and nutrient contents. For example, the best harvesting stage of maize silage is generally the yellow ripening stage, while the best harvesting time of forage is generally the pre-flowering stage.

3 Moisture Adjustment

Theoretically, the optimum moisture content for silage preparation is 60–70%, within which lactic acid fermentation is promoted and butyric acid fermentation is inhibited (Du et al. 2023). If the moisture is too low, the water activity of LAB is restricted, which is not favorable for silage fermentation (Passot et al. 2012). In addition, low moisture content makes compaction to expel trapped air in the cellar difficult, and this enhances the activities of aerobic bacteria and mold to cause deterioration of the feed. Therefore, delayed harvesting of forage should be avoided, and if silage is made from forage with lower moisture, pure water should be added to adjust the moisture content. Excessive moisture also makes compaction and clumping easy but facilitates the actions of *clostridia* and leads to nutrient loss. In the use of high-moisture forage silage modulation, the moisture adjustment method of wilting is often used by exposing the harvested forage to the sun. If climatic conditions, such as rain, are encountered and it is not possible to achieve suitable moisture for silage by wilting, the moisture content can be adjusted by adding dry materials such as hay and agricultural by-products. Microbiological additives such as LAB and cellulolytic enzymes can also be used for high-moisture material to ensure the fermentation quality of the silage (Okoye et al. 2023; Muck et al. 2018; Oladosu et al. 2016).

4 Chopping and Compaction of Material

The length of the chopped material significantly influences the anaerobic conditions within the silage silo and the quality of fermentation (Sun et al. 2020). Usually, the silage material is chopped to a length of approximately 1–2 cm before packing into the silo (Fig. 7.2). This practice helps improve the density of the silage, facilitates compaction, and ensures effective removal of air from the silo. Mechanical compaction methods, such as shovels and tractors, are used to ensure that the compaction density of the silo is high enough to prevent crack spaces and residual air. Compaction

Fig. 7.2 Harvesting and cutting of grass. (Photo by Cai)

can ensure maximum anaerobic conditions in the silo and promote lactic acid fermentation (Dunière et al. 2013). Therefore, chopping and compaction directly affect the fermentation quality of silage.

5 Sealing and Care of Silage

Silage material needs to be sealed promptly after filling and compacting to maintain anaerobic conditions and prevent the loss of water and nutrients. If the silo is not sealed promptly, lactic acid fermentation will be disrupted, leading to an increase in oxygen, moisture, and temperature, which makes the silage more vulnerable to mold. As shown in Fig. 7.3, silos are usually covered with plastic film and then compacted with waste tyres or heavy stones, which effectively improve the sealing of silage (Dunière et al. 2013). After the silo is sealed, the top and periphery of the silo should be well drained to avoid rainwater flowing into the silo. Protective fences should also be set up to avoid animals trampling or damaging the sealing plastic film, etc. At the same time, regular inspection should be carried out, and once the sealing film is found to be broken, it should be repaired in time.

Fig. 7.3 Silo sealing and silage storage. (Photo by Cai)

6 Opening the Silo for Use

Silage is typically ready for use after being stored for 30 days or longer. The silo should be opened only when there is a shortage of forage, allowing the livestock to access high-quality silage. Before opening the silo, it is essential to strictly check the fermentation quality of the silage. This can be done through sensory evaluation, focusing on both color and odor. High-quality silage will have a pleasant sour aroma, while poor-quality silage will emit a moldy or strong odor. If the silage has moldy parts, it needs to be cleaned up in time. When feeding poor-quality silage, the health of livestock needs to be carefully monitored daily and stopped as soon as any abnormality is detected. To ensure the quality of the silage when it is opened, a "one-day pick-up" approach should be adopted, avoiding "one pick-up for multiple days" or "multiple pick-ups for a day." These two incorrect silage pick-up methods tend to cause aerobic spoilage. After picking up the silage, the picking area should quickly be covered with the original plastic film to prevent excess air from entering the silo (Fig. 7.4).

Therefore, the technology and quality management of the silage modulation process is the key to the production of high-quality silage (Ferraretto et al. 2018). Adhering strictly to the technical guidelines for silage preparation and feeding livestock while using the correct methods can enhance the benefits and value of silage. This approach supports the sustainable development of the livestock industry.

Fig. 7.4 Opening of the bunker silo. (Photo by Cai)

References

Bilal AK, Adnan M, Rehman FU, Hasnain A, Usman M, Javed MS, et al. Role of silage in agriculture: a review. Green Rep. 2021;2:9–12.

Coblentz WK, Akins MS. Silage review: recent advances and future technologies for baled silages. J Dairy Sci. 2018;101(5):4075–92.

Du Z, Yang F, Fang J, Yamasaki S, Oya T, Nguluve D, et al. Silage preparation and sustainable livestock production of natural woody plant. Front Plant Sci. 2023;14:1253178.

Dunière L, Sindou J, Chaucheyras-Durand F, Chevallier I, Thévenot-Sergentet D. Silage processing and strategies to prevent persistence of undesirable microorganisms. Anim Feed Sci Technol. 2013;182(1–4):1–15.

Dvorachek HE. Silos and silage in Colorado. Agricultural Experiment Station of the Agricultural College of Colorado, Fort Collins, Colorado, USA; 1914.

Ferraretto LF, Shaver RD, Luck BD. Silage review: recent advances and future technologies for whole-plant and fractionated corn silage harvesting. J Dairy Sci. 2018;101(5):3937–51.

Franco RT, Buffière P, Bayard R. Ensiling for biogas production: critical parameters. A review. Biomass Bioenergy. 2016;94:94–104.

Grant RJ, Ferraretto LF. Silage review: silage feeding management: silage characteristics and dairy cow feeding behavior. J Dairy Sci. 2018;101(5):4111–21.

Hill OJ, Spielman AA. Stack-trench silos. Agriculture Extension Service, State College of Washington, Pullman, WA, USA; 1943.

Hofmann J. Design of food storage facilities. In: Hygienic design of food factories. Cambridge: Woodhead Publishing; 2011. p. 623–46.

Huhnke RL. Bunker silo sizing and management. Oklahoma State University, Stewart, Oklahoma, USA; 2004.

Kim DH, Lee KD, Choi KC. Role of LAB in silage fermentation: effect on nutritional quality and organic acid production—an overview. AIMS Agric Food. 2021; 6(1): 216–34.

Kolli VS, Garg S, Shirkole SS. Silos and bins. In: Transporting operations of food materials within food factories. Cambridge: Woodhead Publishing; 2023. p. 61–93.

Kung L Jr, Muck RE. Silage harvesting and storage. In: Beede DK, editor. Large dairy herd management. Champaign: American Dairy Science Association; 2017. p. 723–38.

Muck RE, Nadeau EMG, McAllister TA, Contreras-Govea FE, Santos MC, Kung L Jr. Silage review: recent advances and future uses of silage additives. J Dairy Sci. 2018;101(5):3980–4000.

Okoye CO, Wang Y, Gao L, Wu Y, Li X, Sun J, et al. The performance of lactic acid bacteria in silage production: a review of modern biotechnology for silage improvement. Microbiol Res. 2023;266:127212.

Oladosu Y, Rafii MY, Abdullah N, Magaji U, Hussin G, Ramli A, et al. Fermentation quality and additives: a case of rice straw silage. Biomed Res Int. 2016;2016:7985167.

Passot S, Cenard S, Douania I, Trelea IC, Fonseca F. Critical water activity and amorphous state for optimal preservation of lyophilised lactic acid bacteria. Food Chem. 2012;132(4):1699–705.

Savoie P, Jofriet JC. Silage storage. Silage Sci Tech. 2003;42:405–67.

Sun Z, Jia T, Gao R, Xu S, Wu Z, Wang B, et al. Effects of chopping length and additive on the fermentation quality and aerobic stability in silage of *Leymus chinensis*. PRO. 2020;8(10):1283.

Chapter 8
Fermented TMR Processing Technology

Hajime Kumagai

Abstract Fermented total mixed rations (TMR) preparation technology enables the preparation of fermented feeds for livestock according to the nutritional requirements of ruminants, utilizing various agricultural by-product resources. This is important in reducing the cost of feed production, improving livestock productivity, and promoting sustainable livestock production. This chapter introduces the role played by the TMR center in effectively utilizing regional feed resources, preparation and storage, production and distribution, etc. It also introduces the role of TMR in maintaining the nutritional balance of livestock, promoting enterprise stability, expanding production scale, improving production environment, improving livestock feeding technology, and increasing farmers' income. Introducing the TMR systems effectively enlarged the farm scale and shortened working time. It is necessary to understand that the chemical composition of by-products, self-produced silage, and imported hay included in TMR have wide varieties of chemical composition. Periodical monitoring of each TMR source is important to utilize the ingredients. The nutritional characteristics of typical food by-products, i.e., soybean curd residue, brewer's grain, sake lees, winery grape pomace and sediment, noodle waste, and potato waste, as ingredients of TMR were discussed. Previous studies on the merits of feeding fermented TMR to lactating dairy cows and fattening beef cattle were proposed.

Keywords Cattle · Fermented total mixed ration · Food by-products · Total mixed ration center

Fermented total mixed rations (TMR) preparation technology enables the preparation of fermented feeds for livestock according to the nutritional requirements of ruminants, utilizing various agricultural by-product resources. This is important in

---

H. Kumagai (✉)
Kyoto University, Kyoto, Japan
e-mail: kumagai.hajime.7n@kyoto-u.ac.jp

Y. Cai, K. Ataku (eds.), *Cultural History and Modern Production Technology of Silage*, https://doi.org/10.1007/978-981-96-5787-2_8

reducing the cost of feed production, improving livestock productivity, and promoting sustainable livestock production. This chapter introduces the role played by the TMR center in effectively utilizing regional feed resources (formulation design of fermented TMR), preparation and storage, production and distribution, etc. It also introduces the role of TMR in maintaining the nutritional balance of livestock, promoting enterprise stability, expanding production scale, improving production environment, improving livestock feeding technology, and increasing farmers' income.

1 Role Played by the TMR Center in Effectively Utilizing Regional Feed Resources

1.1 Importance of Development of Local TMR Centers

The TMR feeding is a system mixing all feeds, roughage, concentrate, minerals, vitamins etc., and feeding to meet livestock requirements. The system is useful for avoiding feed selection, free feeding, quantitative nutrition design, labor saving in feeding, and stability of rumen fermentation. Though the system requires cost and labor for roughage chopping and all feed mixing, it contributes to large-scale livestock farming especially. The roles played by TMR centers are collecting and mixing the local feed resources to produce and distribute the TMR for dairy and beef farmers. Table 8.1 shows the effects of introducing TMR systems on some cows and working time for feed preparation and feeding in dairy farms from Tokachi District, Hokkaido, Japan (Abe 2000c). The system enlarged farm scale and shortened

Table 8.1 Effects of the introduction of TMR systems on working time for feed preparation and feeding in dairy farms

	Before introduction of TMR systems		After introduction of TMR systems	
Farms	Number of multiparous cows	Working time (min/day)	Number of multiparous cows	Working time (min/day)
A	25	230	40	39
B	52	175	60	138
C	60	210	148	222
D	47	235	101	90
E	39	100	52	64
F	48	250	52	85
G	52	340	59	167
H	53	278	71	174
I	70	240	70	130
J	60	110	90	33
K	78	225	102	82
L	23	60	75	65

Division of animal production, extension workers of Tokachi Branch 1996

Table 8.2 Introduction of TMR feeding and free stall in dairy farms (%)

	Number of cows	TMR feeding	Free stall
Hokkaido Division	Total	11.7	13.5
	30–39	2.6	3.3
	40–49	3.4	4.2
	50–74	10.8	12.5
	75–99	39.9	44.9
	100–149	69.5	74.6
	>150	84.8	92.6
Honshu, Shikoku, and Kyushu Divisions	Total	8.1	4.7
	20–29	3.2	1.3
	30–39	8.6	3.1
	40–49	14.8	5.3
	50–74	30.5	18.9
	75–99	53.9	48.9
	100–149	69.8	72.5
	>150	82.9	82.9

Data from Japan Dairy Council 2009

working time. Table 8.2 presents percentages of dairy farms introducing TMR systems and free stalls (Abe 2000c). Forty percent or more farms raising more than 75 cows introduced TMR systems and free stalls.

In building a TMR center, different industries in a region should work together to form a regional industrial complex. Abe (2000b) proposed the following necessities to build a TMR center in a region:

- Use of unused food manufacturing by-products
- Use of self-supplied roughage
- Establishment of agricultural systems based on regional recycling material and development of sustainable agriculture
- Improvement of food self-sufficiency
- Promote understanding of agriculture among the residents

These will form the basis for diverse agriculture and livestock farming.

The following concepts are advocated by Abe (2000b) in the construction of the TMR center.

- A TMR center should be built in a region. In addition to various food manufacturing by-products being brought in, self-supplied roughage and concentrated feed ingredients are used for the TMR materials.
- The TMR center regularly designs feed formulation and monitors the quality of the feed ingredients for TMR.
- The TMR prepared at the TMR center is delivered to dairy and fattening farms in the region by local transport companies.

- Livestock waste is returned to the fields of livestock farmers and crop farmers by feed production contractors. The feed production contractors cultivate pasture and feed crops, prepare silage, and prepare and deliver the compost from livestock farmers.

For beef cattle fattening systems, Oishi et al. (2011) proposed an application of the modified feed formulation to optimize economic and environmental criteria with food by-products, and the use of the by-products possibly reduces costs and environmental impact.

1.2 Types of TMR: Dry TMR, Fresh TMR, and Fermented TMR

Three types of TMR are used in farms. They include dry TMR, which is made by mixing several ingredients of dry concentrate and roughage sources and is distributed to farms. Other concentrate or roughage sources are often fed together with TMR to animals in order to meet their nutrient requirements. Fresh TMR, this is made by mixing several high-moisture food processing by-products and silages with dry concentrate and roughage sources. This TMR is easily damaged by aerobic deterioration from mold and yeast because of the high moisture content. Therefore, TMR should be prepared daily and fed as early as possible after preparation. Fermented TMR, this type of TMR was developed to improve upon the preservation capacity of fresh TMR by anaerobic fermentation in sealed containers. The TMR is often degassed and packed in flexible intermediate bulk containers with plastic inners or packed and wrapped with plastic films by cutting round balers and bale wrappers. The fermented TMR can be distributed in containers or bales from TMR centers to farmers.

1.3 Merits of Fermented TMR

First, the fermented TMR can be preserved for a long period because of anaerobic fermentation during the preservation; at least 2 months in flexible intermediate bulk containers. The TMR can be preserved longer in cutting round balers with bale wrappers. Second, food and agricultural by-products regionally generated with high moisture can be easily utilized as ingredients of fermented TMR at a low price. Third, low-palatable feed resources can be incorporated into the TMR by mixing and fermenting with palatable resources: concentrates or high-quality roughage sources. Fourth, the TMR is less damaged by aerobic deterioration after opening than fresh TMR. The property is an advantage for utilizing the TMR in the summer.

2 Preparation and Storage of Regional Feed Resources and Production and Monitoring of Processed TMR

2.1 Characteristics of TMR Sources

Most of the unutilized- or low-utilized by-products from food processing factories have high moisture: 70–80% or more moisture (Abe 2000d). The characteristics are negative for transport and storage. For TMR ingredients, the moisture should be reduced to around 60% by screw-pressing and/or electro-osmosis, and the ingredients are packed in flexible intermediate bulk containers with plastic inners necessary to promote lactic fermentation. Self-produced grass silage and maize silage included in TMR have wide varieties of moisture content.

2.2 Variation of Chemical Composition

Differences in resources and processes cause variations in chemical composition. Therefore, the variations are large among the food processing factories and small among the daily by-products in the same factories. It is important to understand the chemical composition of by-products for utilization. Self-produced grass silage, maize silage, and imported hay included in TMR have wide varieties of chemical composition.

2.3 Carbohydrate Contents

The digestibilities of sugar, starch, and organic acids are almost 100% in ruminants. The high contents of the carbohydrates in by-products indicate the characteristics of concentrate feed. The digestibility of NDF in by-products influences the energy utilization of by-products. In general, the NDF contents in food by-products are lower than those in agricultural by-products but higher than those in grains.

2.4 Crude Protein (CP) Contents

The CP contents of several food by-products are high. However, there are large variations in the CP contents of food by-products from different factories. Thus, it is important to understand the chemical composition of the by-products for utilization.

2.5 Fat Contents

Several food by-products have high ether extract (EE) contents. The high EE contents increase energy contents but suppress the growth of ruminal microorganisms and depress protein digestibility in the rumen. The EE content should be less than 6% of the total feed.

3 Characteristics of Typical Food By-Products as Ingredients of TMR

3.1 Soybean Curd Residue

Soybean curd residue is produced in factories making soybean curd and soy milk, where soybeans are soaked in water, mashed, heated, and churned, soy milk is separated by squeezing, and soybean curd residue is produced. Differences in the resources of soybeans and processes cause large variations in the chemical composition of soybean curd residue. It is easily damaged by aerobic deterioration from mold and yeast because of moisture. Therefore, soybean curd residue should be preserved in anaerobic conditions. Soybean curd residue has relatively high CP, EE, and NDFom contents, averaging 26.1%, 11.5%, and 33.6%, respectively, on a DM basis (NARO 2010) compared with grains and beans. The digestibilities of CP, EE, and total fiber are also high, showing 85 and 84% (NARO 2010) and 82.5% (Abe 2000a), respectively, and the TDN content is 91.1% (NARO 2010).

3.2 Brewer's Grain

Brewer's grain is produced in breweries making beer, where derooting malt is milled and mixed with rice or starch in hot water, the starch in the mixture is saccharified, wort is separated, and brewer's grain is produced. The moisture content is around 80%. Recently, the residue is dewatered, mixed with molasses and lactic acid bacteria and preserved in anaerobic conditions. Differences in barley breed and process cause variations in chemical composition. Brewer's grain has relatively high CP and EE contents, averaging 24.8% and 10.1%, respectively, and considerably high NDFom content, 67.2%, on a DM basis (NARO 2010). The digestibilities of CP and EE are 67% and 88%, respectively, and the TDN content is 79.0% (NARO 2010). The digestion speed of fiber fraction in brewer's grain is slower than that of soybean curd residue.

3.3 *Sake Lees*

Sake lees are produced in breweries that make Japanese rice wine, i.e., sake, where polished rice is brewed. Two types of process are adopted for breweries. In a conventional brewing process, steamed rice is saccharified by malted rice and fermented by sake yeast. Sake is separated, and then lees are produced. In a new brewing process, rice is liquefied by adding α-amylase to increase the amount of sugar utilized by the sake yeast. Thus, the liquefied sake lees contain more yeast and yeast metabolites than those obtained through the conventional brewing process (Ishida et al. 2013; Izu et al. 2019). Sake lees have high CP contents compared with the other food by-products. However, differences in the process cause variations in chemical composition. Lees produced by the conventional process have an average of 36.9% CP on a DM basis (NARO 2010). Liquefied sake lees have a high CP content, 64.0% on a DM basis, on average (Katsumata et al. 2021).

3.4 *Winery Grape Pomace and Winery Sediment*

There are various kinds of winery wastes: stalks, winery grape pomace, and winery sediment. Winery grape pomace is produced by the process of crushing the grapes to extract their juice after fermentation. Winery sediment is produced by removal with fining agents such as bentonite, albumens, or gelatins. Winery grape pomace has relatively low CP and high EE and NDFom contents, averaging 10.4%, 10.4%, and 50.6%, respectively (NARO 2010), and 9.5%, 8.5%, and 47.4%, respectively (Ishida et al. 2015), on a DM basis because of the inclusion of stalks, skins, and seeds of grapes. Winery sediment has relatively high CP and crude ash (CA) and low EE and NDFom contents, 19.8%, 21.4%, 2.8%, and 6.4%, respectively, on a DM basis (Ishida et al. 2015). Both by-products have high polyphenol contents and DPPH radical scavenging activities, showing 10.0 and 12.7 mg CAE/g and 22.5 and 33.4 umol/TE/g, respectively, for winery grape pomace and winery sediment (Ishida et al. 2015).

3.5 *Noodle Waste*

Noodle waste is waste from producing commercial noodles, udon noodles, buckwheat noodles, Chinese noodles, etc. Noodle waste has substantially high nonfibrous carbohydrates, averagely 83.2% on a DM basis (Kitagawa et al. 2009); thus the TDN, 80.5%, is comparable to barley and oat grains, 84.0 and 80.1% (NARO 2010), respectively.

3.6 Potato Waste

Potato waste is the surface of raw potato including potato skin produced from factories making potato snacks. It has relatively high nonfibrous carbohydrate and TDN contents, 64.6 and 71.3% on a DM basis, and low CP, 6.9% (Kitagawa et al. 2009).

4 Remarks for Formulation Design of Fermented TMR in the TMR Centers

As mentioned in "Sect. 2.1 Characteristics of TMR sources," TMR sources, food by-products, pasture silage, and hay, have large varieties in moisture and chemical compositions. Periodical monitoring of each TMR source is important to utilize the ingredients. For the formulation design of fermented TMR, indices of nutritional management for the production stages of animals and monitoring the chemical composition and fermentation quality of the TMR and production of animals are required.

5 Studies of Formulation Design, Chemical Composition of Fermented TMR and Animal Performance

Original articles concerning the formulation design, chemical composition of fermented TMR, and animal performance have been published: for lactating dairy cows (Inoue et al. 1998; Yani et al. 2012; Yao et al. 2020) and fattening beef cattle (Fujita and Tomiya 1995; Ishida et al. 2012a, b). Outlines of several case studies on fermented TMR using local food by-products, soybean curd residue, potato waste, noodle waste, soy sauce cake etc. are described as follows.

6 Fermented TMR for Lactating Dairy Cows

6.1 Effects of Feeding Fermented TMR Including Soybean Curd Residue for Dairy Cows

Two types of fermented TMR, mixing 30.0% soybean curd residue, 27.8% maize silage, 14.1% hays, and 27.7% commercial concentrate ingredients and mixture (treatment: T), and mixing 38.1% maize silage, 11.2% hays, and 35.2% commercial concentrate ingredients and mixture one conventional (control: C) were prepared on a fresh matter basis (Inoue et al. 1998). The water content of the TMR was the same: 50% and the CP, NDF and TDN contents of the 2 types of TMR were the same (Table 8.3). Cows were assigned to a double reversal trial experimental design to

Table 8.3 Effects of TMR feeding mixed with soybean curd residue on dairy cows

Feed ingredients[a]	T ($n = 9$)	C ($n = 9$)
Soybean curd residue	30.0	—
Maize silage	27.8	38.1
Sudan grass hay	6.4	10.6
Alfalfa pellet	7.7	0.6
Wheat bran	8.2	10.7
Wheat bran with flour	9.5	—
Soybean meal	–	2.3
Commercial concentrate	10.0	22.2
Water	0.4	15.5
Chemical composition[b]		
Crude protein	16.6	16.6
Ether extract	4.0	2.8
Neutral detergent fiber	36.0	36.0
Acid detergent fiber	21.0	19.4
Total digestible nutrients	74.0	74.0
Animal performance		
Dry matter intake (kg/day)	16.6	17.9
Milk yield (kg/day)	20.0	22.5
Fat (%)	4.44	4.33
Protein (%)	3.42	3.56
Lactose (%)	4.68	4.59
Solid nonfat (%)	9.15	9.20
Total solid (%)	13.63	13.52

[a]% on a fresh matter basis
[b]% on a dry matter basis

measure feed intake, milk yield, and milk composition. No significant differences were observed between the treatments. The feed cost was 659 and 972 yen/day/head for T and C, respectively.

6.2 *Effects of Utilization of Local Food By-Products as Fermented TMR on Fermentation Quality, Digestibility and Rumen Condition in Sheep, and Performance in Dairy Cows*

Four wethers were used in a 4 × 4 Latin square design experiment to evaluate in vivo digestibility of total mixed ration (TMR) silage with food by-products for dairy cows, and the ruminal condition and nitrogen (N) balance were examined (Yani et al. 2015). Five by-products (i.e., potato waste, noodle waste, soybean curd residue, soy sauce cake, and green tea waste) were obtained. Four types of TMR silage were used: control (C) containing roughage and commercial concentrate, T1:20% and T1:40% containing the five by-products replacing 20% and 40% of the commercial concentrate on a dry matter (DM) basis, respectively, and T2:40% containing three by-products (potato waste, noodle waste, and soybean curd residue) replacing 40% of the commercial concentrate on a DM basis. Water was added for C to adjust moisture content to the same as T1:40% and T2:40%. The ingredients were mixed and preserved in oil drum silos for 4 months. The TMR silages showed 4.02–4.44% and 1.75–2.19% for pH and lactic acid contents, respectively. The digestibility of DM and neutral detergent fiber and total digestible nutrient content were higher ($p < 0.05$) for T2:40% feeding than for C feeding. Urinary nitrogen excretion tended to be lower ($p = 0.07$) for T2:40% than for C. The results suggested that 40% replacement of commercial concentrate with the three food by-products can be most suitable for TMR silage.

Six primiparous Holstein cows were assigned to a double reversal trial experimental design to measure feed intake, milk yield, and milk composition fed with fermented TMR of treatment (T, T2:40% in the sheep experiment) or control (C) (Yani et al. 2012). The T diet contained 42.8% roughage, 34.4% commercial concentrate, and 22.9% food by-products (11.5% potato waste, 7.9% soybean curd residue, and 3.3% noodle waste) on a DM basis, while the C diet contained 42.8% roughage and 57.3% commercial concentrate without food by-products (Table 8.4). The feed intake and adequacy of TDN were higher in the cows fed T than C ($P < 0.01$), and the CP intake was higher in the cows fed T than C ($P < 0.05$). No significant differences were observed for milk yield and milk composition between C and T groups, while the cows fed C had higher milk urea nitrogen content than those fed T ($P < 0.05$). with the values of 10.4 vs. 9.6 mg/dL. Due to high intake and TDN adequacy in T, the utilization of food by-products can be more suitable for replacing conventional concentrate as a TMR ingredient for dairy cows in a viewpoint of nitrogen utilization.

7 Fermented TMR for Fattening Beef Cattle

7.1 *Effects of Feeding Fermented TMR Including Soybean Curd Residue for Fattening Japanese Black Steers*

Two types of fermented TMR, mixing 15.4 and 19.2% soybean curd residue with maize silage, soybean stem and pods in early and late fattening period, respectively, on a fresh matter basis (treatment: T), mixing large amount of maize silage 63.0 and 50.0% in early and late fattening period, respectively, on a fresh matter basis (control 1: C1), and one conventional feed with grains and rice straw (control 2: C2) were prepared (Fujita and Tomiya 1995). The CP and TDN contents of the three types of feed were the same. T and C1 had lower sugar, starch, and organic acid contents and higher NDF contents than C2 (Table 8.5).

The higher DM intake (DMI), daily gain, carcass weight rib eye area and beef marbling score in steers fed T might have been due to higher digestible NDF from soybean curd residue resulting in higher DMI in steers.

Table 8.4 Effects of utilization of local food by-products as fermented TMR on dairy cows

	T ($n = 9$)	C ($n = 9$)
Feed ingredients[1]		
Maize silage	20.7	20.7
Alfalfa hay	10.9	10.9
Timothy hay	11.2	11.2
Commercial concentrate	34.4	57.3
Potato waste	11.5	—
Soybean curd residue	7.9	–
Noodle waste	3.3	—
Chemical composition[1,2]		
Dry matter	41.2	43.2
Crude protein	14.9	15.0
Ether extract	3.7	3.6
ANDFom	45.5	44.6
ADFom	27.3	26.1
Total digestible nutrients	69.8[a]	63.7[b]
Nutrient adequacy[3] (%)		
Dry matter	107.7	105.4
Crude protein	117.8	116.8
Total digestible nutrients	108.3[c]	97.7[d]
Animal performance		
Dry matter intake (kg/day)	25.3	24.9
Crude protein intake (kg/day)	3.79[a]	3.73[b]
Total digestible nutrients intake (kg/day)	17.7[c]	15.8[d]
Milk yield (kg/day)	31.1	30.9
4% fat collected milk yield (kg/day)	32.0	31.6
Fat (%)	4.26	4.21
Protein (%)	3.66	3.63
Lactose (%)	4.40	4.41
Solid nonfat (%)	9.06	9.05
Somatic cell count (10^4/mL)	6.08	5.36
Milk urea nitrogen (mg/mL)	9.61[a]	10.35[b]

[1]% on a fresh matter basis
[2]% on a dry matter basis except for dry matter
[3]Compared with NARO (2007)
[abcd]Means in the same row with different superscripts differ significantly (ab: $P < 0.05$, cd: $P < 0.01$)

7.2 *Effects of Feeding Fermented TMR Including Local Food By-Products for Fattening Japanese Black Heifers*

Fermented TMR was produced by mixing 38.6% noodle waste, 6.8% soybean curd residue, 5.6% soy sauce cake, 2.2% potato waste, 38.0% commercial concentrate mix for beef fattening, and 8.9% wheat straw on a dry matter (DM) basis (Ishida et al. 2012b). The amount of food by-products in the TMR silage was chosen to be similar to CP, TDN, Ca, and P contents in the concentrate mix. After mixing all

Table 8.5 Effects of TMR feeding mixed with soybean curd residue on fattening Japanese Black steers

	T1 ($n = 5$)	C1 ($n = 4$)	C2 ($n = 5$)
Feed ingredients[a]			
Maize grain	25.0	25.0	45.5
Barley grain	16.5	12.5	27.3
Wheat bran	7.5	11.5	16.4
Salt and minerals	1.0	1.0	1.8
Soybean curd residue	19.2	—	–
Maize silage	25.0	50.0	–
Soybean stem and pods	2.9	—	–
Wheat stem	2.9	–	–
Rice straw	—	—	9.0
Chemical composition[b]			
Crude protein	12.0	11.9	11.1
Ether extract	3.7	3.2	3.2
Sugar, starch, and organic acids	54.2	56.2	60.0
Neutral detergent fiber	25.2	23.7	20.2
Total digestible nutrients	81.0	81.0	80.3
Growth performance			
Dry matter intake (kg/day)	8.1	7.0	7.2
Initial body weight (kg)	319	301	306
Final body weight (kg)	719	624	647
Total gain (kg)	400	323	341
Daily gain (kg)	0.82	0.66	0.70
Ruminal fermentation characteristics[c]			
Acetic acid	49.3	48.5	45.0
Propionic acid	28.8	30.9	37.5
Carcass traits			
Carcass weight (kg)	439	377	384
Rib eye area (cm^2)	46.6	42.0	43.6
BMS No.	7.2	6.8	6.0
Beef marbling grade	4.2	4.3	4.0
BCS No.	4.6	5.0	4.6
Firmness and texture grade	4.2	3.8	3.6

[a]In the late fattening period, % on a fresh matter basis
[b]% on a dry matter basis
[c]At 64 weeks of fattening. % of total short-chain fatty acids

ingredients, the materials were ensiled in flexible container bag silos for 1 month. A switchback experiment was conducted with six 31.3-month-old Japanese Black fattening heifers. The animals were divided into two groups with similar averaged BW between the groups and were allotted to one of the two treatments: (1) control (C: feeding commercial concentrate at 1.4% of BW and wheat straw at 0.3% of BW on a DM basis) and (2) experimental treatment (T: replacing 40% of commercial concentrate in C with food by-products in the TMR silage, i.e., feeding food by-products at 0.56% of BW on a DM basis).

The fermented TMR had relatively lower CP, EE, aNDFom, and ash contents, while higher OM and ADFom compared with the commercial concentrate mix (Table 8.6). Propionic acid or butyric acid was not detected in the TMR silage, and its V-score was 92.6. The DM, CP, and OM digestibility of the TMR silage were numerically lower than those of the commercial concentrate mix determined in the companion study (83.61, 80.31 and 86.47%, respectively) (Ishida et al. 2012a). The nutrient consumption of the heifers in the feeding experiment was not different between the treatments. The BW, daily gain, or feed conversion ratio did not significantly differ between the treatments.

Rumen pH before and after morning feeding was similar in CC and TMR, and there are no significant differences in pH changes between the treatments. Any measured blood components were different between the treatments.

The present results showed that the TMR silage, including noodle waste, soybean curd residue, soy sauce cake, and potato waste, was well preserved with high fermentation quality and had no adverse effects on growth performance, blood components, or ruminal fermentation in heifers. It may be concluded that the TMR silage, including various food by-products, can be used as a substitute for a barley-and-corn-based concentrate for beef fattening.

Table 8.6 Effects of feeding fermented TMR, including local food by-products for fattening Japanese Black heifers

	T (*n* = 6)	C (*n* = 6)
Feed ingredients[a]		
Commercial concentrate	0.44	1.40
Wheat straw	0.21	0.30
Fermented TMR[b]	1.05	—
Chemical composition[c]		
Dry matter	42.3	85.8
Crude protein	12.5	13.4
Ether extract	3.5	7.3
ANDFom	17.8	19.9
ADFom	10.4	7.4
Total digestible nutrients	79.9	83.5
Growth performance		
Dry matter intake (kg/day)	7.62	7.37
Initial body weight (kg)	582	581
Final bodyweight(kg)	602	607
Total gain (kg)	21	26
Daily gain (kg)	0.49	0.62
Ruminal pH		
Before morning feeding	8.06	7.78
Four hours after morning feeding	7.46	7.27

[a] % of body weight on a dry matter basis, daily
[b] 38.6% noodle waste, 6.8% soybean curd residue, 5.6% soy sauce cake, 2.2% potato waste, 38.0% commercial concentrate mix for beef fattening and 8.9% wheat straw on a dry matter basis
[c] % in the fermented TMR for T and commercial concentrate for C on a dry matter basis except for dry matter

References

Abe R. Feed characteristics of food by-products and TMR materials. In: Utilization of food by-products and TMR center (ed). Research and Development Center for Dairy Farming. Sapporo: Sanyo Printing Co. Ltd.; 2000a. pp. 43–73. (in Japanese).

Abe R. Problems in building a TMR center. In: Utilization of food by-products and TMR center (ed). Research and Development Center for Dairy Farming. Sapporo: Sanyo Printing Co. Ltd.; 2000b. pp. 25–42. (in Japanese).

Abe R. TMR and TMR centers. In: Utilization of food by-products and TMR center (ed). Research and Development Center for Dairy Farming. Sapporo: Sanyo Printing Co. Ltd.; 2000c. pp. 3–24. (in Japanese).

Abe R. Un-utilized organic feed resources and characteristics. In: Handbook of usage of un-utilized organic resources for feeds (ed). Science Forum. Tokyo: Tokyo Bunkyudo Co. Ltd.; 2000d. pp. 43–5. (in Japanese).

Fujita K, Tomiya T. Feeding experiment of TMR including soybean curd residue, maize silage, and agricultural by-products. In: Effects of feeding condition on feed intake, ruminal characteristics, body weight gain, carcass traits and cost of production in Japanese black steers, Memoirs of National Institute of Animal Industry (ed), National Institute of Animal Industry, Ministry of Agriculture, Forestry and Fisheries. Tsukuba: Kohmura Printing Co. Ltd.; 1995. pp. 71–7. (in Japanese).

Inoue T, Nagai S, Nakazato B, Yoshida T. TMR silage utilizing soybean curd residue on lactation performance. Bull Nagasaki Prefectural Livestock Exp Stn. 1998;7:6–8. (in Japanese).

Ishida K, Yani S, Kitagawa M, Oishi K, Hirooka H, Kumagai H. Effects of adding food by-products mainly including noodle waste to total mixed ration silage on fermentation quality, feed intake, digestibility, nitrogen utilization, and ruminal fermentation in wethers. Anim Sci J. 2012a;83:735–42.

Ishida K, Yani S, Kitagawa M, Okano K, Oishi K, Hirooka H, et al. Effects of adding food by-products mainly including noodle waste to total mixed ration silage on its feed characteristics, growing performance, blood metabolite concentrations, and ruminal fermentation in Japanese Black fattening heifers. Bull Beef Cattle Sci. 2012b;93:10–8. (in Japanese).

Ishida T, Nishiumi S, Tanahashi T, Yamasaki A, Yamazaki A, Akashi T, et al. Linoleoyl ethanolamide reduces lipopolysaccharide-induced inflammation in macrophages and ameliorates 2, 4-dinitrofluorobenzene-induced contact dermatitis in mice. Eur J Pharmacol. 2013;699:6–13.

Ishida K, Kishi Y, Oishi K, Hirooka H, Kumagai H. Effects of feeding polyphenol-rich winery wastes on digestibility, nitrogen utilization, ruminal fermentation, antioxidant status, and oxidative stress in wethers. Anim Sci J. 2015;86:260–9.

Izu H, Yamashita S, Arima H, Fujii T. Nutritional characterization of sake cake (sake-kasu) after heat-drying and freeze-drying. Biosci Biotech Bioch. 2019;83:1477–83.

Katsumata S, Yamanaka S, Yamauch M, Hirooka H, Kumagai H. Evaluation of the chemical and biological characteristics of sake lees. XXIV international grassland congress and XI international rangeland congress 2021. Online virtual meeting (Nairobi, Kenya). 2022.

Kitagawa M, Kumagai H, Okano K, Murakami T, Goda S, Azumai S. A survey on abundance of food by-products produced in Nantan District, Kyoto. Proceedings of Kansai Animal Science Society conference. 2009. pp. 24. II-9. (in Japanese).

National Agriculture and Food Research Organization, NARO. Japanese feeding standard for dairy cattle (2006). Tokyo: Japan Livestock Industry Association; 2007.

National Agriculture and Food Research Organization, NARO. Standard tables of feed composition in Japan (2009). Tokyo: Japan Livestock Industry Association; 2010.

Oishi K, Kumagai H, Hirooka H. Application of the modified feed formulation to optimize economic and environmental criteria in beef cattle fattening systems with food by-products. Anim Feed Sci Technol. 2011;165:38–50.

Yani S, Ishida K, Goda S, Azumai S, Murakami T, Kitagawa M, et al. Evaluation of food by-products mainly including potato waste and soybean curd residue as total mixed ration silage for lactating dairy cows. Evolution and future challenges of grasslands and grassland agriculture in East Asia. Proceedings of the 4th Japan-China-Korea grassland conference. 2012. pp. 200–1.

Yani S, Ishida K, Goda S, Azumai S, Murakami T, Kitagawa M, et al. Effects of utilization of local food by-products as total mixed ration silage materials on fermentation quality and intake, digestibility, rumen condition, and nitrogen availability in sheep. Anim Sci J. 2015;86(2):174–80.

Yao KY, Wei ZH, Xie XX, Wang DM, Liu HY, Fang D, et al. Lactation performance and nitrogen utilization of dairy cows on diets including unfermented or fermented yellow wine lees mix. Livest Sci. 2020;236:104025.

Chapter 9
Silage Additives

Qinhua Liu and Jianguo Zhang

Abstract Silage additives are vital in livestock production. This chapter reviews them, categorizing them into biological and chemical groups, among others. It discusses their effects on fermentation, aerobic stability, and livestock nutrition. Lactic acid bacteria, enzymes, and other additives improve silage quality and animal performance. The text explores recent findings on silage additives since 2003, including new additives from 2020. The aim is to enhance the understanding and use of additives, preserving feed nutritional value and reducing losses in livestock production.

Keywords Silage additives · Lactic acid bacteria · Fermentation quality · Aerobic stability · Animal performance

Silage additives have assisted in forage preservation for many years. Silage additives typically belong to one of the four main groups according to their impact on the conservation of silage: (1) stimulants for fermentation, (2) inhibitors of fermentation, (3) inhibitors of aerobic degradation, and (4) nutrients and absorbents (McDonald et al. 1992; Kung et al. 2003). With the development of biotechnology and silage technology, new silage additives have emerged, and they play a crucial role in the livestock production process. The objective of this study is to compile the latest findings on silage additives that have been published since the assessment reported by Kung et al. (2003), from new insight to highlight their advantages and utilization. Furthermore, additives in their infancy or on the horizon from 2020 are examined in more detail. Historically, silage additives have been known to operate through multiple mechanisms, drawing from the four aforementioned categories.

Q. Liu (✉)
Yunnan Agricultural University, Kunming, China
e-mail: liuqinhua@njau.edu.com

J. Zhang
South China Agricultural University, Guangzhou, China
e-mail: zhangjg@scau.edu.cn

Y. Cai, K. Ataku (eds.), *Cultural History and Modern Production Technology of Silage*, https://doi.org/10.1007/978-981-96-5787-2_9

However, the above classification is relatively complex and difficult for beginners, masters, and producers to merit their use. Given this, silage additives are divided into only two categories—biological and chemical additives—and further explained by detailed application of additives belonging to each category, as well as the latest combined application of biological and chemical silage additives. For each additive, we will discuss the effects on silage fermentation, aerobic stability, and livestock intake and utilization. We hope that this chapter improves the understanding and cognition of readers to meet their needs from the views of keeping good feed nutritional value and reducing losses.

1 Biological Additive

1.1 Lactic Acid Bacteria (LAB)

1.1.1 Homo-Fermentative LAB for Improving Silage Fermentation

The primary consideration in making silage is to improve its fermentation quality, which is naturally linked to homofermentative LAB that ferments one molecule of glucose into two molecules of lactic acid. The homofermentative LAB such as *Pediococcus (P.), Streptococcus, Enterococcus (E.)*, and *Lactiplantibacillus (L.)* are frequently applied in silage production (Fig. 9.1). However, certain species, which are classified above the genus level, are taxonomically recognized as facultatively heterofermentative LAB rather than obligate homofermentative species. *L. plantarum, Lactobacillus casei, Enterococcus faecium*, and various *Pediococcus* species were commonly found as additives. Silage inoculated with these LAB exhibited homolactic fermentation with lower pH value, contents of acetic acid and ammonia-N, higher lactic acid content, and a higher ratio of lactic acid to acetic acid (>3) compared with the untreated silages. The inoculation of *P. pentosaceus* in silage resulted in a higher dry matter digestibility than *L. plantarum* (Kung et al. 2003). Li et al. found that *Enterococcus faecalis* JF85 and *E. faecium* Y83 isolated from the

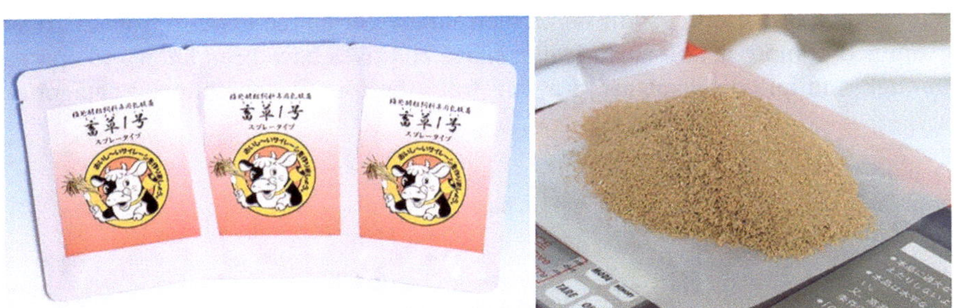

Fig. 9.1 Commercially available lactic acid bacteria inoculant (Snow Brand Seed Co., Ltd., Sapporo, Japan) and freeze-dried powder. (Photos by Cai)

rumen could degrade lignocellulose and improve the fermentation quality of *Pennisetum sinese*. Also, many species of *Pediococcus* have been recommended for probiotic usage because of their antioxidant, anti-inflammatory, detoxifying, and lipid-lowering qualities (Jiang et al. 2021). Additionally, homofermentative LAB have been reported to improve animal performance by 3–5% due to the production of minimal concentrations of ethanol, acetic, and butyric acids (Muck et al. 2018). Nevertheless, this effect is not constant and is also affected by the silage materials. The DM loss of some silage is not low.

1.1.2 Heterofermentative LAB for Improving Aerobic Stability of Silage

Heterofermentative (facultative and obligate) LAB are members of the *Lactobacillaceae* family and encompass genera such as *Lactobacillus, Oenococcus, Leuconostoc, and Weissella*. Four common groups that belonged *to Lentilactobacillus* are *Lentilactobacillus buchneri, L. plantarum, L. reuteri* group, and *L. casei*. Both the *Lentilactobacillus buchneri* and *L. reuteri* groups contain obligate heterofermenters, but the former has been the sole source of obligate heterofermentative strains studied as potential silage inoculants. The *Lentilactobacillus buchneri* group includes species *Lentilactobacillus buchneri, L. diolivorans, L. farraginis, L. hilgardii, L. kefiri, L. kisonensis, L. otakiensis, L. parabuchneri, L. parafarraginis, L. parakefiri, L. rapi, L. sunkii, L. ferintoschensis, L. senioris, L. kunkeei, L. homohiochii, L. fructivorans, L. sanfranciscensis, L. lindneri, L. spicheri, L. namurensis, L. zymae, L. acidifarinae, L. hammesii*, and *L. brevis*. However, only a few species from the *Lentilactobacillus buchneri* group have been evaluated for their effects on silage fermentation, including primarily *Lentilactobacillus buchneri* and much less commonly *L. brevis, L. diolivorans, L. hilgardii, L. kefiri*, and *L. parafarraginis*. In heterofermentative LAB, both *Lentilactobacillus buchneri* and *L. rhamnosus* can enhance the aerobic stability of silage fermentation; however, silage inoculated with *L. rhamnosus* tends to maintain a high pH for an extended period after opening. Weissella is a genus that contains obligate heterofermentative species, among which *W. cibaria, W. confusa,* and *W. paramesenteroides* have shown good aerobic stability in cactus-mixed silage. During the high moisture ensiling process, *Weissella* bacteria are crucial as they produce a significant amount of acetic acid through their metabolism, thereby maintaining the stability of silage quality. *Pediococcus acidilactici* plays a significant role in the aerobic stability of silage. Studies have shown that the addition of *P. acidilactici* not only increases the lactic acid content in silage but also produces acetic acid to further enhance the aerobic stability of the silage. During the production process of total mixed ration (TMR) silage, the use of *Lentilactobacillus buchneri* and *P. acidilactici* has been proven to significantly improve the aerobic stability of silage. These two microorganisms, through their unique fermentation processes, not only accelerate the accumulation of lactic acid and lower the pH value of the silage but also, during the later stages of silage, *Lentilactobacillus buchneri* converts lactic acid into acetic acid, further enhancing the resistance of the silage to aerobic spoilage. This synergistic effect enables TMR silage to preserve its freshness and nutritional value for an extended period post-opening, minimizing losses from aerobic exposure. Consequently, it enhances the overall quality of the feed and the efficiency of animal feeding. The presence of these metabolites can create an environment that is less conducive to the growth of spoilage organisms, thus prolonging the silage's shelf life. Additionally, the balance between homolactic and heterolactic fermentation can be influenced by factors such as the type of substrate available, the pH of the environment, and the presence of other microorganisms. Understanding these dynamics is crucial for optimizing the use of L. brevis as a silage additive to achieve the desired level of aerobic stability. Efficacy of six LAB strains as silage inoculants in forages with different dry matter and water-soluble carbohydrate content. The study also observed that the addition of *Lentilactobacillus buchneri* resulted in a reduction of yeast and mold counts in the silage, which is another indicator of improved aerobic stability. This reduction is likely due to the inhibitory effect of acetic acid on these aerobic microorganisms. Furthermore, the treated silage exhibited a lower pH compared to untreated silage, which can also contribute to the suppression of spoilage organisms. Overall, the findings suggest that *L. buchneri* could be a valuable additive for enhancing the quality and extending the shelf life of corn silage.

1.1.3 Non-LAB

Non-LAB bio-additives are other bacteria and are designed to improve the fermentation quality of silage, enhance its nutritional value, increase its aerobic stability, and inhibit the growth of harmful microorganisms. Among these, *Propionibacterium acidipropionici* can produce propionic acid during ensiling, an organic acid that enhances the aerobic stability of silage. The production of propionic acid helps to inhibit the growth of yeasts and molds, which are often the main factors causing spoilage of silage in aerobic conditions. In some studies, silage treated with *P. acidipropionici* has shown lower dry matter (DM) losses, suggesting its potential advantages in maintaining the nutritional value of silage (Muck et al. 2018). *Streptococcus bovis*, a species found in the rumen, has attracted attention due to its rapid growth characteristics. It can quickly lower the pH value during ensiling, thus acting as an initiating strain. Additionally, this strain has shown the potential to increase crude protein content and reduce gas and effluent losses. Fermentation quality and chemical composition of elephant grass silage were improved by inoculating with rumen strains of *S. bovis*. Other bacteria, for example, the Bacillus genus, such as *Bacillus subtilis*, may inhibit the growth of yeasts and molds by producing bacteriocins, thereby enhancing the aerobic stability of silage feed (Muck et al. 2018). *B. subtilis* can serve as an effective microbial additive in silage fermentation, especially in enhancing aerobic stability and controlling the growth of undesirable microorganisms. However, its effect on improving the nutritional quality of silage may be limited. The use of *Saccharomyces cerevisiae* as a silage inoculant has been shown to improve the nutritional quality of the feed. It can enhance the digestibility of organic matter and increase the concentration of protein. Moreover, the presence of this yeast can reduce the incidence of mycotoxin-producing fungi, which are a significant concern in silage management due to their potential toxicity to livestock. By promoting a healthier microbial environment in the silage, Saccharomyces cerevisiae not only extends the feed's shelf life but also supports the overall health and productivity of the animals consuming it.

2 Enzyme

2.1 Carbohydrate Degrading Enzyme

In silage additives, a variety of carbohydrate-degrading enzymes are present, including but not limited to cellulases, xylanases, pectinases, and amylases. These enzymes play a crucial role in the degradation of cellulose, hemicellulose, pectin, and other plant cell wall components in silage. As a result of their action, soluble carbohydrates are generated and provide abundant substrates for the growth and fermentation of LAB. Simultaneously, these enzymes also contribute to the disruption of plant cell structure while enhancing the palatability and nutritional value of silage. Moreover, they significantly improve feed digestibility when used as silage additives in animal husbandry practices. Therefore, carbohydrate-degrading enzymes have extensive applications in improving feed utilization efficiency and enhancing animal production performance.

2.1.1 Cellulase

Cellulase consists of complex enzymes that degrade cellulose, which includes three primary components: endoglucanase, exoglucanase, and β-glucosidase. Endoglucanase cleaves internal bonds in the cellulose polymer, producing shorter chains. Exoglucanase works from the ends of these chains, releasing cellobiose, while β-glucosidase converts cellobiose into glucose. This multienzyme system is crucial for breaking down the rigid cellulose structure in plant cell walls. The main function of cellulase is to hydrolyze cellulose into glucose, which is a fermentable sugar that enhances microbial fermentation during ensiling. By degrading the plant cell wall, cellulase helps improve the digestibility of crude fiber in silage, permitting better nutrient absorption in animals (Muck et al. 2018). This breakdown increases the energy value of silage, which is especially important in high-fiber feeds like corn stover and straw. Cellulase is widely used in silage production to improve the digestibility of fibrous materials, particularly in crops like corn and grasses. Its usage reduces the cellulose content of the silage, promoting fermentation and increasing the energy yield of the feed (Kung and Shaver 2001).

2.1.2 Xylanase

Xylanase is the hemicellulose-degrading enzyme targeting xylan, which is a major structural component of plant cell walls. The enzyme contains domains that cleave the β-1, 4-xylosidic linkages in xylan, resulting in xylooligosaccharides. There are two main types of xylanase: endoxylanase, which acts on internal bonds, and exoxylanase, which breaks down oligosaccharides into xylose (Shallom and Shoham 2003).

Xylanase facilitates the breakdown of hemicellulose, loosening the cell wall structure and increasing access to cellulose and other polysaccharides. By reducing the rigidity of plant fibers, xylanase enhances the overall digestibility of silage, improves fermentation efficiency, and helps release more fermentable sugars (Kung and Shaver 2001).

Xylanase is especially useful in silages with high hemicellulose content, such as wheat or barley. Its usage accelerates the degradation of resistant plant fibers, increasing lactic acid production and reducing the time needed for proper fermentation. The enzyme also helps in reducing dry matter losses during the ensiling process, which contributes to better feed preservation (Rooke and Hatfield 2003).

2.1.3 Pectinase

Pectinase refers to a group of enzymes that degrade pectin, a polysaccharide found in plant cell walls. Pectinase includes enzymes such as pectin lyase, pectin esterase, and polygalacturonase. These enzymes hydrolyze pectin by breaking down the complex polysaccharides into simpler sugars, such as galacturonic acid (Jayani et al. 2005). Pectinase breaks down the pectin network in plant cell walls, enhancing the accessibility of other enzymes, such as cellulase and xylanase, to act on the remaining plant fibers. This act of loosening the plant structure improves the overall efficiency of fermentation and reduces the amount of trapped oxygen, promoting the anaerobic conditions necessary for proper ensiling (Muck et al. 2018; Weinberg and Muck 1996). Pectinase is particularly effective in silage from fruits, vegetables, and crops with high pectin content. Its usage facilitates better compaction of silage, which allows more efficient fermentation. It is more beneficial in crops like alfalfa or clover, where pectin content is high, enhancing the fermentation process and improving the overall texture of the silage (Jayani et al. 2005; Weinberg and Muck 1996).

2.1.4 Amylase

Amylase comprises two main types: α-amylase and β-amylase. α-amylase cleaves internal α-1,4-glycosidic bonds in starch, producing dextrin and maltose. β-amylase, on the other hand, acts on the nonreducing ends of the starch chains, sequentially releasing maltose.

The primary function of amylase is to degrade starch into maltose and glucose, which are fermentable sugars. This process increases the availability of sugars for lactic acid fermentation, helping to lower the pH of silage more rapidly and thus preventing spoilage (Muck et al. 2018; Kleinschmit and Kung 2006). Starch hydrolysis also increases the energy content of the silage, enhancing its nutritional value for livestock. Amylase is more useful in silage made from crops with high starch content, such as corn and cereal grains. It enhances the efficiency of fermentation, reduces the loss of starch during ensiling, and improves the energy density of the silage. This particularly benefits high-energy-demanding animals like dairy cows and beef cattle (Muck et al. 2018).

2.2 Fatty-Related Enzymes

In addition to carbohydrate-degrading enzymes, fat-related enzymes play a crucial role in silage additives. These include lipase, phospholipase, and cholesterol esterase, which are essential for the fat metabolism of feed. These enzymes effectively break down fats and phospholipids into smaller-molecular-weight fatty acids and glycerol, facilitating enhanced animal absorption. Consequently, this not only improves the energy utilization rate of feed but also enhances its palatability and nutritional value while promoting animal health and production performance.

2.2.1 Lipase

Lipase is a hydrolase enzyme responsible for the breakdown of triglycerides into free fatty acids and glycerol. It typically contains a catalytic triad of serine, histidine, and aspartic acid, which work in synergy to hydrolyze the ester bonds in fats. Lipases can be classified based on their origin (microbial, plant, or animal), with microbial lipases being widely used in various industrial applications due to their stability and efficiency (Jaeger and Eggert 2002). In silage, lipase plays a crucial role in breaking down fat molecules, reducing the overall fat content, and releasing free fatty acids that can enhance fermentation by acting as an energy source for microorganisms. The degradation of fats also helps to prevent rancidity in the silage and maintain its quality over time. Lipases are particularly effective in silages derived from oilseed crops, where they help reduce fat levels, improve fermentation, and enhance the digestibility of lipids in the feed (Geron et al. 2010). Additionally, lipase supplementation is beneficial for dairy cattle and other livestock with high energy demands, as it increases the availability of digestible fatty acids (Jaeger and Eggert 2002).

2.2.2 Phospholipase

Phospholipase refers to a family of enzymes that hydrolyzes phospholipids, which are major components of cell membranes. Depending on the type, phospholipases target different bonds within the phospholipid molecules. For instance, phospholipase A2 cleaves the fatty acid from the second carbon of glycerol, while phospholipase C removes the phosphate head group (Dennis 1997). This hydrolysis releases fatty acids and other components that can be metabolized or used as signaling molecules in cellular processes. In silage, phospholipase breaks down the phospholipid content of cell membranes, which improves the overall quality of fermentation by releasing fatty acids that can serve as substrates for microbial metabolism. The release of fatty acids and their subsequent metabolism helps maintain the stability of the silage by preventing lipid peroxidation, which could otherwise lead to spoilage (Dennis 1997). Phospholipase is particularly effective in silages made from legumes and oilseeds, which tend to have higher phospholipid content. Its use in silage production contributes to improved energy density and better preservation of nutrients.

2.2.3 Cholesterol Esterase

Cholesterol esterase is an enzyme that hydrolyzes cholesterol esters into free cholesterol and fatty acids. It contains a catalytic triad similar to that of lipase (serine, histidine, and aspartic acid), which is essential for the hydrolysis of ester bonds. Cholesterol esterase plays a key role in lipid metabolism by enabling the breakdown of cholesterol esters stored in tissues, thereby allowing the release of free cholesterol for cellular functions or further metabolism. In silage production, cholesterol esterase helps to manage lipid content, particularly in silages that contain animal by-products or high-fat feedstocks. By breaking down cholesterol esters, this enzyme facilitates the digestion and absorption of lipids by animals, improving nutrient utilization and preventing the accumulation of cholesterol and fatty acids in the feed. Additionally, cholesterol esterase can help mitigate the risk of oxidation and spoilage by reducing the levels of reactive lipids in the silage, ensuring better preservation of feed quality over time (Geron et al. 2010).

2.3 *Protein-Related Enzymes*

Protein-degrading enzymes also play an important role in silage additives. By promoting protein decomposition in feed, they convert complex protein molecules into small peptides and amino acids that are more easily absorbed by animals, thus significantly improving the nutritional value and utilization of feed. This conversion process not only improves the digestibility of the feed but also enables the animals to make full use of the nutrients in the feed, thereby promoting their growth and development and improving their health.

2.3.1 Protease

Protease is a broad category of enzymes that catalyze the hydrolysis of peptide bonds within proteins, breaking them down into smaller peptides or amino acids. The two main types of proteases in biological systems are endopeptidases, which cleave internal peptide bonds, and exopeptidases, which remove terminal amino acids from the protein chain. Proteases are often classified based on their catalytic mechanism, which includes serine proteases, aspartic proteases, cysteine proteases, and metalloproteases. Each of these has a different mechanism for breaking peptide bonds (Rawlings et al. 2014).

In silage, proteases play a crucial role in the degradation of plant and microbial proteins into free amino acids and small peptides, which are more easily absorbed and utilized by livestock. The breakdown of proteins also releases nitrogenous compounds that can be used by LAB, enhancing the fermentation process and improving silage preservation. Protease supplementation is particularly beneficial in legume silages, which tend to have higher protein content. By breaking down these proteins during ensiling, proteases help reduce the risk of protein degradation by undesirable microbes and increase the nutritional quality of the silage.

2.3.2 Peptidase

Peptidases are enzymes that break down peptides into individual amino acids. They are classified into two main groups: exopeptidases, which cleave amino acids from the ends of peptides, and endopeptidases, which cleave within the peptide chain. Peptidases are essential in protein metabolism because they release free amino acids that are easily absorbed by animals. The catalytic mechanism of peptidases varies, but they commonly utilize metal ions or catalytic residues (such as serine, threonine, or cysteine) in their active sites to break down peptides. In the context of silage, peptidases help to further break down the peptides produced by proteases, ensuring that more nitrogen is available in the form of free amino acids. This contributes to the fermentation process by providing additional nutrients for the growth of beneficial microorganisms like lactic acid bacteria. Peptidases are more important in high-protein crops, such as alfalfa, where the breakdown of proteins can enhance the overall digestibility and improve the utilization of nitrogen by ruminants. The presence of free amino acids in silage also promotes fast fermentation and lowers the pH more quickly, which improves the preservation quality of the silage.

2.3.3 Aminopeptidase

Aminopeptidase is a type of exopeptidase that specifically cleaves amino acids from the amino (N-) terminus of peptides. They play a key role in the final stages of protein digestion, releasing individual amino acids that are critical for various metabolic processes. Aminopeptidase typically contains a metal ion in its active site, often zinc, which is essential for its catalytic activity (Taylor 1993). They are involved in both intracellular and extracellular protein turnover, ensuring the continuous supply of amino acids for cellular functions. In silage, aminopeptidases contribute to the release of free amino acids from peptides, making them readily available for microbial metabolism and absorption by livestock. This enzymatic activity is particularly beneficial in improving the nutritional quality of high-protein silages, such as those derived from leguminous crops. Aminopeptidase activity ensures a rapid reduction in protein content and promotes efficient fermentation, leading to a more stable and nutritious silage. Furthermore, it helps improve the amino acid profile of the silage, making it more suitable for high-performance animals like dairy cows and beef cattle.

2.3.4 Carboxypeptidase

Carboxypeptidases are exopeptidases that cleave amino acids from the carboxyl (C-) terminus of peptide chains. They are metalloenzymes, often requiring zinc or other metal ions for their activity. Carboxypeptidases are crucial in the final stages of protein breakdown, converting peptides into free amino acids that can be absorbed by the animal's digestive system. There are two main types of carboxypeptidases: those that remove neutral or hydrophobic amino acids (carboxypeptidase A) and those that cleave basic amino acids (carboxypeptidase B). In silage, carboxypeptidases act on the protein fragments generated by other proteolytic enzymes, ensuring a complete breakdown of proteins into free amino acids. This enzymatic activity is especially important in silages with high protein content, such as legume-based silages, where the breakdown of protein can influence the fermentation process and the overall quality of the silage. Carboxypeptidase activity helps to reduce nitrogen loss, improve the protein content of the silage, and enhance the availability of amino acids for ruminants. Its role in increasing the levels of free amino acids also aids in faster acidification of the silage, thereby improving preservation.

2.4 *Other Enzymes*

2.4.1 Ligninase (Lignin-Degrading Enzymes)

Ligninases are a group of enzymes capable of degrading lignocellulose, which includes lignin peroxidase, manganese peroxidase, and laccase. Lignin is a crucial component of plant cell walls, and its complex structure makes it difficult to break down. Ligninase enzymes oxidatively cleave lignin into smaller molecules, making cellulose and hemicellulose more accessible for further enzymatic degradation (Leonowicz et al. 2001).

In silage, ligninase reduces the rigidity of plant cell walls by breaking down lignin, thereby facilitating the breakdown of cellulose and hemicellulose. This improves the digestibility of high-lignin crops, such as grasses and straws, which enhances the energy and nutrient availability in the feed (Fig. 9.2). Ligninase is particularly useful when ensiling high-lignin materials like straw, improving fermentation efficiency and animal digestion.

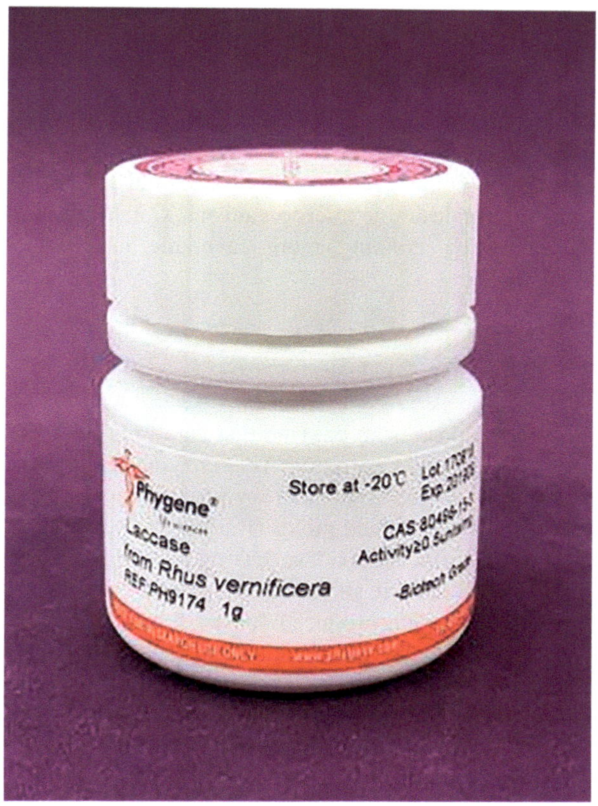

Fig. 9.2 Laccase. (Phygene Biotechnology Co., Ltd., Fuzhou, China; photo by Liu)

2.4.2 Phytase

Phytase is an enzyme that hydrolyzes phytic acid (phytate), the primary storage form of phosphorus in plant seeds. It breaks down phytic acid into inositol and free phosphate, thereby making phosphorus available for animal absorption. Phytase is typically produced by microorganisms and is widely used in animal feed additives (Mullaney and Ullah 2003).

Phytase increases the bioavailability of phosphorus, especially in grain-based silages. Since phytate-bound phosphorus is poorly digestible by monogastric animals like pigs and poultry, adding phytase can significantly improve phosphorus utilization, reduce waste, and minimize environmental pollution (Lei and Porres 2003). Although ruminants can degrade some phytate in their rumen, supplemental phytase can further enhance phosphorus absorption.

2.4.3 Catalase

Catalase is an enzyme that catalyzes the decomposition of hydrogen peroxide into water and oxygen. It contains either iron or manganese at its active site and prevents oxidative stress by eliminating hydrogen peroxide, a harmful byproduct of cellular metabolism (Chelikani et al. 2004).

Besides, catalase helps reduce oxidative stress during the early stages of fermentation by quickly eliminating residual oxygen. This promotes anaerobic conditions, allowing beneficial bacteria like lactic acid bacteria to thrive while inhibiting the growth of undesirable microorganisms. Catalase enhances the quality of silage fermentation by creating more favorable anaerobic conditions (Da Silva and Kung 2022).

2.4.4 Glucose Oxidase

Glucose oxidase is an enzyme that catalyzes the oxidation of glucose to gluconic acid and hydrogen peroxide. This enzyme reduces the available glucose in silage and lowers oxygen levels, thereby suppressing the growth of unwanted microorganisms such as yeasts and molds (Bankar et al. 2009).

In silage, glucose oxidase acts by reducing oxygen and glucose content, which helps inhibit aerobic microorganisms and mold growth, preventing spoilage. The hydrogen peroxide produced further inhibits the growth of harmful microorganisms, improving the preservation and long-term quality of the silage. This enzyme is particularly effective when ensiling high-sugar crops like sugar beets or sugarcane (Gomes et al. 2005).

2.4.5 Lysozyme

Lysozyme is an enzyme that breaks down the peptidoglycan layer of bacterial cell walls, mostly effective against Gram-positive bacteria. It cleaves the β-1,4-glycosidic bonds in the peptidoglycan, leading to bacterial cell lysis and death (Callewaert and Michiels 2010). Lysozyme serves as a natural antimicrobial agent, inhibiting the growth of harmful bacteria that may proliferate during the ensiling process. It helps maintain the hygiene and safety of the silage, improving the quality of the feed, particularly in high-value production systems such as dairy and meat industries (Wang et al. 2006).

3 Chemical Substance

3.1 Mineral and Organic Acids

Common inorganic acids used as silage additives include hydrochloric acid, sulfuric acid, phosphoric acid, etc. These acids can be added to the feed either individually or in mixtures. The addition of inorganic acids to silage can rapidly lower the pH, thereby killing undesirable fermentation bacteria. Historically, the use of inorganic acids in silage fermentation has been more popular in Europe than in North America, but their use has declined due to their corrosive nature (Kung et al. 2003). Inorganic acids, such as hydrochloric acid, orthophosphoric acid, and sulfuric acid, do not have antimicrobial properties and serve only as acidifying agents (Woolford 1978) and are very effective in improving silage fermentation. Commonly used organic acids include formic acid, propionic acid, lactic acid, and acetic acid, among many others. Organic acids can be used either individually or in mixtures as silage additives. Studies have shown that formic acid and mixtures of partially neutralized aliphatic organic acids can both effectively preserve forages, but the combination of acids is more effective in inhibiting fermentation (O'Kiely and Moloney 1994). Additionally, propionic acid can enhance the aerobic stability of silage (Yitbarek and Tamir 2014). The aerobic stability of silage made from formic acid is relatively poor (Bolsen et al. 1996).

3.2 Sugars

The sugar content in forage has a significant impact on the success of ensiling. The main water-soluble carbohydrates in forage crops are glucose, fructose, sucrose, and fructans. Commonly used sugar additives include molasses, fruit sugar, and glucose. Molasses have been proven to be an effective silage additive in numerous experiments, promoting lactic acid fermentation and inhibiting clostridia and proteolysis (Yitbarek and Tamir 2014). Alfalfa, which belongs to the leguminous forage with high buffer and low sugar, is difficult to ensile conventionally. However, adding sugar additives has resulted in satisfactory ensiling outcomes (Seale et al. 1986). Studies have indicated that the addition of molasses to ensiled straw not only ensures the stability of fermentation but also enhances the digestibility in animals and decreases the emission of rumen gases (Zhao et al. 2019).

3.3 *Chemical Microbiocide*

3.3.1 Formic Acid

Formic acid, as a silage additive, provides an effective means for the preservation of silage and the improvement of animal production performance by lowering the pH value, increasing nitrogen utilization, and improving fermentation quality and aerobic stability. In the silage of a mixture of perennial ryegrass and hybrid ryegrass, formic acid treatment increased nitrogen retention, improved silage fermentation quality, and kept the digestibility of dry matter, organic matter, and total energy unchanged in feeding experiments (Rooke et al. 1988). In alfalfa silage treated with formic acid, the formic acid treatment effectively improved the utilization rate of nutrients in the alfalfa silage by lactating dairy cows (Nagel and Broderick 1992).

Formic acid has an antibacterial effect on many kinds of bacteria, including lactic acid bacteria. Therefore, it leads to limited fermentation and a reduction in organic acids, but the silage has a higher content of WSC, which is a better energy source for rumen microbes compared to lactic acid (Rowghani and Zamiri 2009). According to Chamberlain's calculations, using silage with a high content of fermented acid, the energy yield of rumen microorganisms may be much lower than using silage with restricted fermentation. In terms of animal performance, total dry matter (DM) intake and live weight gain were significantly improved compared with untreated or inoculant-treated silage, and formic acid-treated silage had higher content.

3.3.2 Propionic Acid

Propionic acid as a silage additive improves the fermentation quality and aerobic stability of silage through various mechanisms, thereby enhancing the nutritional value of the silage and the production performance of animals. Research has indicated that propionic acid not only prevents spoilage in feed bins but also effectively reduces heating in corn silage and total mixed rations (TMR) (Kung et al. 1998). In silage treated with chemical additives, the concentration of propionic and butyric acids is linearly related to the calculated hydrolysis of the additive, indicating that no clostridia activity occurred. Chemical additives of a mixture of propionic acid and butyric acid can reduce the number of yeasts and improve aerobic stability (Ferrero et al. 2019).

When corn silage is exposed to oxygen in the silo or feed compartment, aerobic microorganisms can cause the temperature to rise and cause spoilage. Lactic acid-metabolizing yeasts are the main spoilage microorganisms of corn silage, although acetic acid bacteria and other microorganisms can also cause spoilage. The deterioration of silage due to exposure to air is irreversible due to the loss of nutrients and the potential negative impact on animal production performance. Various chemical preservatives are used to prevent silage from spoiling in the air, thereby improving the aerobic stability of silage (Fig. 9.3). Among short-chain fatty acids, propionic acid has the greatest antifungal activity. In the past, adding large amounts of propionic acid (1–2% of DM) to silage has improved aerobic stability, but in these cases, a high proportion of acid usually limits fermentation. Many products that are currently added to feed during the silage process to improve aerobic stability contain several active ingredients. Propionic acid usually makes up the largest proportion of these ingredients. The low amount of propionic acid-based chemical preservatives

Fig. 9.3 Propionic acid. (Sigma Merck, Milwaukee, USA; photo by Liu)

does not affect fermentation but can effectively reduce the temperature rise of corn silage and mixed diets, thus reducing the possibility of silage deterioration (Kung et al. 1998).

4 Nutrient Enhancer

4.1 Agricultural By-Products

Agricultural by-products have the potential to enhance nutritional value, improve fermentation quality, and increase palatability as silage additives. In addition to these benefits, they can also promote agricultural recycling and improve the economic benefits of agricultural production. Agricultural by-products, which are rich in dietary fiber exceeding 50%, also contain a large amount of polyphenols (such as flavonoids) and other functional metabolic products. These components can promote the health of animals. Different agricultural by-products have varying impacts on the silage process when used as silage additives.

4.1.1 Sugar Beet Pulp

Beet pulp, a by-product of the sugar industry, has approximately 500,000 tons of dry matter available for feed in the UK each year. This dried molasses beet pulp is a high-quality feed for ruminants, providing 12.5 MJ of metabolizable energy and 79 g of digestible crude protein per kilogram of dry matter. Although it has a high crude fiber content, it is quite digestible. Compared to high-grain diets, diets with a higher proportion of beet pulp have a lesser inhibitory effect on the digestion rate of roughage and do not pose a risk of acidosis. For dairy cows, this highly digestible fiber has a positive impact on milk fat content. Furthermore, due to its high content of water-soluble carbohydrates, beet pulp can also increase the total dry matter intake of animals (Kelly 1983). Replacing some corn silage (CS) with a mixture of beet pulp and straw for silage (BRMS) can increase the dry matter intake and milk production of dairy cows, and as the proportion of BRMS increases, the production of milk protein also increases linearly (Wang et al. 2022a). Sugar beet pulp (SBP), which is rich in pectin and a significant amount of sugars, provides easily fermentable carbohydrates for silage fermentation, thereby enhancing the value of the silage (Gruber et al. 2024).

4.1.2 Rice Bran

Rice bran as a silage additive can enhance feed nutrition and fermentation quality, improve animal production performance, reduce costs, and be environmentally friendly and sustainable. The use of rice bran as an additive increases the recovery rate of digestible dry matter in pearl millet silage and enhances the fermentation and nutritional value of the silage. Rice bran, when used as an additive, can improve the recovery rate of digestible dry matter and enhance the fermentation quality and nutritional value of pearl millet silage. It has been shown to have a positive impact on the fermentation quality of the forage by altering the composition of bacterial communities (You et al. 2022). Adding 10% and 20% rice bran to switch grass silage can increase the content of protein and carbohydrates, as well as improve the rumen degradation kinetics parameters of dry matter, crude protein, and neutral detergent fiber (Negrão et al. 2014).

4.1.3 Wheat Bran

Wheat bran as a silage additive can enhance the nutritional value and fermentation quality, boost animal production performance, reduce costs, decrease agricultural waste, and promote sustainable development. Additionally, its rich dietary fiber and non-starch polysaccharides contribute to the gut health of livestock and poultry, increase nutrient utilization rates, and improve the physical structure of feed. Adding wheat bran to citrus pulp silage can enhance the nutritional value of the citrus pulp silage without causing any adverse effects on this by-product (Kordi and Naserian 2012). Adding wheat bran and cornmeal to the silage of lemongrass can increase the content of dry matter and crude protein and also reduce the levels of ADF and NDF. Considering both the quality of the silage and the cost of the additive, wheat bran is a better choice. This is because wheat bran not only improves the nutritional content of the silage but also does not negatively affect this by-product. Moreover, the addition of wheat bran can enhance the fermentation process and the nutritional value of the silage, making it a more economical and effective additive for improving the quality of lemongrass silage (Cheng et al. 2001).

4.1.4 Edible Fungi Substrate

Edible fungi substrate residue, such as mushroom and tremella substrate residue (Fig. 9.4) as a silage additive, can enhance the nutritional value and fermentation quality of feed, improve animal production performance, reduce costs, decrease agricultural waste, promote sustainable development and its rich prebiotic components contribute to the gut health of livestock and poultry, increase nutrient utilization rates, and improve the physical structure of the feed. By adding 15% molasses to the silage process of mushroom residues, their nutritional content is notably

Fig. 9.4 Mushroom (left) and tremella (right) substrate residue. (Photos by Cai and Liu)

increased, making them a cost-efficient feed by-product for ruminants (Zaboli et al. 2023).

4.2 Food Processing By-Products

4.2.1 Pear Residue

The prickly pear residue, which has similar nutritional components to general fruit residues, can enhance the utilization efficiency of plasma protein and almost does not negatively affect the growth performance, blood metabolites, and rumen fermentation parameters of cattle. Therefore, it can be used as a new resource for cattle feed (Song et al. 2024). The soluble dietary fiber in prickly pear residue has the potential to be used as a modulator to help alleviate ulcerative colitis (Li et al. 2024). Additionally, prickly pear residue is rich in nutritional components, such as total polyphenols, vitamin C, vitamin E, phosphorus, calcium, iron, zinc, etc. These nutrients can enhance the nutritional value and aerobic stability of silage, making it an excellent additive for silage.

4.2.2 Soybean Residue

Soybean curd residue, as a resource abundant in fiber, protein, and lipids, has demonstrated potential as an additive for silage fermentation. Mixing corn stalks or straw with soybean curd residue can increase the levels of lactic acid and protein while simultaneously reducing the pH value, the content of neutral detergent fiber (NDF), acid detergent fiber (ADF), and ammonia nitrogen concentration. It also decreases the bacterial diversity in the silage feed mixture, thereby enhancing the nutritional value and fermentation quality of these mixtures (Wang et al. 2022a, b).

Dried tofu residue has shown the potential to be an efficient desiccant, enabling the production of high-quality silage without wastewater discharge (Yin et al. 2021).

4.2.3 Distiller's Grains and Brewer's Grains

Baijiu distillers' grains, which are rich in crude protein, digestible fiber, ether extractables, and minerals, have the potential to be used as a good silage additive. The addition of distillers' grains promotes an increase in the content of dry matter, water-soluble carbohydrates, and crude protein during the fermentation process of straw silage. At the same time, it also reduces the content of crude fiber and ammonia nitrogen concentration and decreases the bacterial diversity in the silage feed mixture, thereby improving the nutritional value and fermentation quality of these mixtures (Yuan et al. 2016). Brewer's spent grains (BSG) can also be used as a silage additive. Studies have shown that when BSG is used with other feedstuffs in the total mixed ration (TMR) silage, it has the potential to improve the aerobic stability of the silage (Nishino et al. 2003).

4.3 Others

4.3.1 Grains and Cereals

Corn, barley, oats, and wheat: These high-energy crops are often included in silage as a direct source of fermentable starch and carbohydrates. Their addition significantly enhances the energy density of the silage and promotes a more efficient fermentation process (Weinberg et al. 2003). The high starch content of these feed ingredients provides the necessary fermentable carbohydrates for microbial fermentation in the silo, promotes the rapid production of lactic acid, and establishes a stable anaerobic environment. Starch-rich grains provide a quick source of fermentable carbohydrates that support the rapid fermentation of lactic acid. This rapid acidification process not only stabilizes silage but also inhibits the growth of decaying organisms such as *Clostridium*.

Grains and cereals are mainly composed of carbohydrates, especially starch, whose dry matter content can vary between 60% and 70%, depending on the specific grain. They also contain small amounts of protein, fiber, and fat. The high starch content of grains, such as corn and barley, provides easily fermentable carbohydrates, which the lactic acid bacteria can quickly convert into lactic acid, lowering the pH of the silage.

The cell walls of grains are mostly made up of cellulose and hemicellulose and are less lignified than grass, which makes carbohydrates easier for livestock to ferment and digest. The addition of grain increased the availability of fermentable sugar, promoted the rapid fermentation of lactic acid, and improved the quality of feed. This is especially important for silage made from low-energy feed, which is beneficial for high-yielding animals such as dairy and beef cattle.

The presence of starch in grains can also improve the digestibility of ruminants to silage by providing a more balanced energy-to-protein ratio, which is essential for maximizing milk production in cows (Shipandeni et al. 2023). In addition, although the grain fiber content was not as significant as that of feed, it could promote rumen function and digestion efficiency.

The inclusion of grains boosts the availability of fermentable sugars, fostering rapid lactic acid fermentation and improving feed quality. This is particularly important for silage made from low-energy forages and is beneficial for high-production animals like dairy cows and beef cattle (Kung et al. 2003; Weinberg et al. 2003).

4.3.2 Molasses

Molasses is a by-product of sugar processing and is a widely used silage additive that promotes fermentation and increases the energy content of feed. Its high content of easily fermentable sugars, such as sucrose and glucose, makes it an ideal additive for promoting the rapid production of lactic acid, which is essential for stabilizing silage during fermentation (Shipandeni et al. 2023).

Molasses is mainly composed of monosaccharides (sucrose, glucose, and fructose), accounting for about 50–60% of its total dry matter. It also contains small amounts of protein, minerals, and vitamins, especially calcium, potassium, and iron, which can provide some additional nutritional value. The high sugar content in molasses makes it a quick source of energy for microorganisms in silage and for livestock that eat it (Henderson 1993).

Molasses act as a fermentation stimulator by providing easily available sugar to lactic acid bacteria, which produces lactic acid and reduces the pH of the silage. This rapid acidification helps inhibit the growth of harmful microorganisms. Molasses are usually added at a ratio of 3–5% of the total weight of the feed, and by improving the fermentation efficiency, molasses improve the preservation of nutrients and reduce the loss of dry matter, thus improving the silage quality. In addition, the use of molasses as an additive has been shown to increase the intake of animal silage due to its sweet taste and improved palatability while also improving milk production and feed efficiency.

Molasses is often added to grass, corn, and bean silage to balance the carbohydrate content, especially if the feed is low in fermentable sugars. Its use is particularly beneficial when ensiling high-moisture crops or feeds naturally low in sugar, as it compensates for the lack of readily available carbohydrates and ensures a faster fermentation process (Henderson 1993).

4.3.3 Beet Pulp

Beet pulp is a common additive in silage, it is valued for its high energy content and digestibility. It is a by-product of sugar beet processing, consisting primarily of fiber and carbohydrates, such as pectin, hemicellulose, and cellulose. Due to its high soluble fiber content, beet pulp is particularly effective at improving the energy density of the feed and enhancing the efficiency of microbial fermentation in the rumen (Khattab et al. 2013). The inclusion of beet pulp in silage provides a slow-release energy source, promoting better rumen health and stability compared to rapidly fermentable carbohydrates like starch.

Beet pulp is composed of about 15–20% crude fiber, mainly pectin, which is highly digestible by ruminants. Pectin is a non-cellulosic polysaccharide that is easily broken down in the rumen by microbial action, providing a steady supply of energy (Abdel-Rahman et al. 2011). The moderate protein content and low fat make beet pulp a well-balanced additive that supports both energy supply and proper rumen function.

Beet pulp functions as a highly digestible fiber source, which helps in stimulating rumen motility and fermentation. The soluble fiber content encourages the growth of beneficial bacteria, leading to increased production of volatile fatty acids (VFAs), which are crucial for energy metabolism in ruminants. Additionally, beet pulp's low lignin content improves the digestibility of the entire feed mix, making it easier for the animal to utilize the nutrients in the silage.

In practice, beet pulp is often used in silages made from grasses, maize, or other high-fiber crops. Its high moisture absorption capacity makes it particularly useful in balancing silages that are too wet or too dry, thereby enhancing their overall physical structure (Khattab et al. 2013). It also helps in stabilizing the fermentation process by ensuring an adequate supply of fermentable carbohydrates which accelerates lactic acid production, leading to a faster pH drop in the silage. As a result, beet pulp contributes to better preservation of silage nutrients and reduces the risk of spoilage. Additionally, when beet pulp is included in dairy cow rations, it has been shown to increase milk production due to its energy density and digestibility.

4.3.4 Legume Meals

Legume meals, such as soybean meal, alfalfa meal, and lupin meal, are valuable protein sources commonly used to enhance the nutritional quality of silage. Legumes are particularly rich in protein and fiber, making them ideal for improving the protein content of silage and contributing to more balanced diets for livestock. Their use in silage formulations has been increasingly explored due to their ability to improve both the fermentative characteristics and nutritional profile of the feed.

The high protein content of legumes, especially from species like soybeans and alfalfa, helps to elevate the crude protein concentration in silage. This, in turn, contributes to increased nitrogen availability for ruminants, leading to better animal growth and production performance (Khattab et al. 2013). Additionally, legume meals supply essential amino acids, particularly lysine, which are often deficient in cereal-based silages. This makes legume meals an essential component for balancing the amino acid profile of the overall diet (Hoffman et al. 2009).

Legume meals also positively influence the fermentation process of silage. Due to their moderate-to-high buffering capacity, legumes can help stabilize silage pH, creating more favorable conditions for lactic acid bacteria to thrive. This contributes to improved silage preservation and reduces the risks of spoilage. Moreover, legumes are known to contain secondary compounds such as saponins and tannins, which, though typically present in small quantities, can exert antimicrobial properties that limit the growth of undesirable microorganisms during fermentation (Muck et al. 2018). These compounds help maintain silage quality over time.

Research has demonstrated that adding soybean meal to silage improves not only the protein content but also the overall energy availability for ruminants. Soybean meal, for instance, is rapidly fermented in the rumen, providing a quick source of energy and protein for microbial growth. This enhances fiber digestion and increases microbial protein synthesis, which is crucial for the overall efficiency of nutrient utilization in dairy cows and beef cattle (Nadeau et al. 2018).

In summary, legume meals such as soybean and alfalfa provide crucial benefits when used as additives in silage production. They enhance the protein content, improve fermentation characteristics, and contribute to better nutrient utilization by livestock. The presence of essential amino acids and favorable fermentative qualities make legume meals an effective means of boosting both the nutritional value and preservation of silage.

4.3.5 Urea and Ammonia-Based Additives

Urea and ammonia-based additives are commonly utilized in silage production to enhance nitrogen content and improve the fermentation process. These additives are especially beneficial when forages are low in protein, as they can help optimize microbial protein synthesis in the rumen, ultimately improving the nutritional quality of the silage.

Urea is a nonprotein nitrogen source that can be rapidly converted to ammonia by ureolytic bacteria in the rumen. When added to silage, urea increases the nitrogen availability for rumen microbes, promoting their growth and activity. This is crucial for the effective fermentation of fibrous plant materials as it helps balance the carbon-to-nitrogen (C: N) ratio, facilitating more efficient degradation of fibrous components (Muck 1987; Birkelo et al. 1999). However, careful management of urea levels is essential to prevent ammonia toxicity, which can occur if excessive amounts are added.

Ammonia-based additives, including ammonium sulfate and ammonium hydroxide, serve a similar purpose. They provide a source of nitrogen for rumen microbes and enhance the protein content of silage. Research indicates that adding ammonium compounds can improve fermentation efficiency and reduce dry matter loss during ensiling (Muck et al. 2018). Additionally, ammonia treatment of low-quality forages can improve their digestibility by breaking down the lignocellulosic structure, making nutrients more accessible to ruminants.

Both urea and ammonia-based additives can also positively affect the silage fermentation profile. They help lower the pH by increasing lactic acid production, which inhibits the growth of undesirable microorganisms and enhances the preservation of the silage (Muck et al. 2018). The inclusion of these additives has been linked to increased animal performance, particularly in dairy and beef cattle, by providing a more balanced and nutrient-rich diet.

4.3.6 Fat and Oil Supplements

Fat and oil supplements are commonly used in silage to increase the energy density of the feed, which is particularly beneficial for high-producing dairy cows and beef cattle. Fats provide more than twice the energy per unit weight compared to carbohydrates and proteins, making them a concentrated source of calories for animals. These supplements are especially useful in enhancing the overall caloric intake without significantly increasing feed volume, which can help in optimizing the performance of animals with higher energy demands (Palmquist and Jenkins 2003).

Fat supplements in silage are typically derived from vegetable oils (such as soybean oil, canola oil, and sunflower oil) or animal fats (like tallow and lard). These fats are not only a dense energy source but can also have specific roles in improving the digestibility and palatability of silage. For example, the inclusion of vegetable oils has been shown to improve the digestibility of dry matter and fiber in high-forage diets, enhancing overall feed efficiency (Chilliard et al. 2009). Additionally, oils such as soybean oil are rich in unsaturated fatty acids, which may contribute to better animal health by improving the fatty acid profile of milk or meat produced from livestock (Bauman et al. 2006).

However, there are also challenges associated with the inclusion of fats and oils in silage. High levels of fat can inhibit fiber digestion by interfering with microbial fermentation in the rumen, particularly for unsaturated fats, which can be toxic to rumen microbes (Palmquist and Jenkins 2003). Therefore, the application of fats in silage must be carefully managed to avoid negative impacts on fermentation and rumen health. Encapsulated fats or calcium salts of fatty acids are often used to mitigate these issues, as they are less likely to disrupt microbial activity in the rumen while still providing the desired energy boost.

In a study by Wu and Huber (1994), the addition of calcium salts of long-chain fatty acids to corn silage-based diets led to increased milk yield and improved body condition in lactating cows. Similarly, the use of protected fats in silage has been found to reduce energy losses during digestion, particularly in animals under high metabolic demand (Chilliard et al. 2009).

Furthermore, oils, particularly those rich in linoleic acid, have been found to enhance the stability of silage by reducing aerobic spoilage. This is partly due to the antimicrobial properties of certain fatty acids, which can inhibit the growth of undesirable bacteria and fungi during the fermentation process (Bauman et al. 2006). This effect is beneficial in preserving the quality of silage during storage and after silo opening, reducing spoilage and nutrient loss.

In conclusion, the strategic use of fat and oil supplements in silage can significantly enhance the energy content of the feed, improve animal performance, and contribute to better silage preservation. However, the amount and type of fat used must be carefully managed to avoid negative effects on fiber digestibility and fermentation efficiency. Advances in fat supplementation techniques, such as the use of protected fats, continue to improve the applicability of fats in modern silage systems.

4.3.7 Mineral Supplements

Mineral supplements are essential in silage management, and it is vital that mineral supplements are added to silage during the silage process or as part of the total mixed diet (TMR), especially in areas where the soil is mineral-deficient, to correct defects and imbalances that may occur in a feed-based diet. These additives ensure that livestock receive sufficient levels of essential macro and trace minerals to support critical physiological functions such as bone development, enzyme activity, and immune function. Insufficient mineral content in silage can lead to poor performance, metabolic diseases, and reduced yields in animals, especially dairy cows and high-yielding livestock.

Mineral supplements typically include major minerals, such as calcium, phosphorus, magnesium, and potassium, as well as trace elements, such as zinc, copper, manganese, and selenium. They are available in different forms, such as mineral salts or chelated minerals, which improve their bioavailability (National Research Council et al. 2001). Calcium is essential for bone structure and milk production, while phosphorus is important for energy metabolism. Trace minerals such as zinc and selenium play a vital role in antioxidant protection and immune response. For example, selenium supplementation can help prevent reproductive disorders and reduce the incidence of diseases such as mastitis (Weiss and Socha 2005).

The main effect of adding minerals to silage is to make up for the mineral deficiencies often found in feed crops. For example, gramineous silage may have low phosphorus content, while leguminous silage tends to be deficient in magnesium. Proper mineral supplementation ensures that livestock maintain optimal growth, reproductive performance, and milk production.

4.3.8 Yeast and Probiotics

Yeast (*Saccharomyces cerevisiae*) and probiotics (e.g., *Lactobacillus* spp.) are the active microorganisms added to silage (Kleinschmit and Kung 2006). These biological additives not only improve fermentation quality by promoting microbial balance but also improve the nutritional quality of silage. It can also improve the digestibility of feed by creating a good fermentation environment. It indirectly improves the production performance and maintains the health of animals. Due to its role in stabilizing the fermentation process and improving animal health, it is becoming more and more popular in silage production.

Yeast, particularly *Saccharomyces cerevisiae*, is one of the most commonly used microbial additives in silage. It plays a vital role in promoting lactic acid bacteria (LAB) growth and improving aerobic stability. Yeast ferments simple sugars, releasing carbon dioxide and small amounts of ethanol, which help in maintaining anaerobic conditions in silage (Weinberg et al. 1993). Studies have shown that *Saccharomyces cerevisiae* supplementation increases silage dry matter retention, enhances digestibility, and boosts milk yield in dairy cows. Furthermore, yeast supplementation can lower the growth of undesirable microorganisms like molds and spoilage bacteria during the ensiling process, thus improving the feed's stability during feed-out (Muck et al. 2018).

Probiotics, which include a range of beneficial bacteria like *Lactobacillus* and *Enterococcus* species, are also effective in optimizing fermentation. These bacteria outcompete harmful microorganisms, contributing to a more efficient lactic acid fermentation and a quicker reduction in silage pH. This rapid pH drop minimizes the risk of nutrient loss and spoilage (Guedes et al. 2008). Probiotic additives have been shown to improve fiber degradation, which enhances the digestibility of the feed for ruminants, leading to better overall performance (Ávila and Carvalho 2020). Furthermore, the use of probiotics in silage has been linked to improved animal health, particularly in enhancing rumen microbial activity and boosting immune responses (Elferink et al. 2000).

References

Abdel-Rahman MA, Tashiro Y, Sonomoto K. Recent advances in lactic acid production by microbial fermentation processes. Biotechnol Adv. 2011;29(6):902–10.

Ávila CLS, Carvalho BF. Silage fermentation—updates focusing on the performance of micro-organisms. J Appl Microbiol. 2020;128(4):966–84.

Bankar SB, Bule MV, Singhal RS, Ananthanarayan L. Glucose oxidase—an overview. Biotechnol Adv. 2009;27(4):489–501.

Bauman DE, Mather IH, Wall RJ, Lock AL. Major advances associated with the biosynthesis of milk. J Dairy Sci. 2006;89(4):1235–43.

Birkelo CD, Burch KD, Lindgren RW. Urea and ammonia effects on fermentation characteristics and nutrient composition of low-quality forage silage. J Anim Sci. 1999;77(8):2182–8.

Bolsen KK, Ashbell G, Weinberg ZG. Silage fermentation and silage additives-review. Asian-Australas J Anim Sci. 1996;9(5):483–93.

Callewaert L, Michiels CW. Lysozymes in the animal kingdom. J Biosci. 2010;35(1):127–60.

Chelikani P, Fita I, Loewen PC. Diversity of structures and properties among catalases. Cell Mol Life Sci. 2004;61(2):192–208.

Cheng YK, Chen CS, Peng PW. Effects of different additives on silage quality of napier-grass. 2001.

Chilliard Y, Ferlay A, Mansbridge RM, Doreau M. Ruminant milk fat plasticity: nutritional control of saturated, polyunsaturated, trans and conjugated fatty acids. Ann N Y Acad Sci. 2009;1190:114–26.

Da Silva ÉB, Kung L Jr. A meta-analysis of the effects of a chemical additive on the fermentation and aerobic stability of whole-plant maize silage. Agriculture. 2022;12(2):132.

Dennis EA. Phospholipases. Encycl Life Sci. 1997;4(1):1–5.

Elferink SJWHO, Driehuis F, Gottschal JC, Spoelstra SF. Silage fermentation processes and their manipulation. FAO plant production and protection papers; 2000. pp. 17–30.

Ferrero F, Tabacco E, Borreani G. Effects of a mixture of monopropionine and monobutyrin on the fermentation quality and aerobic stability of whole crop maize silage. Anim Feed Sci Technol. 2019;258:114319.

Geron LJV, Takahashi H, Alves T. Use of enzymes in silage preparation: advances and perspectives. J Anim Sci. 2010;88(1):118–25.

Gomes E, de Souza SR, Grandi RP, Silva RD. Production of glucose oxidase by *Aspergillus niger* in different aeration conditions using wheat bran. Braz J Microbiol. 2005;36(4):414–9.

Gruber T, Fliegerová K, Terler G, Resch R, Zebeli Q, Hartinger T. Mixed ensiling of drought-impaired grass with agro-industrial by-products and silage additives improves the nutritive value and shapes the microbial community of silages. Grass Forage Sci. 2024;79(2):179–97.

Guedes CM, Gonçalves D, Rodrigues MAM, Dias-da-Silva A. Effects of a Saccharomyces cerevisiae yeast on ruminal fermentation and fibre degradation of maize silages in cows. Anim Feed Sci Technol. 2008;145(1–4):27–40.

Henderson N. Silage additives. Anim Feed Sci Technol. 1993;45(1):35–56.

Hoffman PC, Lundberg KM, Bauman LM, Shaver RD. The effect of maturity on NDF digestibility in alfalfa silage. J Dairy Sci. 2009;92(5):2249–54.

Jaeger KE, Eggert T. Lipases for biotechnology. Curr Opin Biotechnol. 2002;13(4):390–7.

Jayani RS, Saxena S, Gupta R. Microbial pectinolytic enzymes: a review. Process Biochem. 2005;40(9):2931–44.

Jiang S, Cai L, Lv L, Li L. *Pediococcus pentosaceus*, a future additive or probiotic candidate. MCF. 2021;20:1–14.

Kelly P. Sugar beet pulp—a review. Anim Feed Sci Technol. 1983;8(1):1–18.

Khattab IM, El-Kady RI, El-Hassanein EE. Utilization of beet pulp as a silage additive for improving the nutritional value and fermentation characteristics of maize silage. Anim Biosci. 2013;7(2):65–74.

Kleinschmit DH, Kung L Jr. A meta-analysis of the effects of *Lactobacillus buchneri* on the fermentation and aerobic stability of corn and grass silages. J Dairy Sci. 2006;89(10):4005–13.

Kordi M, Naserian AA. Influence of wheat bran as a silage additive on chemical composition, in situ degradability and in vitro gas production of citrus pulp silage. AJB. 2012;11(63):12669–74.

Kung L Jr, Shaver RD. Interpretation and use of silage fermentation analysis reports. Focus Forage. 2001;3(13):1–5.

Kung L Jr, Sheperd AC, Smagala AM, Endres KM, Bessett CA, Ranjit NK, et al. The effect of preservatives based on propionic acid on the fermentation and aerobic stability of corn silage and a total mixed ration. J Dairy Sci. 1998;81(5):1322–30.

Kung L Jr, Stokes MR, Lin CJ. Silage additives. Silage Sci Technol. 2003;42:305–60.

Lei XG, Porres JM. Phytase enzymology, applications, and biotechnology. Biotechnol Lett. 2003;25(23):1787–94.

Leonowicz A, Matuszewska A, Luterek J, Ziegenhagen D, Wojtaś-Wasilewska M, Cho NS, et al. Biodegradation of lignin by white rot fungi. Fungal Genet Biol. 2001;27(2–3):175–85.

Li L, Zhang X, Wang L, Gao M, Wang Y, Zhang Z, et al. Protective effect of soluble dietary fiber from *Rosa roxburghii* Tratt residue on dextran sulfate sodium-induced ulcerative colitis by regulating serum metabolism and NF-κB pathway in mice. J Sci Food Agr. 2024;104(12):7258–70.

McDonald P, Henderson AR, Heron SJE. The biochemistry of silage. 2nd ed. Exp Agric. 1992;28:125.

Muck RE. Urea in silage. J Dairy Sci. 1987;70(3):499–507.

Muck RE, Nadeau EMG, McAllister TA, Contreras-Govea FE, Santos MC, Kung L Jr. Silage review: recent advances and future uses of silage additives. J Dairy Sci. 2018;101(5):3980–4000.

Mullaney EJ, Ullah AH. The term phytase comprises several different classes of enzymes. Biochem Biophys Res Commun. 2003;312(1):179–84.
Nadeau EMG, Buxton DR, Russell JR, Allison MJ, Young JW. Silage fermentation and additive effects on aerobic stability. J Dairy Sci. 2018;101(8):1231–41.
Nagel SA, Broderick GA. Effect of formic acid or formaldehyde treatment of alfalfa silage on nutrient utilization by dairy cows. J Dairy Sci. 1992;75(1):140–54.
National Research Council, Committee on Animal Nutrition, Subcommittee on Dairy Cattle Nutrition. Nutrient requirements of dairy cattle: 2001. Washington, DC: National Academies Press; 2001.
Negrão FDM, Zanine ADM, Cabral LDS, Souza ALD, Alves GR, Ferreira DDJ, et al. Fractionation of carbohydrates and protein and rumen degradation kinetic parameters of Brachiaria grass silage enriched with rice bran. Rev Bras Zootec. 2014;43:105–13.
Nishino N, Harada H, Sakaguchi E. Evaluation of fermentation and aerobic stability of wet brewers' grains ensiled alone or in combination with various feeds as a total mixed ration. J Sci Food Agr. 2003;83(6):557–63.
O'Kiely P, Moloney AP. Silage characteristics and performance of cattle offered grass silage made without an additive, with formic acid or with a partially neutralised blend of aliphatic organic acids. Ir J Agric Food Res. 1994;33:25–39.
Palmquist DL, Jenkins TC. Challenges with fats and fatty acid methods. J Anim Sci. 2003;81(12):3250–64.
Rawlings ND, Waller M, Barrett AJ, Bateman A. MEROPS: the database of proteolytic enzymes, their substrates and inhibitors. Nucleic Acids Res. 2014;42(D1):D503–9.
Rooke JA, Hatfield RD. Biochemistry of ensiling. Silage Sci Technol. 2003;42(1):95–139.
Rooke JA, Maya FM, Arnold JA, Armstrong DG. The chemical composition and nutritive value of grass silages prepared with no additive or with the application of additives containing either *L. plantarum* or formic acid. Grass Forage Sci. 1988;43(1):87–95.
Rowghani R, Zamiri MJ. The effects of a microbial inoculant and formic acid as silage additives on chemical composition, ruminal degradability and nutrient digestibility of corn silage in sheep. Iran J Vet Res. 2009;10(2):110–8.
Seale DR, Henderson AR, Pettersson KO, Lowe JF. The effect of addition of sugar and inoculation with two commercial inoculants on the fermentation of lucerne silage in laboratory silos. Grass Forage Sci. 1986;41(1):61–70.
Shallom D, Shoham Y. Microbial hemicellulases. Curr Opin Microbiol. 2003;6(3):219–28.
Shipandeni MNT, Paula EM, Esposito G, Faciola AP, Raffrenato E. Effects of starch sources varying in particle sizes on ruminal fermentation, nutrient flow, starch digestibility, and lactation performance of dairy cows. J Anim Sci. 2023;101:147.
Song X, Yang Y, Wang C, Zhu W, Wu W. *Rosa roxburghii* Tratt residue can be used as a novel feed resource for cattle. 2024.
Taylor A. Aminopeptidases: structure and function. FASEB J. 1993;7(2):290–8.
Wang J, Zhao G, Zhang X, Zhang S, Li S. Lysozyme: a review of biochemical and pharmacological aspects. Food Bioprod Process. 2006;84(4):301–8.
Wang X, Song J, Liu Z, Zhang G, Zhang Y. Fermentation quality and microbial community of corn stover or rice straw silage mixed with soybean curd residue. Animals. 2022a;12(7):919.
Wang Y, Xia K, Wang XN, Lin X, Liu J, Li YJ, et al. Improvement of feed intake, digestibility, plasma metabolites, and lactation performance of dairy cows fed mixed silage of sugar beet pulp and rice straw inoculated with lactic acid bacteria. J Dairy Sci. 2022b;105(1):269–80.
Weinberg ZG, Muck RE. New trends and opportunities in the development and use of inoculants for silage. FEMS Microbiol Rev. 1996;19(1):53–68.
Weinberg ZG, Ashbell G, Hen Y, Azrieli A. The effect of applying lactic acid bacteria at ensiling on the aerobic stability of silages. J Appl Microbiol. 1993;75(6):512–8.
Weinberg ZG, Shatz O, Chen Y, Yosef E, Nikbahat M, Ben-Ghedalia D, Miron J. Effect of lactic acid bacteria inoculants on in vitro digestibility of wheat and corn silages. J Dairy Sci. 2003;86(5):1556–60.

Weiss WP, Socha MT. Dietary manganese for dry and lactating dairy cows. J Dairy Sci. 2005;88(1):2517–23.

Woolford MK. Antimicrobial effects of mineral acids, organic acids, salts and sterilizing agents in relation to their potential as silage additives. J Brit Grassland Soc. 1978;33(2):131–6.

Wu Z, Huber JT. Relationship between dietary fat supplementation and milk protein concentration in lactating cows: a review. Livest Prod Sci. 1994;39(2):141–55.

Yin X, Wu J, Tian J, Wang X, Zhang J. Dried soybean curd residue: a promising absorbent for cleaner production of high-quality silage. J Clean Prod. 2021;324:129300.

Yitbarek MB, Tamir B. Silage additives. Open J Appl Sci. 2014;4(5):17.

You J, Zhang H, Zhu H, Xue Y, Cai Y, Zhang G. Microbial community, fermentation quality, and in vitro degradability of ensiling caragana with lactic acid bacteria and rice bran. Front Microbiol. 2022;13:804429.

Yuan X, Dong Z, Desta ST, Wen A, Zhu X, Rong T, Shao T. Adding distiller's grains and molasses on fermentation quality of rice straw silages. Ciênc Rural. 2016;46(12):2235–40.

Zaboli K, Kalvandi S, Malecky M, Nasrabadi M. Nutritional value of spent mushroom (*Agaricus bisporus*) compost silage treated with different levels of molasses in sheep feeding. Iran J Appl Anim Sci. 2023;13(1):57–65.

Zhao J, Dong Z, Li J, Chen L, Bai Y, Jia Y, et al. Effects of sugar sources and doses on fermentation dynamics, carbohydrates changes, in vitro digestibility and gas production of rice straw silage. Ital J Anim Sci. 2019;18(1):1345–55.

Chapter 10
Silage Safety

Xianjun Yuan and Lin Sun

Abstract Silages constitute a component of dairy and beef rations worldwide. They are required to fulfill the demands of both the animal and the ultimate consumer of animal products, particularly with respect to safety. Silage is susceptible to the introduction of hazards throughout the entire production chain, from field to feed. Ensiled forages may contain various mycotoxins, stemming from preharvest contamination by *Fusarium* and *Aspergillus* species and/or from postharvest contamination by toxigenic molds, such as *Aspergillus* and *Penicillium*. Silage has also been detected to be contaminated with a diverse range of pathogens and chemical hazards. Most of these hazards are associated with poorly fermented silages, which can be avoided by enhancing lactic acid fermentation and preventing aerobic deterioration. This chapter summarizes types, prevalence, and levels of mycotoxins, chemical hazards, and pathogens in ensiled forages, describes their detrimental impacts on the health of ruminants, and outlines effective prevention and mitigation strategies.

Keywords Chemical hazard · Mycotoxin · Pathogen · Silage safety

Silage is a crucial component of dairy and beef cow diets worldwide. The presence of microbiological or toxicological contaminants in silage not only compromises animal health, productivity, and welfare but also poses significant threats to food safety. Ensuring the safety of silage is a fundamental requirement for the production of safe and nutritious food products derived from animals intended for human consumption. Poor-quality silage may be susceptible to contamination from various microbiological hazards (Fink-Gremmels 2012; Ogunade et al. 2018; Queiroz et al.

X. Yuan (✉)
Nanjing Agricultural University, Nanjing, China
e-mail: yuanxianjun@njau.edu.cn

L. Sun
Inner Mongolia Academy of Agriculture and Animal Husbandry Sciences, Hohhot, China

Y. Cai, K. Ataku (eds.), *Cultural History and Modern Production Technology of Silage*, https://doi.org/10.1007/978-981-96-5787-2_10

2018), chemical hazards (Driehuis et al. 2018), and natural toxins (Monbaliu et al. 2012).

Almost all forages are susceptible to fungal infestation, either during the growing season or within the stage of preservation, transportation, and storage. Consequently, mycotoxins represent a considerable threat to both human and animals. Silages frequently become contaminated with pathogens, such as when slurry is utilized as a fertilizer on fields or when forages are contaminated with soil-borne pathogens during the harvesting process. Chemical hazards associated with undesirable silage fermentation cause a variety of adverse health effects to livestock and human. Additionally, some plant toxins present in standing crops at harvest time can persist through the ensiling process, potentially causing poisoning in both livestock and humans.

For ensure safety and efficacy of silage, it is crucial to understand the origins of these hazardous factors in silages, to recognize their adverse effects, and to implement effective detection and control strategy. Therefore, the present review is to give an up-to-date summary of the adverse effects of hazard factors and most recent developments in mycotoxin-mitigation during silage production (Table 10.1).

1 Toxigenic Fungi and Mycotoxins in Silage

Microbial hazards in silage predominantly originate from the fermentation process where improper ensiling techniques facilitate the proliferation of pathogenic bacteria, molds, and yeasts. These microorganisms can multiply rapidly under the anaerobic conditions of silage, potentially leading to the production of mycotoxins and other harmful substances.

Molds present in silage can be categorized based on their origins. Firstly,molds related to crop infestations during field growth, such as *Aspergillus flavus*, *Aspergillus parasiticus*, and species from *Fusarium* and *Alternaria* genera. Secondly, endophytic symbiotic molds found existing in grasses or cereals, such as *Claviceps* and *Neotyphodium* species, are also part of the classification. Finally, molds emerged in poor silages, such as *Penicillium roqueforti*, *Penicillium paneum*, *Aspergillus fumigatus*, *Monascus ruber*, *Byssochlamys nivea* (Driehuis et al. 2018). With the exception of *R. nigricans* and *C. sitophila*, all the aforementioned molds possess the ability to produce mycotoxins (Driehuis 2013).

1.1 Fusarium and Fusarium Toxin

Fusarium species are regarded as field fungi, owing to their capacity to thrive during the growth and maturation phases of plants. These molds are widespread and pose a contamination risk to field crops in both temperate and warm climatic regions. The proliferation of *Fusarium* species is particularly favored humid environments (>

Table 10.1 The effects of hazards in common silage on ruminants

Types of toxins		Summary of effects
Mycotoxin	Aflatoxins	Reduced milk production in dairy cows
		Decreased milk quality and safety due to carry-over of toxin from contaminated feed
		Decreased feed efficiency and rate of gain in beef cows
		Compromised immune and ruminal functions
		Liver malfunctions
	T-2	Immunosuppression in cattle because of reduced antibody production, neutrophil function, and lymphocyte blastogenesis
		Bovine infertility and abortion in late gestation
		Extent of transfer into the milk of ruminants is negligible
	Deoxynivalenol	Gastrointestinal problems and reduced performance probably due to feed refusal
		No evidence of carry-over into milk of ruminants
	Zearalenone	Infertility, reduced milk production, and hyperestrogenism in cattle
		Extent of transfer into milk of ruminants is negligible
	Fumonisins	Reduced performance probably due to feed refusal
		Mild liver disease
		Carry-over of the toxin into the milk of ruminants is negligible
	Ochratoxins	No significant toxicity to cattle when fed alone in naturally occurring doses
		Carry-over of the toxin into milk is minimal
	Roquefortine	Feed refusal, reproductive disorders, and paralytic effects
		No evidence of transfer into milk of ruminants
	Mycophenolic acid	Research data on the effects in cattle are lacking
		No evidence of transfer into milk of ruminants

(continued)

Table 10.1 (continued)

Types of toxins		Summary of effects
Pathogenic bacteria	*Salmonella typhimurium*	Led to sepsis, which is characterized by abortion, enteritis, and hemorrhagic pneumonia
	Listeria monocytogenes	Caused encephalitis and meningitis in ruminants, as well as miscarriage and diarrheal disease
	Escherichia coli	The production of enterotoxin can result in the development of severe diarrhea, sepsis, and even meningeal bleeding
	Clostridium perfringens	The production of exotoxins leads to the development of gas gangrene and necrotizing enteritis
	Clostridium botulinum	Affected the nervous system of ruminants, resulting in muscular paresthesia, dyspnea, cerebral congestion, and edema
Pathogenic fungi	*Aspergillus flavus*	The bacterium produces enterotoxins, which result in severe diarrhea, and even aflatoxins
	Fusarium graminis	Produced emetic toxin and F-2 toxin, which disrupts the nervous system and exhibits carcinogenic and teratogenic properties
	Fusarium moniliforme	The production of fumonitoxin, a group 2B carcinogen
	Fusarium pseudocladoides	The production of T2 toxins induces symptoms such as emesis, diarrhea, and potential organ dysfunction
	Penicillium viridis	Produced ochratoxin, resulting in hepatotoxicity, nephrotoxicity, damaged to other organs, exhibited teratogenic effects
	Penicillium islandum	The production of island-penicillium toxin and flavin has been associated with carcinogenic effects
	Yeast	Excessive consumption can lead to diarrhea and potentially result in gastrointestinal and respiratory disorders
Chemical toxins	Nitrite	The condition led to the development of methemoglobinemia
		The dilation of blood vessels results in peripheral circulatory failure
		Facilitated the compensatory response of the thyroid gland and contributed to the synthesis of carcinogenic N-nitroso compounds
	Total arsenic	The nervous system, digestive system, cardiovascular system, hepatic system, renal system, and other organs sustain damage
		The teratogenic effect on embryonic development is clearly evident
	Lead	Induced neurotoxicity in ruminants, impaired hematopoiesis and coagulation processes
		Inflicted renal damage, disrupted reproductive functions, as well as exhibited teratogenic and carcinogenic properties

	Mercury	Led to the manifestation of severe symptoms such as profuse diarrhea, intense abdominal pain, significant dehydration, sudden collapse, and subsequent shock
	Cadmium	Led to damage to vascular endothelial cells, internal organs, and interference with the absorption of calcium, phosphorus, iron, and other elements
		Additionally, it will affect the reproductive function and immune response in ruminants
		Potentially causing carcinogenesis, teratogenicity, and mutagenicity
	Chromium	The application of high concentrations of chromium on the skin, mucosa, and digestive tract can cause stimulation and corrosion, leading to skin congestion, erosion, ulcers, etc., which may potentially result in the development of skin cancer
		Chromium has the ability to accumulate in the liver, kidney, and lung; particularly hexavalent, chromium exhibits the highest level of toxicity
	Fluorine	The joints experience swelling, the length of hoof nails increases, and the bones become lax, leading to paralysis
	Polychlorinated biphenyls	The consumption of this substance can result in bone and tooth damage, as well as potentially leading to chronic carcinogenesis and genetic variation
		Ruminants exposed to this substance may experience disorders in their endocrine system and an increased risk of developing malignant tumors
	Hexachlorane	Caused acute poisoning characterized by respiratory symptoms. Its chronic toxicity manifests as hepatomegaly, hepatocyte degeneration, and damage to the central nervous system and kidneys
	DDT	Impacted the nervous system of ruminants, resulting in nerve conduction disorders and reproductive system impairment. Function in ruminants
		DDT has the potential to induce hepatic tumors and significantly compromise liver
	Hexachlorobenzene	The exposure can result in the development of endocrine disorders, reproductive and immune disorders, neurobehavioral and developmental disorders, cancer, and other diseases

70%), particularly where there are significant temperature fluctuations between day and night. In contrast, ensiling conditions are generally unfavorable for *Fusarium* species, as they cannot tolerate low pH and anaerobic environments. Most *Fusarium* toxins are chemically and thermally stable during ensiling. Hence, the concentrations of *Fusarium* toxins detected in silage may serve as an indicator of the Fusarium toxins contaminated extent at harvest time (Driehuis et al. 2008a). Fusarium toxins present in forages encompass trichothecenes, fumonisins, and zearalenone.

Trichothecenes constitute a group of sesquiterpenoid metabolites, which can be categorized into two principal classes: macrocyclic and nonmacrocyclic. Macrocyclic trichothecenes are characterized by an ester-ether bridge connecting carbon-4 and carbon-15, whereas nonmacrocyclic trichothecenes lack this specific structural trait. The nonmacrocyclic trichothecenes are predominantly produced by *Fusarium* species and are further classified into type A and type B. Type A trichothecenes are characterized by a hydroxyl or acyl group at the carbon-8 position and include compounds such as T-2 toxin, HT-2 toxin, T-2 triol, T-2 tetraol, scirpentriol, and diacetoxyscirpenol. In contrast, type B trichothecenes are identified by a keto functional group at the carbon-8 position and comprise deoxynivalenol (DON), nivalenol, acetylated deoxynivalenol, fusarenon X, etc.

DON, frequently referred to as vomitoxin, is a mycotoxin that can cause a range of adverse effects in animals, including vomiting, feed refusal, diarrhea, reproductive issues, and death. Ruminants display greater resistance to DON, as the rumen microflora can detoxify mycotoxins through enzyme synthesis. Previous studies indicated that DON is the most frequently detected mycotoxins in silages. A survey by Driehuis et al. (2008a) revealed that corn and wheat silages sampled from the Netherlands are major sources of DON in dairy cattle diets, with average concentrations of 854 and 621 μg/kg, respectively, and maximum concentrations reaching 3142 and 1165 μg/kg. In a recent multiyear mycotoxin survey in Poland by Kosicki et al. (2016), DON was present in 86% of 143 corn silage samples, with average and maximum concentrations of 223 μg/kg and 7860 μg/kg, respectively.

Zearalenone (ZEA), a macrocyclic β-resorcyclic acid lactone, is an estrogenic metabolite produced by several *Fusarium* species. Because of its structural similarity to estrogen, ZEA trigger a variety of reproductive problems, particularly in swine, such as hyperestrogenism, vaginitis, and mammary gland enlargement. In contrast, ruminants are less susceptible than swine to ZEA as ruminal microbiota, especially protozoa, can convert ZEA into its hydroxylated metabolites α-zearalenol and β-zearalenol. Notably, α-zearalenol has a higher affinity for estrogen receptors than ZEA itself.

Several studies have indicated that ZEA is one of the most common mycotoxins in silages, particularly in ensiled forages contaminated with *Fusarium* spp. A survey of corn, alfalfa, and sorghum silage samples from dairy farms in Argentina revealed that 45.5% of contaminated corn silages, 45.5% of contaminated alfalfa silages, and 60% of contaminated sorghum silages contained ZEA levels exceeding 300 μg/kg (Ogunade et al. 2018). The maximum concentrations detected were 1774.8 μg/kg for corn silage, 1599.3 μg/kg for alfalfa silage, and 595.3 μg/kg for sorghum silage, respectively.

Fumonisins are produced by several species of *Fusarium* genus, which includes *Fusarium verticillioides*, *Fusarium proliferatum*, *Fusarium anthophilum*, and *Fusarium nygamai*. Over 28 distinct forms of fumonisins have been identified and categorized into A-series, B-series, C-series, and P-series. Fumonisin B, especially fumonisin B1, is regarded as the most toxic in feed contamination. Among mycotoxins, fumonisins pose the most threats to livestock. This is evidenced by a 64% occurrence frequency and an average concentration of 1965 μg/kg within the 7049 livestock feed samples from the Americas, Europe, and Asia. In Brazil, approximately 48.6% of 327 corn silage samples were contaminated with fumonisin B1, exhibiting an average concentration of 369 μg/kg (Schmidt et al. 2015). González Pereyra et al. (2008) reported that fumonisin in corn silage ranged from 340 to 2490 μg/kg of dry matter (DM). Significantly higher concentrations were commonly witnessed in samples procured from the upper layer and side walls of silos. These particular regions are more prone to air incursion and are characterized by poor fermentation quality.

1.2 Aspergillus and Aspergillus Toxin

Aspergillus is one of the most important fungal genera in the spoilage of foods and animal feeds, particularly in warm-temperate climates and the tropics. Some species, such as *Aspergillus flavus*, have the potential to contaminate crops and generate aflatoxins in the field when exposed to high temperature conditions exceeding 32 °C, elevated humidity levels above 80%, or during periods of drought stress. Insect-induced damage to forages can render crops more susceptible to infection or intensify preexisting infections. The most common mycotoxins produced by *Aspergillus* species encompass *aflatoxins*, *ochratoxins*, *gliotoxins*, *fumonisins*, *sterigmatocystin*, and *patulin*.

Aflatoxins constitute a collection of difuranocoumarin derivatives that possess toxicity, mutagenicity, and carcinogenicity. These are synthesized by certain strains of *Aspergillus flavus*, *Aspergillus parasiticus*, and *Aspergillus nominus*. The most common and significant aflatoxins include B_1, G_1, B_2, and G_2 and are frequently detected in contaminated food and feed. *Aspergillus* species are mainly prevalent in tropical or subtropical climatic regions (Cheli et al. 2013).

Aflatoxin B_1 (AFB_1) is recognized as the most potent carcinogen found in nature and produced by *Aspergillus* species. Compared with other mycotoxins, the prevalence and amount of aflatoxin are relatively low in well-preserved silages. However, substantial aflatoxin can be detected in poorly fermented silages or in silages prepared with diseased corn (Queiroz et al. 2012). In addition, higher occurrences of aflatoxins have been documented in trench silos compared to silo bags as trench silos are often constructed on concrete bases or bare ground, making them more susceptible to environmental conditions (Alonso et al. 2013). As much as 51 μg/kg of DM of AFB_1 was detected in corn silage collected from a trench silo in France (Richard et al. 2009).

Chronic exposure to aflatoxins within the diet of lactating cattle has been correlated with multiple negative consequences. These involve a decline in overall well-being and productivity, a detriment to liver functionality, a dampening of the immune system's reaction, and a heightened susceptibility to illnesses, even when vaccinations have been administered. High-dose exposure to AFB_1 may cause acute aflatoxicosis, characterized by inappetence, ataxia, rough hair coats, or hepatomegaly (Ogunade et al. 2018).

After ingestion by dairy cows, a portion of AFB_1 is transformed by certain ruminal microbes into aflatoxicol, while the remaining AFB_1 proceeds to the intestine, where it is rapidly absorbed and transported via the portal bloodstream to the liver. In humans and susceptible animals, aflatoxins, especially AFB_1, are transformed by cytochrome P-450 enzymes into aflatoxin-8, 9-epoxide. This compound is extremely noxious, mutagenic, and carcinogenic, and it induces chromosomal harm by binding and forming complexes with DNA. Additional metabolic routes for AFB_1 in the liver involve processes such as O-dealkylation to AFP_1, ketoreduction to aflatoxicol, and hydroxylation to aflatoxin M_1 (AFM_1; highly toxic and carcinogenic), AFQ_1, or AFB_{2a}. Aflatoxin B_1 is excreted in dairy milk as AFM_1 (Battacone et al. 2005). Contaminated milk and dairy items present a substantial safety hazard because AFM_1 is also categorized as a Group 1 human carcinogen. The transfer rate of aflatoxin from feed to milk varies, ranging from 1% to 6%. Aflatoxin is the only mycotoxin for which there are governmental regulations regarding its maximum allowable amounts in feed and milk products. The limits set by the US FDA are 0.5 for liquid whole milk and 20 μg/kg for feed ingredients provided to dairy cattle. In Europe, the maximum recommended levels are 0.05 and 5 μg/kg for liquid whole milk and feed ingredients, respectively.

Ochratoxins are a major class of fungal secondary metabolites produced by *Aspergillus ochraceus*, *Aspergillus niger*, *Aspergillus carbonarius*, or *Penicillium verrucosum* (Keller et al. 2013). Ochratoxin A (OTA) is known as a potent immunotoxic, teratogenic, and carcinogenic mycotoxin. OTA is regarded as the most toxic among ochratoxins. The incidence and concentration of OTA in silage are typically lower compared to other major mycotoxins. This relatively low incidence might be ascribed to the incapacity of the fungi responsible for producing this toxin to endure high levels of acetic acid and CO_2 (Gallo et al. 2015b). For example, only 1% of 124 corn silage samples were contaminated with OTA with a maximum concentration of 3 μg/kg. Ruminants are much less sensitive to OTA than non-ruminants, as the ruminal microbes, with protozoa being a central group, can degrade the mycotoxin extensively into less toxic or nontoxic product in the rumen (Table 10.2).

1.3 Penicillium and Penicillium Toxin

Penicillium molds are capable of thriving in typical silage environments characterized by water activity levels ranging from 0.79 to 0.83 aw, which are comparatively lower than those conducive to the growth of other fungal organisms. *Penicillium*

Table 10.2 Regulatory levels of common mycotoxins in ruminant feeds and human foods

Common mycotoxin	Agency	Intended maximum use	
		Ruminant feeding	Human consumption
Aflatoxins	US Food and Drug Administration	20 μg/kg for dairy animals 100 μg/kg for breeding cattle 300 μg/kg for finishing beef cattle	20 μg/kg in all foods intended for human consumption 0.50 μg/kg in milk (aflatoxin M1)
	European Commission	5 μg/kg for dairy animals 20 μg/kg for all other ruminants	4 μg/kg in all cereals and cereal products with the exception of corn and rice subjected to physical treatment (10 μg/kg) 0.1 μg/kg in foods for infants and young children 0.05 μg/kg in milk (aflatoxin M1)
	China Commission	30 μg/kg for all ruminants (aflatoxin B1)	20 μg/kg in corn and its peoduct, 10μg/kg in rice, 5μg/kg in wheat (aflatoxin B_1), 0.5μg/kg in milk (aflatoxin M_1)
Deoxynivalenol	US Food and Drug Administration	10 mg/kg for ruminating beef and feedlot cattle (>4 mo old) 5 mg/kg for ruminating dairy cattle older than 4 mo	1 mg/kg in wheat products intended for human consumption
	European Commission	5 mg/kg for ruminants	0.75 mg/kg in cereals intended for direct human consumption 0.2 mg/kg in foods for infants and young children
	China Commission	5 mg/kg for ruminants	1 mg/kg in grains and their products
Total fumonisins	US Food and Drug Administration	15 mg/kg for breeding ruminants 30 mg/kg for ruminants at least 3 mo old raised for slaughter	2 mg/kg in degermed dry milled corn products 3 mg/kg in popcorn 4 mg/kg in whole or partially degermed dry milled corn products and dry milled corn bran
	European Commission	50 mg/kg for adult ruminants (>4 mo old)	1 mg/kg in corn intended for direct human consumption 0.2 mg/kg in foods for infants and young children
	China Commission	60 mg/kg for ruminants	Currently not stipulated

(continued)

Table 10.2 (continued)

Common mycotoxin	Agency	Intended maximum use	
		Ruminant feeding	Human consumption
Zearalenone	US Food and Drug Administration	Currently not stipulated	Currently not stipulated
	European Commission	0.5 mg/kg for ruminants	0.075 mg/kg in cereals intended for direct human consumption 0.02 mg/kg in foods for infants and young children
	China Commission	1 mg/kg for ruminants	60 μg/kg in grains and their products
Ochratoxins	US Food and Drug Administration	Currently not stipulated	Currently not stipulated
	European Commission	Currently not stipulated	0.003 mg/kg in cereals intended for direct human consumption 0.0005 mg/kg in foods for infants and young children
	China Commission	100 μg/kg for ruminants	5 μg/kg in grains and their products
T-2 and HT-2	US Food and Drug Administration	Currently not stipulated	Currently not stipulated
	European Commission	Currently not stipulated	0.05 mg/kg in foods intended for direct human consumption with the exception of corn (0.1 mg/kg) and oats (0.2 mg/kg) 0.015 mg/kg in foods for infants and young children
	China Commission	0.5 mg/kg for ruminants	Currently not stipulated

molds are predominantly observed during the storage phase; they can also develop on plants in the field under excessively wet conditions. Among these molds, *Penicillium roqueforti* is frequently referred to as the "silage mold" due to its acid tolerance and ability to grow in low-oxygen environments.

Penicillium species can produce a diverse array of chemical compounds, including *Penicillium roqueforti* (PR) toxin, mycophenolic acid, and roquefortine. Mycophenolic acid and roquefortines usually have been detected in ensiled forage; however, PR toxin is seldom detected in silages (Gallo et al. 2015a, 2016).

A survey conducted on dairy farms revealed that mycophenolic acid was detected in 13% of 47 corn and grass silage samples and 10% of 29 ensiled by-products, exhibiting an average concentration of 0.52 mg/kg. Roquefortine C was identified in 19% and 7% of samples from silage sample and ensiled by-prodcuts, respectively, with an average concentration of 0.52 μg/kg (Driehuis et al. 2008b). Other *Penicillium*-derived exometabolites were also discovered in silages, such as *andrastin* A, *citreoisocumarin*, *agroclavine*, *festuclavine*, and *marcfortine*.

Toxic symptoms, including reproductive disorders, mastitis, and loss of appetite, were detected in cattle herds fed silages containing 0.2 to 1.5 mg/kg of roquefortine. Mycophenolic acid can block T and B lymphocyte proliferation and clonal expansion and prevent the generation of cytotoxic T cells and other effector T cells; thus, it is classified as an immunosuppressive agent (Eugui et al. 1991). Despite the researches regarding the impacts of mycophenolic acid on cattle are rather scarce, a study carried out by Mohr et al. (2007) ascertained that no harmful impact were observed in sheep, which were subjected to daily dosages of 10 to 300 mg of mycophenolic acid per kilogram of their body weight for 44 d.

2 Foodborne Pathogens in Silage

Poorly fermented silages can harbor pathogens that detrimentally affect the performance of dairy cows, compromising the safety of dairy products and posing risks to consumers' health. Pathogenic bacteria that are frequently or occasionally detected in silage include *Enterobacteria, Listeria, Bacillus* spp., *Clostridium* spp., and *Salmonella* (Queiroz et al. 2018).

2.1 Enterobacteria

Enterobacteria are facultatively anaerobic gram-negative bacteria found in poorly preserved silage. The epiphytic enterobacteria such as *Erwinia herbicola* and *Rahnella aquitilis* dominate in the fresh grass, but they are typically outcompeted by other species, including *Escherichia coli*, *Hafnia alvei*, and *Serratia fonticola*, during ensiling fermentation (Driehuis and Elferink 2000). During the initial stage of ensiling fermentation, the enterobacteria would compete for nutrients with LAB although their growth and viability diminish as the pH decreases. Enterobacteria are involved in the deamination and decarboxylation of amino acids in silages, resulting in the accumulation of nitrogen oxide and dioxide, ammonia, and biogenic amines The retarded initial fermentation or contamination of aerobically exposed silage can create favorable conditions for the proliferation of Enterobacteria (Ogunade et al. 2017).

Escherichia coli O157:H7, a gram-negative bacterium with the capacity to synthesize Shiga toxin, is the most notorious member within the enterobacteria family. It is responsible for hemorrhagic colitis and hemolytic uremic syndrome in humans (Queiroz et al. 2018).

Crops can be contaminated with *E. coli* O157:H7 via manure or irrigation water. And this pathogen has been frequently identified as a part of the epiphytic microbial consortium present on forage crops. Research has demonstrated that a rapid decline of pH could effectively eliminate *E. coli* in silage (Byrne et al. 2002). Pedroso et al. (2010) assessed the efficacy of three inoculants on controlling *E. coli* O157:H7 in

corn silages and found that irrespective of the silage inoculants used, as long as the pH fell below 4.0, the pathogen was eliminated within 3 days of ensiling (Chen et al. 2005). Duniere et al. (2015) also reported that when the pH dropped below 5.0, *E. coli* O157:H7 was eliminated in ensiled artificially contaminated wheat and corn. The elimination of this pathogen in these studies was presumably due to the antimicrobial properties of organic acids at low pH (Bjornsdottir et al. 2006).
Pathogenic *E. coli* may endure throughout the ensiling process if the acidification rate is inadequate (Ogunade et al. 2017). Ogunade et al. (2016) demonstrated that inoculating *L. plantarum* or *L. buchneri* eliminated the *E. coli* O157:H7 by accelerating pH decline. For the corn silage, the lower buffering capacity than alfalfa contributed to the rapid decline in silage pH, resulting in the elimination of the pathogen from both inoculated and non-inoculated silages within 3 d (Ogunade et al. 2017). However, when all corn silages were subsequently reinoculated with *E. coli* after aerobic exposure, the *L. plantarum* and control silages exhibited higher *E. coli* counts and higher pH values than the *L. buchneri*-inoculated silages. The presence of oxygen in the silo may also extend the survival of *E. coli* during ensiling. During the feed-out stage, significant populations of *E. coli* (up to 4000 cfu/g of silage) were detected in the upper corners or shoulders of bunker silos of wheat and corn silages; the high population densities are due to the low density of the silage in these areas, which makes them more prone to air penetration (Weinberg et al. 2004).

2.2 *Listeria*

Listeria represents a group of opportunistic, gram-positive, facultative anaerobic bacteria. These pathogens are capable of inducing a high mortality rate and a diverse array of diseases in both immune-compromised animals and humans (McDonald et al. 1991). 1992). *Listeria* has a particular affinity for the central nervous system, especially the brainstem, with encephalitis being the most frequently observed manifestation. While animals affected by listeriosis may exhibit a range of symptoms, in cattle and sheep, the condition is primarily recognized as a neurological disease or encephalitis, presenting symptoms such as depression, head tilt, and facial muscle paralysis, leading to an inability to eat and potentially death. Other manifestations may include abortions, the birth of weak calves or lambs, or systemic infections resulting in septicemia with miliary abscesses throughout various organs.

Listeria monocytegenes is the principal causative agent of the food-borne life-threatening disease listeriosis and is ubiquitous in dairy operations. The outbreaks of listeriosis are more frequently associated with sheep than cattle because they exhibit less resistance than cattle (McDonald et al. 1991)(1992). Spoiled silage is regarded as the primary source of this pathogen responsible for listeriosis cases and outbreaks in ruminants. *L. monocytogenes* can adapt, survive, and even grow over a wide range of environmental stress conditions, such as refrigeration temperatures, low water activity, and a broad range of pH levels. *L. monocytogenes* is naturally present in the soil and can proliferate rapidly in silage when oxygen is available, particularly in poorly fermented silages that lack sufficient acidity.

Pauly and Tham (2003) identified strong correlations between the presence of *L. monocytogenes* in grass silage with silage pH ($r = 0.92$), lactic acid concentration ($r = -0.80$), and the total amount of undissociated acids ($r = -0.83$). They also demonstrated that low moisture conditions did not inhibit the pathogen because ensiling wilted forage (540 g/kg of DM) did not inhibit the proliferation of *Listeria*. However, inoculation with *L. plantarum* decreased the *Listeria* counts in both direct-cut and wilted grass silage. Wilson et al. (2005) evaluated the anti-listerial activity of cell-free culture filtrates of *L. plantarum* SK1 MRS broth and confirmed that lactic acid could effectively reduce *Listeria* counts.

Bacteriocins or bacteriocin-producing bacteria can be used to prevent *Listeria* in silages. Amado et al. (2012) demonstrated that nisin-producing *Lactococcus lactis* CECT 539 and pediocin-producing *Pediococcus acidilactici* NRRL B-5627 could serve as silage inoculants to control the growth of pathogenic bacteria not only in the silo but also in the gastrointestinal tract of the animals. Amado et al. (2016) evaluated the effect of pediocin alone or combined with an inoculant on *L. monocytogenes* in corn silage and found that *L. monocytogenes* disappeared within 8 d in silages treated with pediocin, with or without the inoculant, while *L. monocytogenes* in untreated silage persisted until d 30 of fermentation.

2.3 Bacilli

Bacillus species are aerobic or facultatively anaerobic, gram-positive, sporulating, rod-shaped bacteria. Their endospores exhibit remarkable resilience to extreme conditions, such as high temperatures, freezing, and drying (Masiello et al. 2014). Typically, *Bacillus* spp. proliferate in silage following the oxidation of lactic acid by yeast under aerobic conditions. Hence, the implementation of measures to prevent silage aerobic deterioration can effectively attenuate the probability of contamination by *Bacillus* spp. Warm summer conditions also promote the proliferation of thermoduric *Bacillus* spp. in corn silage, which may initiate silage spoilage. Buehner et al. (2014) documented a significant increase in spore counts in corn silage, with the count increasing from 3.60 log during the winter period to 6.33 log in the summer, accompanied by a doubling of spore counts in milk sample.

Strategies aimed at reducing *Bacillus* spp. contamination in silage should focus on controlling aerobic and facultative anaerobic bacteria, which are not effectively inhibited by fermentation organic acids or low pH (Woolford 1975; Pahlow et al. 2003). Under optimal ensiling conditions, *Bacillus* spp. would be outcompeted by LAB, and homofermentative LAB inoculants may effectively suppress *Bacillus* spp., thereby promoting efficient silage fermentation. Aerobic *Bacillus* spp. exhibit a greater propensity to colonize the surface and sidewalls of silos, attributable to oxygen infiltration and the low silage density. The application of sorbate and benzoate has proven effective in minimizing the growth of *Bacillus* spp. (Woolford 1975).

Various methods, including heat treatment, UV radiation, high pressure, and pulsed electric fields, have been shown to effectively mitigate *Bacillus* in the food industry. However, these techniques are not feasible for large quantities of silage in farm-scale silos. The use of bacteriocins and bacteriocin-producing inoculants may represent promising strategies for controlling *Bacillus* spores in contaminated silages.

2.4 Clostridia

Clostridium species are gram-positive, sporulating, and predominantly obligate anaerobes. They flourish in low-sugar silages, especially when the plant moisture surpasses 70%, the pH exceeds 4.6, the temperature is higher than 30 °C, and the buffering capacity is elevated. Thus, they often can be found in unwilted legume silages. The presence of *Clostridium* spp. in silage is primarily associated with soil contamination or the application of slurry. The outer layer of baled silos and the top layer of bunker silos are particularly susceptible to *Clostridial* spore formation. The spores present in the feed can endure the digestion process and subsequently enter manure or soil, contaminating the exterior of teats and, consequently, the milk (Vissers et al. 2007). *Clostridia* can be grouped into proteolytic and saccharolytic types based on their substrates. *Saccharolytic clostridia* mainly ferment sugars and organic acids, while proteolytic clostridia principally ferment proteins and amino acids.

The consumption or utilization of lactate during the feed-out stage by yeast and saccharolytic *Clostridium* species present in the silage leads to an elevation in pH. This rise in pH creates a favorable environment for the proliferation of proteolytic *Clostridium* species, which deaminated amino acids to release ammonia. The subsequent increase in ammonia concentration and protein solubility renders the silage protein less suitable for high-producing cattle and results in the nitrogen pollution caused by cows consuming such silages. Additionally, the clostridial proteolysis may produce biogenic amines, including cadaverine, glucosamine, histamine, putrescine, and tyramine in silages. Although these compounds exist in relatively small amounts within all cells and can be substantially degraded in the rumen, biogenic amines possess potential toxicity. Many of these putrefaction-associated compounds emit unpleasant odors and are unpalatable, thereby reducing the feed intake of livestock.

Although *Clostridium* spp. are part of normal flora within ruminants, they can produce potent toxins under dietary stress, injury, management alterations, and parasitism. Those toxins can cause sudden episodes of abdominal pain, diarrhea, ulceration, and even death in calves, which are also prevalent in cows, humans, lambs, and monogastric livestock. Another issue related to *Clostridium* is that *Clostridium* spores transmitted from silage into milk, resulting in the formation of outgrowths or gas pockets.

It is well-known that *Clostridium* spp. are unable to tolerate a low pH (below 4.5); therefore, any additive-enhanced lactic acid production and hastening of the pH drop could inhibit clostridial fermentation in silage. Wilting restricts the growth of *Clostridium* due to the moisture preference of these species. The count of *Clostridium* spores has been shown to decrease as the DM content of grass at the time of ensiling rises from 30% to 45%. Chemical additives can also be effective in reducing *Clostridium* spores in silage. The combination of sodium benzoate, sodium nitrite, hexamine, and sodium propionate has been shown to inhibit *Clostridial* growth in low-wilted grass/clover mixtures.

2.5 *Salmonella*

Salmonella are motile, gram-negative, facultative aerobic bacilli belonging to the Enterobacteriaceae family. *Salmonella enterica* is a significant pathogen responsible for approximately 42,363 cases of foodborne illness annually in the United States (Scallan et al. 2011). *S. enterica* is relatively common and is represented by diverse serotypes on dairy farms (Callaway et al. 2005). Salmonellosis is a disease caused by the *Salmonella* bacteria, it is usually characterized by watery or bloody diarrhea, dehydration, fever, depression, and poor performance. The transmission of *Salmonella* to herds appears to be more closely associated with management practices, as its introduction typically occurs via contaminated animals, tools, and personnel, rather than through feed sources (Adhikari et al. 2009). Weinberg et al. (2004) examined fresh plant and silage samples from sewage-irrigated sorghum and corn fields in Israel and discovered that only one out of five corn silage samples was positive for *Salmonella*. Natural disasters such as storms, hurricanes, flooding, hailstorms, or lodging can result in the contamination of forages by soil sediment, increasing the epiphytic counts of *Salmonella* and other pathogens. However, if the fermentation process is properly established and a low pH (<4) is attained, pathogens like *Salmonella* will not endure. Therefore, the presence of the pathogen in silage indicates poor silage-making procedures or contamination that occurs after ensiling. Generally, the incidence of *Salmonella* contamination of silage samples is relatively low, and contamination typically becomes an issue only when silage fermentation is insufficient. No studies have been found regarding the inhibition of by silage additives or inoculants. Thus, further research is required to investigate the efficacy of silage inoculants and organic acids such as propionic and acetic acid in controlling this pathogen in feed ingredients (Carrique-Mas et al. 2007; Jones 2011).

3 Chemical Hazard

Chemical hazards in silage can stem from various sources, including contaminated feedstock, pesticides, fertilizers, and additives. Residual chemicals from agricultural practices or improper storage can contaminate silage, thereby posing risks to animal health and environmental safety.

3.1 Nitrate, Nitrite, and Oxide Gases of Nitrogen

Nitrate (NO_3^-) is an essential macronutrient, which also serves as signal for plant growth. Although nitrate itself is not toxic to livestock, nitrite (NO_2^-) poisoning can ensue upon the consumption of forages with high concentrations of NO_3^-. The normal conversion pathway of NO_3^- in the rumen involves its reduction to ammonia, with NO_2^- as an intermediary (Table 10.3). However, in cases of excessive NO_3^- intake, NO_2^- can accumulate in the rumen and be absorbed into the blood, where it reacts with hemoglobin to form methemoglobin, which is incapable of transporting oxygen from the lungs to tissues. The condition known as methemoglobinemia can present with varying degrees of severity, including reduced feed intake, abortion, respiratory distress, coma, and even sudden death. Forages with NO_3^- concentrations exceeding 1700 mg of NO_3^--N/kg of DM may cause acute toxicity and should be incorporated into livestock rations in a balanced manner with other feeds to alleviate toxicity.

Nitrate is degraded during early stage of ensiling by enterobacteria and LAB, leading to the production of ammonia and nitrous oxide (N_2O) gas, with nitrogen dioxide (NO_2) and nitric oxide (NO) gases emerging as intermediates (Pahlow et al. 2003; Rooke and Hatfield 2003). The generation of nitric oxide and nitrogen dioxide gases in silage poses a significant threat to both livestock and human health, as well as presenting respiratory hazards. O'Kiely et al. (1999) described an incident in which ten calves and a farmer endured severe respiratory distress 24 h after an adjacent silo was filled with grass, resulting in the appearance of a brown haze of nitrogen dioxide gas in the building adjacent to the silo. Livestock should not be housed and individuals should avoid working near silos emitting nitrogen oxide gases during or immediately after filling.

Table 10.3 Main conversions of nitrate-N occurring during ensiling

Reaction	pH range	Method
1 $NO_3^- + AH2 + H+ \rightarrow NO_2^- + A + H2O$	>(4.5–5.0)	Enterobacteria
	>5.5	Plant enzymes
	>4.0	Lactobacilli
2 $2NO_2^- + 2AH2 + 2H+ \rightarrow N2O + 2A + 3H2O$	>(4.5–5.0)	Enterobacteria
3 $NO_2^- + 3AH2 + 2H+ \rightarrow NH_4^+ + 3A + 2H2O$	>(4.5–5.0)	Enterobacteria
4 $3NO_2^- + 2H+ \rightarrow 2NO + NO_3^- + H2O$	<6	Chemical

AH_2, electron donor; e.g., carbohydrate, NAD

3.2 Biogenic Amines and Ammonia

Biogenic amines are a group of low-molecular weight basic nitrogenous compounds formed mainly by decarboxylation of amino acids or by amination and transamination of aldehydes and ketones. They can be detected in silage and may exert detrimental impacts on feed intake, preference of ruminants, and animal health. The most common biogenic amines found in feed or food are histamine, tyramine, putrescine, cadaverine, spermine, and spermidine. These amines can accumulate in dairy products especially cheese (Linares et al. 2011). Biogenic amines taken in high amounts are known to cause headaches, nausea, vomiting, diarrhea, and hypertension (Gardini et al. 2016).

The production of amines in silage is attributed to slow acidification during ensiling (Van Os et al. 2009). Amines formed in the early stage of ensiling are attributable to the decarboxylation of amino acids by enterobacteria. In contrast, those formed in the later stage of ensiling result from the subsequent growth of proteolytic *Clostridia species* (Pahlow et al. 2003; Rooke and Hatfield 2003). Amines can be detected in both grass and maize silages, with putrescine, cadaverine, and spermidine being the predominant amines (Križek et al. 1993). Although results regarding the effect of amines on silage intake were inconsistent, the accumulation of amines and NH_3 in poorly fermented silages substantially reduced silage intake and utilization by animals (Scherer et al. 2015).

3.3 Butyric Acid

Butyric acid is a volatile fatty acid with a very pungent odor, likened to sweaty feet or a strong cheese-like smell, and an unpleasant taste. Butyric acid in poorly fermented silage is primarily produced by *Clostridia*, accompanied by a range of undesirable by-products, including amines, ammonia, acetic acid, carbon dioxide, and hydrogen. High amounts of butyric acid in silages indicate that the silage has undergone clostridial fermentation and is low in nutritive value, as butyric acid fermentation in silage is associated with the degradation of soluble nutrients. Silage with a high concentration of butyric acid showed less palatability and low intake, adversely affecting animal performance. Feeding silage with high butyric acid content increases the risk of subclinical ketosis (Driehuis et al. 2018).

Prevention of clostridial fermentation in forages necessitates the implementation of sound management strategies across the entire forage production, harvesting, and storage continuum. Firstly, efficient manure management protocols should be adopted to curtail the potential for contamination during the growth and harvesting of forages. Specifically, manure application should be avoided within 28 days prior to forage harvest. Secondly, forages should be harvested at the most suitable growth stage and with an appropriate dry matter content. Using clean equipment and avoiding harvesting in muddy or heavily soiled conditions can effectively minimize soil and manure contamination during forage harvesting. Inoculants are recommended to promote extensive fermentation and a rapid decline in pH, suppressing the growth of harmful clostridia species.

4 Conclusion

The probability of mycotoxins, as well as chemical and microbiological hazards, occurring in silage can be reduced through the implementation of appropriate silage-making practices. Both prior to and during the ensiling process, management strategies that facilitate rapid homolactic fermentations should be assured. During aerobic exposure, practices that guarantee the aerobic stability of silages and preclude pH elevation during feed-out are of utmost importance. The use of antifungal chemicals or inoculants is vital for minimizing mycotoxin contamination in livestock feeds. To prevent the production of *C. botulinum* toxins, butyric acid, or biogenic amines, it is important to ensure a rapid decline in silage pH below 4.5 or to reduce water activity through quick field wilting before ensiling. Sustaining low silage pH levels also restricts the growth of enterobacteria, thus reducing the risk of the production of toxic nitrogen oxide gas from nitrate. Any strategies that prevent aerobic deterioration can effectively inhibit the proliferation of *L. monocytogenes* and molds. Further research is required to transform silage epiphytic microbes and potent yet safe chemicals into innovative silage additives that not only enhance silage fermentation and storage but also effectively mitigate these hazards.

References

Adhikari B, Besser TE, Gay JM, Fox LK, Davis MA, Cobbold RN, et al. Introduction of new multidrug-resistant *Salmonella enterica* strains into commercial dairy herds. J Dairy Sci. 2009;92(9):4218–28.

Alonso VA, Pereyra CM, Keller LA, Dalcero AM, Rosa CA, Chiacchiera SM, et al. Fungi and mycotoxins in silage: an overview. J Appl Microbiol. 2013;115(3):637–43.

Amado IR, Fuciños C, Fajardo P, Guerra NP, Pastrana L. Evaluation of two bacteriocin-producing probiotic lactic acid bacteria as inoculants for controlling Listeria monocytogenes in grass and maize silages. Anim Feed Sci Technol. 2012;175(3–4):137–49.

Amado IR, Fuciños C, Fajardo P, Pastrana L. Pediocin SA-1: a selective bacteriocin for controlling in maize silages. J Dairy Sci. 2016;99(10):8070–80.

Battacone G, Nudda A, Palomba M, Pascale M, Nicolussi P, Pulina G. Transfer of aflatoxin B1 from feed to milk and from milk to curd and whey in dairy sheep fed artificially contaminated concentrates. J Dairy Sci. 2005;88(9):3063–9.

Bjornsdottir K, Breidt F Jr, McFeeters RF. Protective effects of organic acids on survival of *Escherichia coli* O157:H7 in acidic environments. Appl Environ Microbiol. 2006;72(1):660–4.

Buehner KP, Anand S, Djira GD, Garcia A. Corrigendum to 'Prevalence of thermoduric bacteria and spores on 10 Midwest dairy farms'. J Dairy Sci. 2014;97(12):8009–16.

Byrne CM, O'Kiely P, Bolton DJ, Sheridan JJ, McDowell DA, Blair IS. Fate of *Escherichia coli* O157:H7 during silage fermentation. J Food Prot. 2002;65(12):1854–60.

Callaway TR, Keen JE, Edrington TS, Baumgard LH, Spicer L, Fonda ES, et al. Fecal prevalence and diversity of Salmonella species in lactating dairy cattle in four states. J Dairy Sci. 2005;88(10):3603–8.
Carrique-Mas JJ, Bedford S, Davies RH. Organic acid and formaldehyde treatment of animal feeds to control Salmonella: efficacy and masking during culture. J Appl Microbiol. 2007;103(1):88–96.
Cheli F, Campagnoli A, Dell'Orto V. Fungal populations and mycotoxins in silages: from occurrence to analysis. Anim Feed Sci Technol. 2013;183(1–2):1–16.
Chen Y, Sela S, Gamburg M, Pinto R, Weinberg ZG. Fate of *Escherichia coli* during ensiling of wheat and corn. Appl Environ Microbiol. 2005;71(9):5163–70.
Driehuis F. Silage and the safety and quality of dairy foods: a review. Agric Food Sci. 2013;22(1):16–34.
Driehuis F, Elferink SJWHO. The impact of the quality of silage on animal health and food safety: a review. Vet Quart. 2000;22(4):212–6.
Driehuis F, Spanjer MC, Scholten JM, Giffel MCT. Occurrence of mycotoxins in maize, grass and wheat silage for dairy cattle in The Netherlands. Food Addit Contam B. 2008a;1(1):41–50.
Driehuis F, Spanjer MC, Scholten JM, te Giffel MC. Occurrence of mycotoxins in feedstuffs of dairy cows and estimation of total dietary intakes. J Dairy Sci. 2008b;91(11):4261–71.
Driehuis F, Wilkinson JM, Jiang Y, Ogunade I, Adesogan AT. Silage review: animal and human health risks from silage. J Dairy Sci. 2018;101(5):4093–110.
Duniere L, Jin L, Smiley B, Qi M, Rutherford W, Wang Y, et al. Impact of adding Saccharomyces strains on fermentation, aerobic stability, nutritive value, and select lactobacilli populations in corn silage. J Anim Sci. 2015;93(5):2322–35.
Eugui EM, Almquist SJ, Muller CD, Allison AC. Lymphocyte-selective cytostatic and immunosuppressive effects of mycophenolic acid in vitro: role of deoxyguanosine nucleotide depletion. Scand J Immunol. 1991;33(2):161–73.
Fink-Gremmels J. 1 - Introduction to animal feed contamination. In: Fink-Gremmels J ed., Animal Feed Contamination: Woodhead Publishing 2012: 1–10
Gallo A, Giuberti G, Bertuzzi T, Moschini M, Masoero F. Study of the effects of PR toxin, mycophenolic acid and roquefortine C on in vitro gas production parameters and their stability in the rumen environment. J Agric Sci. 2015a;153(1):163–76.
Gallo A, Giuberti G, Frisvad JC, Bertuzzi T, Nielsen KF. Review on mycotoxin issues in ruminants: occurrence in forages, effects of mycotoxin ingestion on health status and animal performance and practical strategies to counteract their negative effects. Toxins. 2015b;7(8):3057–111.
Gallo A, Bertuzzi T, Giuberti G, Moschini M, Bruschi S, Cerioli C, et al. New assessment based on the use of principal factor analysis to investigate corn silage quality from nutritional traits, fermentation end products and mycotoxins. J Sci Food Agr. 2016;96(2):437–48.
Gardini F, Özogul Y, Suzzi G, Tabanelli G, Özogul F. Technological factors affecting biogenic amine content in foods: a review. Front Microbiol. 2016;7:1218.
González Pereyra ML, Alonso VA, Sager R, Morlaco MB, Magnoli CE, Astoreca AL, et al. Fungi and selected mycotoxins from pre- and postfermented corn silage. J Appl Microbiol. 2008;104(4):1034–41.
Jones FT. A review of practical Salmonella control measures in animal feed. J Appl Poult Res. 2011;20(1):102–13.
Keller LAM, González Pereyra ML, Keller KM, Alonso VA, Oliveira AA, Almeida TX, et al. Fungal and mycotoxins contamination in corn silage: monitoring risk before and after fermentation. J Stored Prod Res. 2013;52:42–7.
Kosicki R, Blajet-Kosicka A, Grajewski J, Twaruzek M. Multiannual mycotoxin survey in feed materials and feedingstuffs. Anim Feed Sci Technol. 2016;215:165–80.
Križek M, Kalač P, Peterka J. Biogenic amines in silage 3. The occurrence of six biogenic amines in farm-scale grass and maize silages. Arch Anim Nutr. 1993;45(2):131–7.
Linares DM, Martín M, Ladero V, Alvarez MA, Fernández M. Biogenic amines in dairy products. Crit Rev Food Sci. 2011;51(7):691–703.

Masiello SN, Martin NH, Watters RD, Galton DM, Schukken YH, Wiedmann M, Boor KJ. Identification of dairy farm management practices associated with the presence of psychrotolerant sporeformers in bulk tank milk. J Dairy Sci. 2014;97(7):4083–96.

McDonald P, Henderson AR, Heron SJE. The biochemistry of silage. 2nd ed. Exp Agric. 1992;28:125.

Mohr AI, Lorenz I, Baum B, Hewicker-Trautwein M, Pfaffl M, Dzidic A, et al. Influence of oral application of mycophenolic acid on the clinical health status of sheep. J Vet Med A. 2007;54(2):76–81.

Monbaliu S, Van Peteghem C, De Saeger S. Detection and determination of natural toxins (mycotoxins and plant toxins) in feed. Animal feed contamination. Woodhead Publishing; 2012. pp. 286–325.

O'Kiely P, Turley T, Rogers PA. Exposure of calves to nitrogen dioxide in silage gas. Vet Rec. 1999;144(13):352–3.

Ogunade IM, Kim DH, Jiang Y, Weinberg ZG, Jeong KC, Adesogan AT. Control of *Escherichia coli* O157:H7 in contaminated alfalfa silage: effects of silage additives. J Dairy Sci. 2016;99(6):4427–36.

Ogunade IM, Jiang Y, Kim DH, Cervantes AAP, Arriola KG, Vyas D, et al. Fate of *Escherichia coli* O157:H7 and bacterial diversity in corn silage contaminated with the pathogen and treated with chemical or microbial additives. J Dairy Sci. 2017;100(3):1780–94.

Ogunade IM, Martinez-Tuppia C, Queiroz OCM, Jiang Y, Drouin P, Wu F, et al. Silage review: mycotoxins in silage: occurrence, effects, prevention, and mitigation. J Dairy Sci. 2018;101(5):4034–59.

Pahlow, G., R. E. Muck, F. Driehuis, S. J. W. H. Oude Elferink, and S. F. Spoelstra. 2003. Microbiology of ensiling. Pages 31–93 in Silage Science and Technology. L. Al-Amoodi, ed. American Society of Agronomy, Crop Science Society of America, and Soil Science Society of America, Madison, WI.

Pauly TM, Tham WA. Survival of Listeria monocytogenes in wilted and additive-treated grass silage. Acta Vet Scand. 2003;44(1–2):73–86.

Pedroso AF, Adesogan AT, Queiroz OC, Williams SK. Control of *Escherichia coli* O157:H7 in corn silage with or without various inoculants: efficacy and mode of action. J Dairy Sci. 2010;93(3):1098–104.

Queiroz OC, Han JH, Staples CR, Adesogan AT. Effect of adding a mycotoxin-sequestering agent on milk aflatoxin M1 concentration and the performance and immune response of dairy cattle fed an aflatoxin B1-contaminated diet. J Dairy Sci. 2012;95(10):5901–8.

Queiroz OCM, Ogunade IM, Weinberg Z, Adesogan AT. Silage review: foodborne pathogens in silage and their mitigation by silage additives. J Dairy Sci. 2018;101(5):4132–42.

Richard E, Heutte N, Bouchart V, Garon D. Evaluation of fungal contamination and mycotoxin production in maize silage. Anim Feed Sci Technol. 2009;148(2–4):309–20.

Rooke, JA., Hatfield, RD., Buxton, DR. (Ed.), Muck, RE. (Ed.), & Harrison, JE. (Ed.) (2003). Biochemistry of ensiling. In DR. Buxton, RE. Muck, & JE. Harrison (Eds.), Silage Science and Technology (pp. 95–139). American Society of Agronomy. Advance online publication.

Scallan E, Hoekstra RM, Angulo FJ, Tauxe RV, Widdowson MA, Roy SL, et al. Foodborne illness acquired in the United States—major pathogens. Emerg Infect Dis. 2011;17(1):7.

Scherer R, Gerlach K, Südekum KH. Review-biogenic amines and gamma-amino butyric acid in silages: formation, occurrence and influence on dry matter intake and ruminant production. Anim Feed Sci Technol. 2015;210:1–16.

Schmidt P, Novinski CO, Junges D, Almeida R, de Souza CM. Concentration of mycotoxins and chemical composition of corn silage: a farm survey using infrared thermography. J Dairy Sci. 2015;98(9):6609–19.

Van Os M, Lassalas B, Toillon S, Jouany JP. In vitro degradation of amines by rumen microorganisms. J Agric Sci. 2009;125(2):299–305.

Vissers MM, Driehuis F, Te Giffel MC, De Jong P, Lankveld JM. Short communication: quantification of the transmission of microorganisms to milk via dirt attached to the exterior of teats. J Dairy Sci. 2007;90(8):3579–82.

Weinberg ZG, Ashbell G, Chen Y, Gamburg M, Sela S. The effect of sewage irrigation on safety and hygiene of forage crops and silage. Anim Feed Sci Technol. 2004;116(3–4):271–80.

Wilson A, Sigee RD, Epton HA. Anti-bacterial activity of *Lactobacillus plantarum* strain SK1 against Listeria monocytogenes is due to lactic acid production. J Appl Microbiol. 2005;99(6):1516–22.

Woolford MK. Microbiological screening of the straight chain fatty acids (C1-C12) as potential silage additives. J Sci Food Agr. 1975;26(2):219–28.

Part III
Silage Production and Livestock Raising in the World

Chapter 11
Silage Production and Livestock Raising in Africa

Ansah Terry, Peter Dele, Delma B. Jethro, and Damião W. Nguluve

Abstract Feed shortage during the dry season is an important factor limiting livestock production in tropical regions such as Africa. To address this problem, several African countries are increasingly adopting silage and TMR fermentation technology to improve the efficiency of feed resource utilization and promote livestock production. Tropical forage and crop straw are important feed resources in Africa, known for their high biomass production capacity. They have enormous production potential as raw materials for silage. Effectively utilizing local feed resources to produce high-quality silage that can be stored for extended periods is crucial for alleviating feed shortages and promoting sustainable development of animal husbandry in Africa. This chapter focuses on the overview of feed resources, silage production, and livestock raising in the African region.

Keywords Africa · Feed resources · Feed shortage · Livestock raising · Silage production

A. Terry (✉)
University for Development Studies, Tamale, Ghana
e-mail: tansah@uds.edu.gh

P. Dele
Federal University of Agriculture, Abeokuta, Nigeria
e-mail: delepa@funaab.edu.ng

D. B. Jethro
Centre National de la Recherche Scientifique et Technologique, INERA, Département Production Animale, Ouagadougou, Burkina Faso

D. W. Nguluve
Agricultural Research Institute of Mozambique, Matola, Mozambique

Y. Cai, K. Ataku (eds.), *Cultural History and Modern Production Technology of Silage*, https://doi.org/10.1007/978-981-96-5787-2_11

1 Silage Production and Livestock Raising in Ghana

1.1 Importance of Livestock and Production System in Ghana

Livestock contributes about 2–4% of the national GDP and 15–20% of the agricultural GDP in Ghana. In Ghana's Savanna agro-ecological zone (SAEZ), with its semi-arid climate, livestock production provides food and employment to most of the population. Around 60% of the rural population is engaged in livestock farming, and this zone hosts over 60% of the country's ruminant livestock.

Smallholder crop-livestock farmers dominate this region, who typically manage 5–30 ruminants and cultivate 1–5 hectares of land annually. Livestock management is primarily extensive or semi-intensive (79% of farmers), with transhumance cattle herders dominating the cattle business with large animal populations. The common ruminant livestock species kept are cattle, sheep, and goats. The popular cattle breeds are the West African Short Horn (WASH), N'Dama, white Fulani, and the offspring of crosses between the WASH or N'Dama and the white Fulani (Sanga). The sheep and goat breeds are primarily dominated by the Djallonké sheep and goats, with a few Sahelian breeds often brought into the country from neighbouring countries during festivities.

1.2 Livestock Feed Resource

Feed sources include natural pasture, on-field crop residues, and agro-industrial by-products. Seasonal changes significantly affect feed quality and availability. During the wet season, extensive land cultivation for food production limits grazing despite abundant pasture, while the dry season provides access to grazing lands but with poor-quality feed, which is often destroyed by bushfires. During the dry season, animals travel 1.8–3.5 km daily in search of food. Crop residues from legumes are usually collected and stored as hay, but due to poor storage conditions, these residues often reduce in quality. This poor-quality feed and improper manure management lead to increased methane and nitrous oxide emissions, affecting livestock growth and productivity. In the wet season, the country experiences sporadic clashes between transhumance herders and crop farmers due to the invasion of crop farms belonging to the herders. Biomass yields from natural pasture are higher in the wet than in the dry season (0.63–13.43 t/ha). Crude protein content drops significantly from about 13% in the rainy season to 5% or less in the dry season. Ruminants in Ghana will require about 2–3% of their body weight in dry matter daily for maintenance and growth. The maintenance energy requirement is estimated to be about 70–80 MJ/100 kg live body weight with an additional 10–15 MJ for growth. A typical smallholder farmer keeping about 1–10 TLU (tropical livestock unit) will require daily metabolizable energy of about 175–1750 MJ for maintenance.

Harvested grass

Chopping of grass for ensiling with a forage chopper

Packing of chopped grass into the bucket silos

Opening of silos for sampling

Estimating Fermentation loss

Silage after 3 months

Fig. 11.1 Silage preparation and analysis of fermentation quality. (Photos by Terry and Cai)

Ensiled grass generally provides about 8–11 MJ/kg of dry matter (DM) in metabolizable energy, while natural pasture in the dry season provides between 6 and 9 MJ/kg of DM. This may be influenced by factors such as grass type, maturity, fermentation quality, and harvesting practices. Therefore, silage can be a significant energy source for livestock in Ghana, especially during long periods of drought. It may be fed as a supplementary feed with natural pasture to meet the nutritional needs of livestock.

Silage use among Ghanaian farmers is low despite the high demand for reliable and nutritious feed sources throughout the year (Fig. 11.1).

1.3 Importance of Silage

Silage is crucial for livestock feed, particularly in Ghana, due to its ability to provide consistent, high-quality nutrition to livestock throughout the year. Ghana's climate, characterized by distinct wet and dry seasons, with about 7-month-long dry seasons, especially in the Savanna agro-ecological zones, presents challenges for maintaining a steady supply of fresh forage. During the dry season, the availability of natural pastures and high-quality feed diminishes, which can lead to nutritional deficiencies in livestock.

Silage addresses this issue by allowing farmers to preserve and store high-moisture forage, such as maize, sorghum, or grasses, when they are at their peak nutritional value. Through fermentation, silage converts these forages into a stable, palatable feed that can be used during periods when fresh forage is scarce. This ensures that livestock receive balanced nutrition year-round, enhancing their health, productivity, and overall performance.

Silage provides a reliable source of high-quality feed throughout the year, especially during periods when fresh forage is unavailable, such as in the dry season. This helps to maintain consistent nutrition for livestock. Silage retains the nutritional value of the forage used, preserving essential nutrients such as energy, protein, and vitamins. This ensures that livestock receive balanced and nutritious feed even when natural pastures are less productive. Through fermentation, silage converts these forages into a stable, palatable feed that can be used during periods when fresh forage is scarce. This ensures that livestock receive balanced nutrition year-round, enhancing their health, productivity, and overall performance. Consistent and high-quality nutrition from silage can enhance livestock health, growth, and productivity. This leads to better weight gain, milk production, and overall performance. Silage production allows farmers to make use of surplus or seasonal forage, optimizing feed resources and reducing feed costs. This can lead to more efficient and profitable livestock operations. Silage can be produced in various types and forms (e.g., corn silage, grass silage) to suit different livestock needs and farming systems. It offers flexibility in feed management and planning.

1.4 Types of Silage

Silage can be made from corn, sorghum, grass, legume, crop residues, or mixed crop silage, depending on the availability of each of these resources. Each of these offers different nutritional benefits regarding energy, fibre, and protein. During the wet season, ensiling grass species such as *Pennisetum*, *Panicum*, and *Brachiaria* in the Savanna agro-ecological zone of Ghana reduced fermentation losses by 8–10%. The CP content decreased from about 90 g/kg DM in fresh grass to 80 g/kg DM in the silage. Ensiling successfully eliminated coliform bacteria and mould from various grass species. Additives such as lactic acid bacteria (LAB) and cellulase inoculants containing *Lactiplantibacillus plantarum* and *Acremonium* species, respectively, have been successfully used to ensiling various grass species in Ghana. Total digestible nutrients (TDN) decreased by about 7% in silage with LAB but increased by approximately 6% when LAB was combined with a cellulose additive.

1.5 Silage Production Process

Forages should be harvested at the right stage of maturity to maximize nutritional value. For most grasses, the ideal time is when they are still in the vegetative or early boot stage. Harvesting should be done when the forage's moisture content is appropriate (typically between 65% and 75%) for grass silage. Avoid harvesting during wet conditions to prevent excessive moisture and spoilage.

Under smallholder farmer conditions, forages may be cut manually using sickles or cutlasses. This method is labour-intensive but suitable for small-scale farming. The harvested forages should be chopped with a forage chopper to a 2–5 cm length. This improves packing and enhances the fermentation processes. Additives such as LAB or cellulase may be included individually or combined to enhance fermentation.

Silage can be prepared in various types of silos, including pits, drums, buckets, and bag silos. The bags, drum, and bucket silos have been used in the Savanna agro-ecological zone of Ghana. The buckets and drums are often lined with a thick polythene sheet before packing the grass. The chopped forage must be immediately transferred into the silos and tightly packed to prevent spoilage and promotes proper fermentation. Properly seal silage with plastic sheeting or covers to prevent air infiltration, which can lead to spoilage and mould growth.

Regularly check the silage for signs of spoilage or inadequate fermentation. Adjust storage practices as needed to maintain quality. Good-quality silage should have a stable and low pH, typically between 3.8 and 4.2, indicating effective fermentation and preservation. The silage should have a pleasant, slightly sour smell. Off-smells or a rancid odour can indicate spoilage. The texture of matured silage should be uniform, without excessive moisture or dry spots. It should be easy to handle and mix with other feeds. Feed out silage from the silos carefully to minimize exposure to air and prevent contamination.

1.5.1 Challenges of Silage Technology Adoption

In Ghana, the adoption of silage technology faces several unique challenges that impact its uptake among farmers.

1.5.2 The Initial and Operational Cost of the Technology

The cost of inputs and equipment required for silage production, such as quality, forage harvesters and choppers, silos, and sealing materials, can be prohibitive for smallholder farmers in Ghana. Access to high-quality inputs, such as silage additives and improved forage seeds, is often limited. Limited access to credit and financial support makes it difficult for farmers to afford these initial investments. Additional expenses related to labour, maintenance, and the purchase of silage additives further strain farmers' finances. Many smallholders prioritize immediate needs over medium- to long-term investment in technology.

1.5.3 Lack of Knowledge and Training

Smallholder farmers are generally unaware of the benefits and techniques of silage production. Educational programs and extension services on silage technology are often insufficient or not widely accessible. Effective silage production requires specific technical skills, including proper harvesting, chopping, fermentation, and storage techniques. The absence of targeted training programs contributes to low adoption rates.

1.5.4 Cultural and Traditional Practices

Many Ghanaian farmers rely on traditional feed preservation methods such as hay-making, which may not be applicable in the wet season when there is an abundance of quality forage. These practices are familiar and may seem more straightforward compared to the new technology of silage production.

Farmers may be reluctant to adopt new technologies due to perceived risks and uncertainties regarding their effectiveness and profitability.

1.5.5 Economic and Market Considerations

Limited access to markets for silage or insufficient demand for preserved feeds can affect farmers' willingness to invest in silage technology. Farmers may not see a clear economic benefit if there is no strong market for silage. Farmers may be sceptical about the returns on investment in silage technology, especially if they have not seen successful case studies or lack information on cost-benefit analysis.

1.6 Potential for Growth and Development in Silage Technology in Ghana

1.6.1 Increasing Awareness and Training

Develop and expand training programs and workshops on silage production techniques. Partner with agricultural extension services and universities to offer practical, hands-on training for farmers. Use media, demonstrations, and farmer-to-farmer networks to share successful silage production practices and case studies. Utilize mobile apps and digital platforms for extension services, market information, and supply chain management.

1.6.2 Reducing Costs Through Innovation

Develop cost-effective, locally manufactured silage equipment and materials. Encourage innovation in low-cost silage production techniques suitable for smallholder farmers. Improve access to credit and subsidies for purchasing silage equipment and inputs. Create financial products tailored to smallholder farmers.

1.6.3 Research and Development

Foster partnerships between research institutions, government agencies, and private sector players to drive silage production and livestock farming innovation. Research climate-resilient livestock management practices and feed production techniques suitable for Ghana's diverse agro-ecological zones.

1.6.4 Enhancing Policy and Support Frameworks

Advocate for policies that support the growth of silage production and livestock farming. This includes subsidies, tax incentives, and research grants. Develop and enforce standards for silage production to ensure quality and safety.

1.7 Conclusion

Economic and Nutritional Importance: Livestock agriculture is vital to Ghana's economy and rural livelihoods, contributing nearly 13% to agricultural GDP. It provides essential protein through meat, milk, and eggs, supports employment, and stimulates rural development. However, it faces challenges such as diseases, inadequate veterinary services, and year-round feed shortages, particularly during the dry season.

Challenges in Forage Availability: Natural pasture availability in Ghana is highly seasonal, with severe forage shortages during the dry season. The wet season offers abundant forage but is often restricted due to land use for crop production. The dry season sees limited grazing lands and poor-quality feed, necessitating alternative feeding strategies like the use of silage.

Importance of Silage: Silage is crucial for maintaining consistent, high-quality nutrition for livestock throughout the year, especially in Ghana's climate with its prolonged dry season. It allows for preserving high-moisture forage, providing a stable, nutritious feed when fresh forage is scarce, thus enhancing livestock health and productivity.

Adoption of silage technology in Ghana is low due to factors such as high costs, limited knowledge, cultural practices, and economic constraints.

(Ansah Terry)

2 Silage Production and Livestock Raising in Nigeria

Nigeria covers a total land area of about 923,768 km^2 and accommodating nearly 230 million people (US Department of State 2023) making it the most populated country in Africa and the sixth most populous in the world. Nigeria shares borders with Benin Republic to the west, Niger to the north, Chad to the northeast, and Cameroon to the east. Nigeria is made up of 36 states plus the Federal Capital Territory, which is home to the nation's capital, Abuja. Nigeria's nominal GDP accounts for 9% of the Africa GDP making it the fourth in Africa after South Africa, Egypt, and Algeria, respectively. In 2023, agriculture was noted to have contributed 24.65% to nominal GDP of Nigeria in 2023 and accounted for more than 35% of total employment in 2020. In terms of Nigerian agricultural GDP, livestock subsector is in a distant second to the crop subsector and is a promising subsector contributing about 5% of Nigeria's GDP and 17% of its agriculture GDP.

In Nigeria, silage production plays a significant role in the husbandry of livestock, especially ruminants like cattle, sheep, and goats. High-moisture forages like Napier grass, Guinea grass, sorghum, and maize are ensiled and utilized during the dry season when fresh forage is scarce because of the variation in the Nigeria climate, with tropical rainforests in the south and desert and semi-arid regions in the north. Both the kinds of forage crops that can be planted and the timing of silage production are impacted by this variability. The period of forage scarcity has been associated with loss of weight and disease pre-disposition in the livestock industry and sometimes death. Silage production has significantly contributed to the sustainability and productivity of livestock farming in Nigeria (Olorunnisomo and Adewumi 2015), and in the future it is believed to have the potential for better sustainability.

2.1 History of Silage Production in Nigeria

Silage has been defined as the fermented and acidic stored feed from agricultural crops (Wilkinson et al. 2003). However, in practice within the Nigeria livestock space, farmers/producers also use agro by-products such as tuber peels, unsalable farm produce (Dele et al. 2013) in the tropics. In Nigeria, silage making has been dated back to the 1950s (Miller and Rains 1963) with the use of forage crops like *Andropogon gayanus*, *Chloris gayana*, *Hyparrhenia rufa*, *Panicum maximum*, *Pennisetum pedicellatum*, *Pennisetum purpureum*, *Sorghum vulgare*, and *Zea mays*. These forage crops were ensiled on experimental basis with different silos in use; the silos ranged from miniature towers 1.83 m tall and 0.76 m wide which contain chopped materials of about 363 kg and pilot-scale pits with 3.66 m depth and 3.66 m diameter and can contain chopped materials of 9000 kg to large-scale trench silos which are 13.72 m long, 3.66 m wide, and 1.83 m deep. The northern Nigeria was profound in forage conservation in the 1950s and 1960s (Miller and Rains 1963; Rains 1963; Akinola 1974), and this might be as a result of the higher number of ruminants in the region (NASS 2011) with over 90% of cattle and 70% of small ruminants.

2.2 Livestock Production in Nigeria

Livestock production is a part of Nigeria's agriculture system. In 2017, the country had approximately 180 million poultry birds, 76 million goats, 43.4 million sheep, 18.4 million cattle, 7.5 million pigs, and 1.4 million equids (FMARD 2017). Nigeria has the largest population of small ruminants (sheep and goat) in Africa and noted to be the sixth in terms of cattle population. Given the importance of beef consumption in the nation, cattle are considered essential livestock in Nigeria. The cattle production in Nigeria has been basically on extensive system of production with about 82% of the 18.4 million, and 16.8% and 1.1% have been reported to be raised under the semi-intensive and intensive system of production (Fig. 11.2), respectively. It has been reported that 99% are indigenous breeds with 1% being exotic breeds. Majority of the cattle and small ruminants' population are concentrated in the northern part of Nigeria (Fig. 11.3) (NASS 2011) and are majorly on the extensive system of production by nomadic herders.

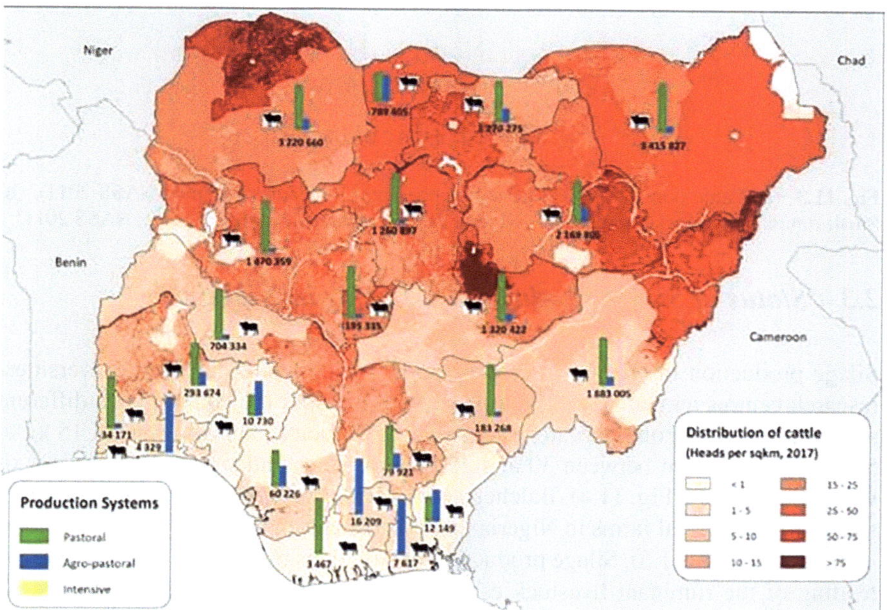

Fig. 11.2 Cattle distribution by region and production system in Nigeria. (Source: GLW 3.0 and FMARD 2017)

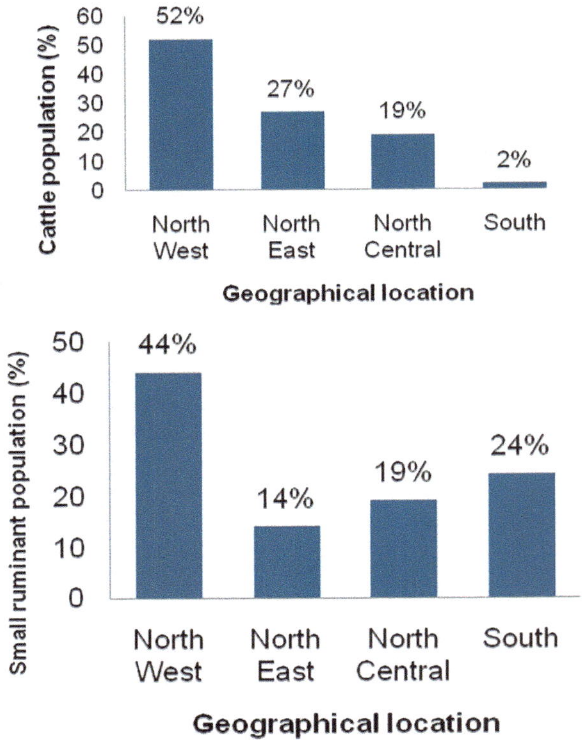

Fig. 11.3 (**a**) Cattle population in Nigeria by geographical location. (Source: NASS 2011). (**b**) Small ruminant livestock population in Nigeria by geographical location. (Source: NASS 2011)

2.3 Status of Silage Production Capacity in Nigeria

Silage production in Nigeria is majorly under experimental farms in universities/research centres and smallholder farms scattered across the country with different silo types ranging from laboratory silos of 1 kg capacity to plastic bags of 15 kg to 50 kg, drum silos of between 30 and 250 kg, pit silos, and small-sized bunkers of varying capacities (Fig. 11.4). Balehegn et al. (2022) have reported the utilization of silage in commercial farms in Nigeria that practise mechanized harvesting and storing of silage (Fig. 11.5). Silage production in Nigeria is majorly centred towards the feeding of the ruminant livestock especially in the dry season when there is the scarcity of fodder with very few towards the poultry and rabbit subsector (Ologhobo and Balogun 1987; Amos et al. 2021). Currently, there are efforts geared towards youths and women empowerment in silage production in different regions of the countries. In the northern region, here is a partnership between Nestle and 2SCALE aiming to encourage the intensification of cattle husbandry in Nigeria by

Fig. 11.4 Filling (**a**), compacting (**b**), sealing (**c**), and storage (**d**) of the silo at the Federal University of Agriculture, Abeokuta, Nigeria. (Photos by Dele)

Fig. 11.5 Silage making process at Ilmar integrated farms subsidiary of Nasrun Nigeria Ltd. Kaduna, Nigeria. (Photos by Dele)

transitioning from pastoral to smallholder dairy farming by training youths in the art of silage making during the wet season when there is abundance of forage materials and later feed dairy cows during scarcity to reduce the fluctuation of milk production (Figs. 11.6 and 11.7).

Chopping process **Compaction process**

Fig. 11.6 Training youths in Southern Nigeria on silage making on graduate farms affiliated to Federal University of Agriculture Abeokuta. (Photos by Dele)

Fig. 11.7 Final year students chopping maize stover for silage making. (Photo by Dele)

2.4 Silage Role in Livestock Production in Nigeria

Inadequate feed is the major challenge in the livestock industry in Nigeria especially cattle production, particularly during the dry season where there is acute shortage of fodder. Farmers are less familiar with silage making as they are familiar with using agricultural by-products as animal feed. Adoption has been relatively low, despite a great deal of study being done on silage production at research centres and universities in Nigeria. This is due to a variety of factors, such as a scarcity of herbage, the perception that silage production is difficult and easily go bad, and a lack of funding for new equipment. Amaranth has a good potential to produce good-quality silage for the sheep (Olorunnisomo and Ayodele 2009; Olorunnisomo 2010). Amuda et al. (2020) also reported that feeding of ensiled maize stover supplemented with protein concentrates improves the performance of west African dwarf sheep with respect to dry matter intake (DMI), nutrient digestibility, and nitrogen utilization. Ensiling sugarcane top with non-protein nitrogen has been reported to improve the digestibility in WAD sheep (Akinbode et al. 2018). Amole et al. (2013) noted that supplemented ensiled maize stover with Lablab improved the total ruminal volatile fatty acid of grazing calves by 68% when compared to calves that are only graze, which implied that the calves benefited from the ensiled maize stover-Lablab silage as a source of energy. Silage-based feeding (legume or grass ensiled with cassava peels) has shown to increase and improve milk yield and quality of Sokoto Gudali dairy cattle in Nigeria (Olorunnisomo 2013), and ensiled *Moringa oleifera* fed to Sokoto Gudali cows has potential to improve local milk production (Olorunnisomo 2014).

2.5 Silage Research Output Published in Nigeria

The research outputs on silage in Nigeria are limited, and this is a reflection of the attention given to silage making and its adoption among livestock stakeholders. The two major outlets of research outputs on livestock production in Nigeria are, namely, Nigerian Journal of Animal Production (NJAP) and Nigerian Journal of Animal Science (NJAS). The 'silage outputs' with these two outlets are dismally low with 31 articles out 2325 articles published in the NJAP from 1974 to 2023 making just 1.33% of the total output. The output in the NJAS was similar to that of the NJAP with 1.15% of the total articles published between 1999 and 2023 (11 articles out of 960); of the outputs in the two journal outlets, the scale of silage production has been experimental with no reference to large-scale silage production or collaboration with between the universities/research centres and commercial silage producers of farms. Going by the statistics, 90% of the silage outputs in NJAP were between 2013 and 2023, and 91% of the silage outputs in NJAS were between 2016 and 2023, which showed that there is a rising interest in silage making in the country though with low adoption, but it is a reflection that silage making is gaining

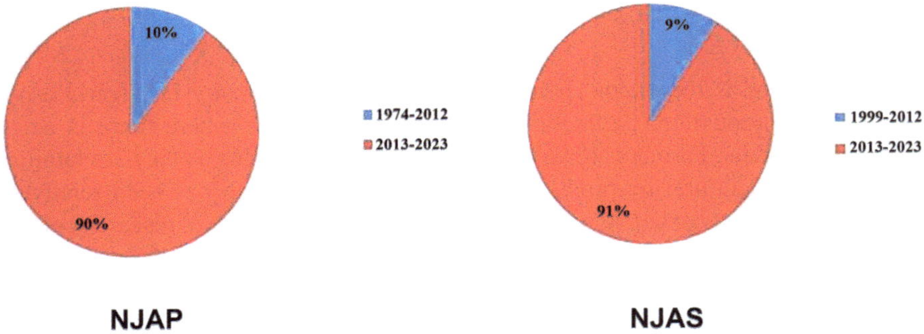

Fig. 11.8 Silage articles published in foremost Nigerian-based Animal Science Journals (NJAP, $n = 2325$; NJAS, $n = 960$)

attention, and there is a prospect (Fig. 11.8). The Silage Group Nigeria was formed in 2023 and is intensifying the awareness campaign of silage production in the country and currently holding a month webinar on silage featuring professionals both in the academia and industry. By the recruitment of open data available at Scival/Scopus (scival.com/benchmarking/analyse) in April 2023, Nussio and Daniel (2023) reported that silage-related articles linked to Nigeria are about 0.5% of the total silage (22,083) articles between 1908 and 2022, which is also a reflection that the adoption of silage in Nigeria is low.

2.6 Challenges Associated with Silage Production in Nigeria

Forage conservation practices (Lukuyu et al. 2011), especially as they relate to silage making, related costs, poor fit into the current farming systems, and lack of simple and appropriate technologies for ensiling (Tufail et al. 2020), are some of the factors contributing to the limited adoption of silage making in sub-Saharan Africa and especially Nigeria. With the pace of research on silage production in Nigeria which is considered low which could be associated to paucity of research grant and more reason why most research are usually on laboratory and experimental level and the involvement of extension workers have been considered low (Shishi and Agbo 2018), which is important in training and dissemination of research outputs to farmers. Comprehensive training programs for farmers on silage production techniques and management procedures are critical. Extension services should focus on providing farmers with the information and skills required to efficiently produce and store silage. Insufficient/inadequate government policies and assistance programs for smallholder farmers may hinder the development of silage production. Farmers' financial burdens can be reduced by making silage-making equipment and materials more accessible through subsidies, cooperatives, or credit facilities. Investment in infrastructure, such as adequate storage facilities, is also required to reduce losses. Government policies should encourage silage production by giving incentives, promoting research and development, and establishing a favourable climate for private sector participation in the supply chain.

2.7 Challenges of Livestock Production in Nigeria

Livestock production is an important and integral economic part of the Nigerian system is but faced with some constraints especially the ruminant subsector. A major limitation in ruminant livestock production is insufficient pasture going by the system of production; the natural pasture has been over relied on, which is affected by the season (Dele 2012). In the raining season, forages are available in excess and in high quality, but at the onset of the dry season, both quality and quantity are affected drastically most especially in the northern part of the country. Therefore, the need to gear up the awareness campaign on forage conservation especially silage making needs to be taken seriously. This seasonality has led to the movement of livestock from the northern part to the southern part (Ayanlade and Ojebisi 2019) which is generally humid and with spared vegetation in the dry season. This movement of livestock to the southern part is not but with its resultant consequences causing clashes between crop farmers and herders. The availability of pasture has also been affected with urbanization drive in most regions of the country (Shabu et al. 2021) where grazing lands and reserves have been encroached. The scarcity of water resources has a substantial impact on ruminant production in Nigeria. This difficulty influences numerous areas of livestock husbandry, including animal health and pasture availability. Water scarcity can cause dehydration and heat stress in ruminants, lowering their general health and production owing to the production system where animals move around (Akinmoladun et al. 2019). Nigerian ruminant production is greatly affected by climate change, which has an impact on a number of factors including animal health, pasture availability, water supplies, and overall productivity (Nwobodo et al. 2022). Unimproved breeding and management practices have greatly impede ruminant livestock productivity in Nigeria leaving our production abysmally low comparing with exotic breeds as most livestock farmers are resource poor and only rely on indigenous breeds with lower genetic potential, for instance, Sokoto Gudali cows have been reported to produce 295 kg for cows on extensive production system to 1300 kg for cows on high nutrient intake per lactation period compared to Holstein Friesian cow which produces 6000–8000 kg per lactation period (Silke et al. 2002). Enhancing ruminant productivity requires improved breeding and management system, availability to superior genetic stock, and awareness campaign.

2.8 Conclusion

Silage production has huge potential to boost livestock production in Nigeria especially the ruminant livestock being a major source of animal protein for the populace, making it more robust and productive. Farmers can significantly enhance their livelihoods and contribute to the country's food security by overcoming current challenges. The way forward is a coordinated effort to educate, equip, and empower farmers, ensuring that silage production becomes a cornerstone of sustainable livestock farming in Nigeria.

(Peter Dele)

3 Silage Production and Livestock Raising in Burkina Faso

In Sahelian countries like Burkina Faso, livestock feed is primarily based on natural grazing with a little supplementation (Sanfo et al. 2020). Dependence on natural grazing exposes the livestock sector to the spatiotemporal insufficiency of forage, which is a real obstacle to livestock productivity in tropical areas like Burkina Faso (Toe et al. 2022; Bilal et al. 2021; Zampaligré et al. 2013). The increased severity of rainfall seasonality in these regions has a direct impact on the livestock systems practised. Indeed, natural pastures vary with climate change and shrink with the increase in livestock numbers and rampant urbanization. Moreover, it is evident that crop residues collected and stored in a peasant manner after harvest are generally of poor quality. There is also the high cost of commercial feeds, which are used as supplements to compensate for the shortages of natural resources (Sanou et al. 2016).

Faced with this situation, the practice of forage cultivation and the practice of appropriate conservation methods would be a better alternative to overcome the shortage of good-quality forage throughout the year and to maintain the production capacity of livestock systems. Silage is a method of preserving wet forage characterized by the fermentation of forage in an anaerobic environment. The use of silage in animal feed during the dry seasons of the year is a means to maintain ruminant herds (Bilal et al. 2021). This practice can also help farms manage ruminant feeding to ensure a continuous supply of meat and dairy products for human consumption.

3.1 Livestock Systems in Burkina Faso

The livestock systems in Burkina Faso are as follows:

(i) The pastoral system is characterized by extensive exploitation of natural resources without the use of zootechnical inputs (RPCA 2010). It is based on mobility to ensure forage and water supply for the herd throughout the dry season.
(ii) The agropastoral system combines agriculture and livestock with the practice of transhumance over short distances. This system combines the use of natural pastures, forage crops, and agro-industrial by-products in animal feed.
(iii) Semi-intensive and intensive livestock systems are characterized by animal housing, a high level of intensification and technicality, and a constant supply of zootechnical inputs to animals (Sanon et al. 2014; Gnanda et al. 2016).

3.2 *The Practice of Silage in Burkina Faso*

In sub-Saharan African countries, particularly in Burkina Faso, silage is mostly produced under experimental conditions in research centres or on semi-intensive or intensive farms, with small farmers in rural areas rarely producing it due to the lack of extension services, inputs, and technical knowledge (Tamboura et al. 2005; Balehegn et al. 2022). Studies have shown that producers primarily use sorghum (*Sorghum bicolor*) and maize (*Zea mays*) for silage, while legumes are rarely used for silage in Burkina Faso. There are several types of silos for silage production in Burkina Faso, and they include open-air silos, vertical silos, pit silos, and bale or sack silos (Fig. 11.9) (Tamboura et al. 2005). However, it is evident that the use of silage as cattle feed in Burkina Faso is rare, with its practice largely found in dairy farming systems, which require significantly higher investment (Gnanda et al. 2016; Vall et al. 2021). In contrast, the conservation of forage through haymaking remains the most practised by Burkinabe farmers (Amole et al. 2022). Although the use of chemical or microbial additives in forage conservation is uncommon, urban dairy producers use additives such as molasses and salt in silage conservation.

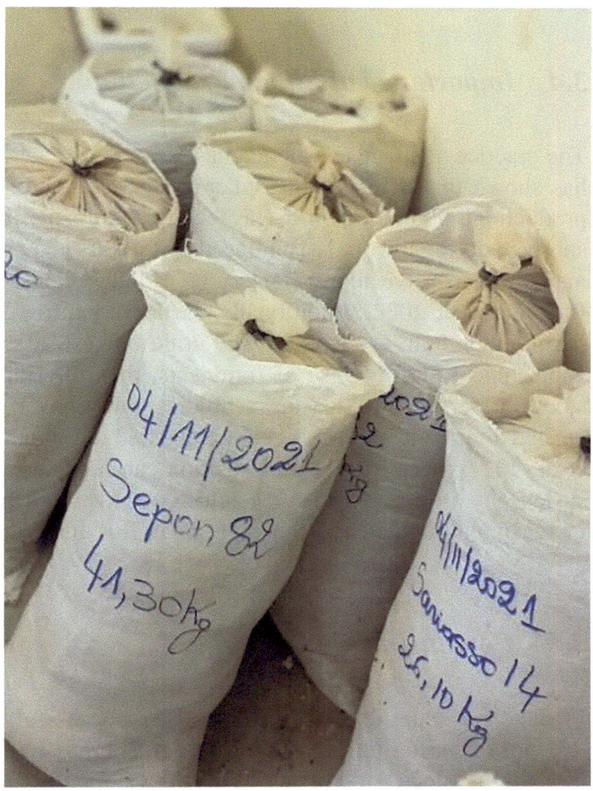

Fig. 11.9 Triple-bottom bags silage. (Photo by Jethro)

Few studies have been conducted on the quality of silage and the use of additives in silage production in Burkina Faso, with the majority of published research being relatively recent.

For instance, Cai et al. (2021) investigated the factors affecting sorghum silage, while Cai et al. (2020a, b) examined the chemical and microbial composition of sorghum and pearl millet stover silage in the region. Furthermore, few research studies have been conducted on silage production at different stages of forage development in Burkina Faso.

3.3 Constraints Related to the Adoption of Silage

The adoption of the practice of forage conservation through silage in Burkina Faso faces various constraints, as is the case in most sub-Saharan countries. These challenges include limited access to inputs (silage equipment, grinders, and silos) and resources (forage seeds), insufficient labour availability, lack of knowledge and skills in silage production, and the socio-cultural factors of producers (Balehegn et al. 2022).

3.4 Importance of Silage Production

The practice of silage production has several advantages in Burkina Faso. Research has shown that using cereal stalks (maize, sorghum, and pearl millet) for silage production optimizes their use as feed for ruminants. This helps address the forage shortage during the dry season faced by farmers and provides a cost-effective alternative to expensive supplementary feeds like cottonseed cake (Cai et al. 2020a, b; Balehegn et al. 2022). Additionally, the use of improved crop varieties, such as dual-purpose sorghum ensiled after the grain harvest, has been shown to enhance the zootechnical performance of ruminants, thus offering economic benefits to livestock farmers in Burkina Faso. Studies have shown that the quality of sorghum ensiled after grain harvest is better compared to sorghum stover stored in heaps (Cai et al. 2021). Finally, silage production can help sustain milk production in dairy systems by providing additional nutrition to livestock during the dry season, thereby increasing milk yield and consumption in Burkina Faso.

3.5 Conclusion

In Sahelian countries like Burkina Faso, successful livestock farming primarily depends on mastering the feeding system. Therefore, the practice of silage is beneficial for livestock enterprises as it allows for year-round feeding management.

Utilizing local crop by-products in silage production is an ideal strategy to alleviate the seasonal feed shortages in Burkina Faso.

(Delma B. Jethro)

4 Silage Production and Livestock Raising in Mozambique

Mozambique is a home for about 33 millions of people, among which 65.3% is living in rural areas being agriculture, forestry, fishery, and livestock the bases for their livelihoods. These subsectors combined contribute 25.9% of national GDP. The smallholder production sector (subsistence livestock production system) alone is holding 89% of national livestock production, particularly cattle (2.1 million), small ruminants (6.7 million), chicken (23.5 million), and pigs (1.4 million) (DNDP 2023). The remaining share is held by commercial and government sectors using natural pasture for livestock grazing both holding 10.9% and 0.1% population of cattle, respectively. The major of cattle producers are in Maputo, Gaza, Inhambane, and Tete holding 73.9% of cattle out of 2.3 million of cattle countrywide. However, livestock production in Mozambique declined through 2015/2016 due to the effect of climatic changes characterized by severe drought and flooding. In response, the Government of Mozambique designed and implemented the so-called Livestock Restocking Program. Assuming that weather conditions were kept favourable for the practice of agriculture, it would have maintained positive performance, supported by investment inputs, towards raising annual growth rates of 4% for cattle, 2% for swine, and 3% for goat industries, from 2022 to 2023 (DNDP 2023). Despite these growth rates, the country is far from a satisfactory response to meat consumer's demand. More than 15% of meat is imported from neighbouring countries. Dairy sector is insignificant (3261 heads), under pasture-based grazing system at low cost, thereby saving financial resources in supplementation, animal health management, and infrastructures.

Based on DNDP reports, poultry industry ranks first (48.3%), followed by, in descending order, goats (15%); ducks (11.8%); pigs (10%) and cattle (with 6.6%). In terms of livestock production, 1.9 million raise chickens; 636,800 goats; 411,990 pigs; 473,803 ducks; and 242,473 cattle. However, the production of red meat tends to grow in favour of cattle. The range is 15.4 million tons, with an annual growth rate of 25.8% from 2016 to 2017; small ruminants 2.4 million tons with growth rate of 12.5%; and pigs with 2.2 million tons of meat and annual growth rate of 23.7%. In 2020, the production of red meat reached 141.0 millions of tons, with all indicators for red meat on the rise. The projection for large ruminants in 2024 is 20.9 million tons.

Mozambique is rich in forage species, diverse and available crop by-products from staple food, industrial crops, fruits brewers, and molasses. Furthermore, the country grows hulls and corn stover from cereals, sugarcane tops and banana steam and leaf, oil crops, coconut residues, peanuts, and soybean, rich in vitamins, minerals, fibre, and proteins. However, most of the products are left-overs burned in the field; others are in huge volumes, with high content of moisture; they constitute a waste that can harm and bring about environmental and economic concerns (Cai et al. 2020a, b), if correct measures of management are not taken into account. The utilization of such resources for animal feed in small-scale silage and diet formulation by farmers still has to be promoted and extensively disseminated among the producers platform, given the fact that feed availability and quality are seasonal in various agro-ecological zones of Mozambique. In such a farming uncertainty, farmers do not predict the growth season for various reasons including the prevailing dry season and severity of weather variations.

The objective of this review is to summarize the production and utilization of silage made from traditional crops, agricultural by-products, and non-conventional biomass, and to explore itsapplication in diet formulation for improved performance in ruminant livestock.

4.1 Ruminants

Mozambique is a tropical country with high potential for ruminant production especially goat, cattle, and sheep. Indigenous breeds of all species are highly adapted to harsher environmental conditions such as high temperatures, drought, and feed scarcity. In response, the annual growth of red meat species 'is increasing at rates of 18.0 % and 11.00%, for bovines and goats, respectively'. The total number of domestic ruminants is about 9.0 million, including water buffalos. Smallholder farmers account to 80% of national livestock, with 95% of indigenous breeds. The majority of cattle are concentrated in the southern part of the country, given the periodical outbreak of tsetse fly in central and northern areas of the country, against the availability of rangeland vegetation, both grasses and browsable shrubs.

There are four well-identified indigenous breeds of cattle in Mozambique, namely, the Angone and the Bovine of Tete (representing 20% of national herd), Landim (70%), and Namaacha Nguni (3%). Angone (7%) is a Zebu type, while the Landim is a Sanga type with a traceable origin. The Bovine of Tete is categorized as a Sanga type, but its origin is not well studied yet. Reports by Madilindi et al. (2019) and Regé and Tawah (1999) indicated Bovine of Tete derives from crossbreeding between the Landim and the Angone breeds. Non-conclusive reports suggest that the Bovine of Tete derives from the Mashona breed (Sanga type), which inhabits at the border between Zimbabwe and Mozambique. Namaacha Nguni is an ecotype from KwaZulu-Natal province of South Africa, belonging to Sanga class (Madilindi et al. 2019).

In addition to indigenous breeds of cattle, in the commercial pasture-based sector, there are some exotic breeds too, both zebuines and Taurinas, pure or crossbred at certain degree. Brahman is spread across different provinces, particularly in Zambezia, Nampula, and southern of the Save River. The Bonsmara is found in Maputo, Gaza, and Inhambane, while Mashona, Friesian, Hereford, and Simmental in Manica, Sofala, and the south of the Save River as well. Jersey and Brown Swiss are in the south of the Save River. Angus can be found in the province of Zambezia. There is also an attractiveness and indiscriminate exchange of exotic breeds between smallholder and commercial farmers for reproduction purpose (Madilindi et al. 2019). Such practices should be discouraged to avoid gene erosion of indigenous breeds of cattle before complete studies on adaptability and their genetic potential are concluded.

According to Martinho (1947), Friesian, Jersey, and Brown Swiss are dairy breeds of cattle early introduced into the country with the goal of improving country's dairy production. These breeds were raised pure or crossbred for the purpose of finding the best crossover levels meeting the country's agro-ecological conditions. Pure Friesian cattle had serious adaptation difficulties at Animal Science Research Station. Animals with highest percentage of exotic blood performed poorly. The 3/4 Friesian had 42% birth rate, while 1/2 Friesian had 52%. Mortality rates followed the same response as birth rates, with higher mortality compared to those with the highest percentage of exotic blood (15% for Friesian and 33% for Jersey).

4.2 Silage Production from Conventional Crops and By-products

The livestock production sector is dominated by smallholder farmers. The farming system is hampered by harsher environmental conditions, weather uncertainty, and climatic variation (droughts, floods, cyclones, and climate change). To cope with these weather events, farmers have been adopting low-input and cost-effective climate-smart agriculture practices, such as small livestock rearing, crop residue management for cattle feed, intercropping, fodder bank establishment, and smart animal waste management. In this type of integrated farming, where crops are not a choice, livestock rising is a smart livelihood alternative option (Madilindi et al. 2019). Feed supply in southern provinces, where there is a majority of livestock populations, is dependent on crop by-products and its supply from the central and northern regions of the country, as the larger grain market suppliers, and from the neighbouring countries (Maciel et al. 2013). Along with these measures, there is a need for intensifying silage making to provide supplemental feeding especially for milk and meat industries, towards minimizing the import of such commodities.

In the recent years, studies on silage making have been undertaken by using Napier grass, maize, and industrial by-products. However, to avoid competition, green maize stover, sugarcane tops, and silvo-pastoral browsing legumes (*Leucaena leucocephala*, *Gliricidia sepium*, *Moringa oleifera*, and *Morus alba*) have been the potential option, because they stand green yearlong, with high content of dry matter and crude protein (Muir 1996). Meanwhile, most by-products are left in the field and burned; others are in huge volumes, with high content of moisture, and constitute a waste that can pose a serious threat to environmental, public health, as well as economic concerns. In order to improve the fermentation ability and quality of silage, it is recurrent to add LAB strains and cellulase into silage, molasses, urea, and common salt. The results confirm that LAB and cellulase are suitable as potential additives for silage preparation and can improve fermentative qualities of Napier grass and other silage making crops, in Mozambique.

To date, studies on non-conventional silage crop making and characterization have been carried on, in the country. LAB (*Lactiplantibacillus plantarum*) was isolated from silage of corn stover and sugarcane tops prepared with LAB and cellulase enzyme, in Mozambique. Information by Du et al. (2020) indicates that aerobic bacteria were dominant in the microbial populations of both crops prior to ensiling, while after ensiling, LAB became the dominant bacteria. The natural LAB community, including some common, such as *Lactiplantibacillus plantarum*, *Lactobacillus brevis*, and *Pediococcus acidilactici*, were present in fresh matter of both crops. *Lactiplantibacillus plantarum* and *Weissella cibaria* were the dominant LAB population in corn stove and sugarcane tops, respectively. Fresh corn stover had high WSC content and LAB abundance and low lactate buffering capacity (LBC), resulting in silages of good quality compared to sugarcane tops, where the control silage had poor quality, because of the high LBC and low WSC content and LAB count. However, when sugarcane top silages were inoculated with LAB and cellulase, the fermentation quality was improved. The study indicates that such fermented crops and their mixtures with other ingredients can be disseminated across small and commercial livestock enterprises, along with addition of LAB. Additionally, such diet components and formulations became a significant tool to improve the quality of by-products by ensiling them for a short period of time followed by a mixture with mineral premix and concentrate feeds. These kinds of diets can fulfill the livestock feed security during the dry season or whenever emergence feeds are needed, for some reason in the farm. On hand total mixed ratio can be an additional farm business when it is prepared and folded in a plastic bag, by selling it to other farms, during the harsh dry season. Similar study concluded that *Lactiplantibacillus plantarum* strain MO1 isolated from local silage crops can modulate the microbial community and metabolic pathways and achieve health fermentation (Cai et al. 2020a, b; Du et al. 2020, 2023). It is of great significance to alleviate feed shortage and promote sustainable production of livestock farming sector, using such low-cost technologies.

4.3 Impact of Silage on Ruminant Performance

Rangeland vegetation in Mozambique will be for long period of time the low-cost feed resource available, since it shares in the ruminant feeding balance for almost 90% of total feeding system. However, its availability and quality are hampered by uncertainty and distribution of rainfall. While other resources are available, famers should look for other feed options. The utilization of silvo-pastoral legumes, ensiled Napier grass and crop residues showed to be solely the promising alternative, for

Table 11.1 Effect of silage or total mixed ration on livestock production in Lao PDR

References	Animal	Feed	Results
Cai et al. (2020b)	–	Lactic acid bacteria and cellulase in Napier grass silage enhance	LAB and cellulase enhanced fibre degradation and increase WSC; convert WSC into lactic acid; reduce pH and preserve silage (improve fermentation quality of Napier grass silage)
Nguluve et al. (2021)	Goats	Sorghum silage combined with maize and wheat bran and 10% of feed mill	Goats fed sorghum silage have highest liveweight gain, carcass quality, and feed conversion ratio
Massango et al. (2007)	Pigs	Woods plants (*Leucaena* and *Moringa*) and QMP	Pigs fed TMR have highest liveweight gain, carcass quality, and feed conversion ratio (feed/gain)
Muir et al. (1992)	Pigs	Woods plants (*L. leucocephala*) and wheat bran	Feed conversion was improved by the addition of the forages by 59%
Du et al. (2020)	Dairy (Jersey)	Napier grass-based silage, mixed with other ingredients as a TMR	Increased ad libitum intake, digestibility, milk yield, and economic benefits

LAB lactic acid bacteria, *WSC* water-soluble carbohydrate, *TMR* total mixed ratio

Fig. 11.10 Corn stover silage making and fermentation quality analysis. (Instituto de investigação Agrária de Moçambique, Matola, Mozambique; photos by Nguluve and Cai)

both smallholder and commercial sectors. Chemical composition of fermented total mixed ration for 10–16 days improved the fermentation characteristics of the mixture, such as crude protein and digestible carbohydrates at 20% and 40%, respectively (Table 11.1).

Du et al. (2020) experimented the utilization of total mixed ration (25% of formula, 12.5% of wheat bran, 12.5% of corn bran, and 50% of fermented Napier grass for 10 days), for Jersey dairy cows, and concluded that feed intake and digestibility were increased by 25.0% and 15.0%, respectively, and milk production by 50.5%, an economic benefit of 40.0%, compared to traditional feed, indicating significant impact on dairy yield. Impact of total mixed ration based on short day's fermentation of fresh maize stover silage at 60% incorporation with maize bran and 30% sugarcane molasses showed that the addition of such ingredients on diets increased milk production by 30% compared to control diet (Fig. 11.10). The above data clearly indicate the reasonable possibility of using fermented silage crops and by-products to formulate dairy cattle diets, because after short fermentation period, they provide highly digestible carbohydrates and proteins.

Study on effects of replacing protein source by *Moringa*, *Leucaena*, and quality protein maize (QPM) on pig diet indicated that *Moringa* leaves, *Leucaena* leaves (at 30% of diet inclusion), and QPM increased the average daily weight gain by 12, 10, and 20%, respectively. Animals consuming diets with such unconventional and wood plants feed ingredients decreased the production cost by 12–37.5%/animal and produced animals with 40% less fat (Massango et al. 2007). Considering the high cost of pig production, based on commercial feeds/concentrate, it is clear that farmer formulated diet and total mixed ration can make a significant difference on the costs and quality of meat, in the pig industry. Muir et al. (1992) reported that feed mill for mixed breeds of pigs can be replaced by a total mixed ration (diet) based on fermented wheat bran + *Leucaena* leaves (WB + L), wheat bran+ forage turnip (*Brassica napus*-WB + B*)*, as well as wheat bran + *Leucaena* + *Brassica napus* (WB + L + B), with an improvement of feed conversion rate of 47%, 25%, and 59%, respectively. The study suggests that the feed mill cost on pig production can be lowered by the addition of forages on them or on wheat bran, with the similar outcomes. Additionally, studies on replacement of carbohydrate sources in TMR, based on sorghum silage, mixed at 30–40% for goats increased liveweight gain by 12% (almost 3 times) compared to goats feeding shopped hay mixed maize bran, the traditional practices adopted by local farmers (Nguluve et al. 2021). The study also reported an increase in hot carcass lent and ad libitum ingestive behaviour when goats were feeding diets of TMR based on sorghum silage.

4.4 Conclusion

The availability of different staple crops, a range of by-products from cash crops and browsing rangeland resources, reveals that there is an opportunity for adopting resilient technologies to cope, in the long run, with climate change uncertainty that negatively impacts livestock productivity in the tropics and Mozambique in particular. Continuous test of other potential and candidate crops and by-products for silage and total mixed ratio may diversify and overcome feed shortage, protein, minerals, and carbohydrates and consequently increase productivity and income.

The recently isolated *Lactiplantibacillus plantarum* MO1 and its possible addition to silage will significantly enhance the fermentative quality and characteristics of silage using local resource to provide highly soluble carbohydrates to livestock and thus reduce the dependence on feed mill or imported feedstuffs for self-sufficiency in the animal farming. Such diverse conventional (Napier grass, corn stover, sugarcane tops) and non-conventional feed resources (silvo-pastoral or multipurpose fodder plants) for animal production, either preserved, as silage or as ingredients for formulating of total mixed ratio, for ruminants and monogastric, can boost animal husbandry, in Mozambique, especially for small and medium-class farmers.

However, efforts have to be done on research investment and technology transfer from research to extension services and then to the producers, so that animal productivity be enhanced through the application of mitigating and suitable technologies to increase animal farming, using those livestock adapted to roughage and coarse feeds. Additionally, research should play an important innovation role to meet smart faming technologies and practices, adapted and resilient to climate change, so that farm productivity and income be generated to improve livelihood of local communities.

(Damião W Nguluve)

References

Akinbode RM, Isah OA, Oni AO, Adebayo KO, Aderinboye RY, Adelusi OO, et al. Nutrient digestibility and blood parameters of West African dwarf sheep fed sugarcane top silage. Niger J Anim Prod. 2018;45(2):304–15.

Akinmoladun O, Muchenje V, Fon FN, Mpendulo CT. Small ruminants: farmers' hope in a world threatened by water scarcity. Animals. 2019;9(7):456.

Akinola JO. Grassland research in Nigeria-a survey with recommendations. Niger J Anim Prod. 1974;1(2):162–8.

Amole T, Augustine A, Balehegn M, Adesogoan AT. Livestock feed resources in the West African Sahel. Agron J. 2022;114(1):26–45. https://doi.org/10.1002/agj2.20955. Epub 2021 Dec 29. PMID: 35910094; PMCID: PMC9303701.

Amole TA, Oduguwa BO, Jolaosho AO, Onifade OS, Arigbede OM, Olanite JA, et al. Effect of sole maize and maize-Lablab silage on the ruminal volatile fatty acids (VFAs) of grazing calves in the dry season. Niger J Anim Prod. 2013;40(1):133–43.

Amos AT, Oso AO, Durojaiye OJ, Agazue K, Obanla AO, Anthony AC, et al. Growth response, economics of production, carcass characteristics and blood profile of weaner rabbits fed ensiled cassava root-leaf blends (ECRLB) as a replacement for maize. Niger J Anim Prod. 2021;48(1):152–65.

Amuda AJ, Onaleye KJ, Babayemi OJ. Dry matter intake, nutrient digestibility and nitrogen balance of WAD sheep fed ensiled maize stover and concentrate supplement. Niger J Anim Prod. 2020;46(2):284–94.

Ayanlade A, Ojebisi SM. Climate change impacts on cattle production: analysis of cattle herders' climate variability/change adaptation strategies in Nigeria. Change Adapt Socio-Ecol Syst. 2019;5(1):12–23.

Balehegn M, Ayantunde A, Amole T, Njarui D, Nkosi BD, Müller FL, et al. Forage conservation in sub-Saharan Africa: review of experiences, challenges, and opportunities. Agron J. 2022;114(1):75–99.

Bilal AK, Adnan M, Rehman FU, Hasnain A, Usman M, Javed MS, et al. Role of silage in agriculture: a review. Green Rep. 2021;2(4):9–12.

Cai Y, Du Z, Yamasaki S, Delma BJ, Nignan M. Chemical composition, characteristics concerned with fermentative quality and microbial population of ensiled pearl millet and sorghum stover in semi-arid West Africa. Anim Sci J. 2020a;91(1):e13463.

Cai Y, Du Z, Yamasaki S, Nguluve D, Tinga B, Macome F, et al. Influence of microbial additive on microbial populations, ensiling characteristics and spoilage loss of delayed sealing silage of Napier grass. Asian Australas J Anim Sci. 2020b;33:1103–12.

Cai Y, Du Z, Delma BJ, Nignan M, Yamasaki S. Analysis of main factors affecting silage fermentation of sorghum prepared with whole crop and stover in semiarid West Africa. Afr J Range Forage Sci. 2021;38(2):169–78.

Dele PA. Evaluation of dry matter yield and nutritive quality of forage, hay and silage produced from three grasses fertilized with animal manures. Ph.D. thesis, Federal University of Agriculture, Abeokuta; 2012. pp. 263.

Dele PA, Jolaosho AO, Arigbede OM, Ojo VOA, Amole TA, Okukenu OA, et al. Chemical composition and in vitro gas production of silage from Guinea grass, cassava peel and cashew apple waste at different periods of ensilage. Pak J Biol Sci. 2013;16(23):1801–5.

Direcção Nacional de Desenvolvimento Pecuário (DNDP). Relatório Anual. Ministério de Agricultura e Desenvolvimento Rural. Maputo. 2023.

Du Z, Yamasaki S, Oya T, Nguluve D, Euridse D, Tinga T, et al. Ensiling characteristics of total mixed ration prepared with local feed resources in Mozambique and their effects on nutrition value and milk production in Jersey dairy cattle. Anim Sci J. 2020;91:13370.

Du Z, Yamasaki S, Oya T, Nguluve D, Euridse D, Tinga B, et al. Microbial network and fermentation modulation of Napier grass and sugarcane top silage in southern Africa. Microbiol Spectr. 2023;12:00520–4.

FMARD. Animal population data. Fed Minist Agric Rural Dev. 2017.

Gnanda BI, Bado D, Reusse S, Pare L. Caractéristiques et typologie des exploitations laitières bovines de la zone périurbaine de Ouagadougou, au Burkina Faso. Renc Rech Ruminants. 2016;23:165–8.

Lukuyu BA, Duncan AJ, Kariuki K, Baltenweck I. Feeds and feeding practices. East Africa Dairy Development Project Baseline Survey Brief No. 3. 2011.

Maciel S, Amimo J, Martins M, Okeyo AM, Scholtz MM, Neser FWC. Feedlot performance of the Nguni ecotypes in southern Mozambique. Livest Res Rural Dev. 2013;25(6).

Madilindi MA, Banga CB, Bhebhe E, Sanarana YP, Nxumalo KS, Taela MG, Mapholi NO. Genetic differentiation and population structure of four Mozambican indigenous cattle populations. Livestock Res Rural Dev. 2019. Vol 31, Retrieved May 7, 2025, from http://www.lrrd.org/lrrd31/4/matom31047.html.

Martinho JC. Anais dos serviços veterinários de Moçambique. Ind Anim. 1947;1:125–51.

Massango Z, Fato P, Chaúque P, Nguluve D, Dias G, Vilanculos J, et al. Efeito da utilização de milho sussuma na engorda de suínos. Relatório Anual. Matola: Instituto de Produção Animal; 2007.

Miller TB, Rains AB. The nutritive value and agronomic aspects of some fodders in northern Nigeria. II. Silages. J Br Grassland Soc. 1963;18:223–9.

Muir JP. Effect of cutting height and frequency on *Leucaena leucocephala* forage and wood production. J Range Forage Sci. 1996;15(1/2):7–10.

Muir JP, Massaete ES, Tsombe HN. Effect of *Leucaena leucocephala* and *Brassica napus* on growth of pigs fed wheat bran diets. Livest Res Rural Dev. 1992;4(2):19.

National Bureau of Statistics/Federal Ministry of Agriculture and Rural Development. Collaborative survey on national agriculture sample survey (NASS), 2010/2011. 2011.

Nguluve D, Macome F, Tinga T, Massaete F, Mussalafo, Mapande D. Silagem de resíduos frescos de mapira e feno no desempenho produtivo de caprinos landim. Animal- DCA-IIAM-Matola. 2021.

Nussio LG, Daniel JLP. Advances and opportunities in silage research. In: Yang F, Guo X, editors. XIX International Silage Conference (ISC); Beijing, China 2023. pp. 1–12: 8–12.

Nwobodo CE, Nwokolo B, Iwuchukwu JC, Ohagwu VA, Ozioko RI. Determinants of ruminant farmers' use of sustainable production practices for climate change adaptation and mitigation in Enugu State. Niger Front Vet Sci. 2022;9:735139.

Ologhobo AD, Balogun AM. The value of dried fish silage for pullets and the layer hen. Niger J Anim Prod. 1987;14:67–76.

Olorunnisomo OA. Nutritive value of conserved maize, amaranth or maize-amaranth mixture as dry season fodder for growing West African dwarf sheep. Livest Res Rural Dev. 2010;22(10):1–10.

Olorunnisomo OA. Milk production in *Sokoto Gudali* cows fed legume or elephant grass ensiled with cassava peel. Livestock Res Rural Dev. 2013;25.

Olorunnisomo OA. Intake and milk yield of Zebu cows fed Moringa forage ensiled with cassava peel. Afr J Livestock Ext. 2014;14:72–6.
Olorunnisomo OA, Adewumi AS. Grass silage as conserved forage for cattle production in the humid parts of Nigeria. In: 1st Biennial Conference: Society for Grassland Research and Development in Nigeria. Federal University of Agriculture, Abeokuta, Nigeria. 2015.
Olorunnisomo OA, Ayodele OJ. Effects of intercropping and fertilizer application on the yield and nutritive value of maize and amaranth forages in Nigeria. Grass Forage Sci. 2009;64(4):413–20.
Rains AB. Grassland research in Northern Nigeria. 1963:56–8.
Regé JEO, Tawah CL. The state of African cattle genetic resources. II. Geographical distribution, characteristics and uses of present-day breeds and strains. Anim Genet Resour Inf. 1999;26:1–25.
RPCA (Réseau de Prévention des Crises Alimentaires). L'élevage au Sahel et en Afrique de l'Ouest. 26ème réunion annuelle. Accra (Ghana). 2010;10:14–6.
Sanfo A, Nouhoun Z, Kulo EA. Analyse des préférences des agropasteurs pour la production et la conservation du fourrage à base de variétés améliorées de cultures à double objectifs dans deux zones agro-écologiques au Burkina Faso. J Anim Plant Sci. 2020;46(3):8318–35.
Sanon HO, Drabo A, Sangaré M, Kiendrebeogo T, Gomgnibou A. Caractérisation des pratiques d'embouche bovine dans l'ouest du Burkina Faso. IJBCS. 2014;8(2):536–50.
Sanou KF, Ouédraogo S, Nacro S, Ouédraogo M, Kaboré-Zoungrana C. Durabilité de l'offre et valeur nutritive des fourrages commercialisés en zone urbaine de Bobo-Dioulasso, Burkina Faso. Cah Agric. 2016;25:25.
Shabu T, Fate S, Ukula MK. Impact of urbanization on agricultural land in Makurdi Local Government Area of Benue State, Nigeria. NASS J Agric Sci. 2021;3(1):21–8.
Shishi CM, Agbo TOO. Silage making to improve the nutrition of milking animals for small scale farmers in Benue State. Int J Agric Home Econ Educ. 2018;5(1):211–20.
Silke V, Diskin MG, Kenny DA, Boland MP, Dillon P, Mee JE, et al. Extent, pattern, and factors associated with late embryonic loss in dairy cows. Anim Reprod Sci. 2002;71:1–12.
Tamboura HH, Bougouma V, Traoré A, Kaboré A, Ouédraogo S de B, Sawadogo L. Technique de conservation des fourrages par voie humide: L'ensilage. CIRDES. 2005;4.
Toe A, Sanon HO, Obulbiga F, Bougouma VMC. Amélioration du disponible fourrager par différents modes de cultures de sorgho et de niébé à double objectif. J Anim Plant Sci. 2022;54(1):9808–21.
Tufail MS, Mbuku S, Mutimura M, Guo X, Piltz J. Utilisation of conserved forage to improve livestock production on smallholder farms in Asia and Africa. 2020.
U.S. Department of State. U.S. Relations with Nigeria. Bilateral relations fact sheet. Bureau of African Affairs; 2023.
Vall E, Sib O, Vidal A, Delma BJ. Dairy farming systems driven by the market and low-cost intensification in West Africa: the case of Burkina Faso. Trop Anim Health Prod. 2021;53(2):288.
Wilkinson JM, Bolsen KK, Lin CJ. History of silage. Madison: Silage Science and Technology. American Society of Agronomy; 2003. p. 1–30.
Zampaligré N, Dossa LH, Schlecht E. Contribution of browse to ruminant nutrition across three agro-ecological zones of Burkina Faso. J Arid Environ. 2013;95:55–64.

Chapter 12
Silage Production and Livestock Raising in Southeast Asia

Smerjai Bureenok, Yuli Yanti, Viengsakoun Napasirth, Pattaya Napasirth, and Yantyati Widyastuti

Abstract Livestock production is on the rise in Southeast Asian countries like Thailand, Laos, and Indonesia, driven by both local consumption and export markets. Roughage sources such as forage crops, agricultural by-products, and waste are the main feed resources in this region. Similar to other tropical regions, fodder production is at its peak during the rainy season and declines during the dry season. Consequently, feed availability for livestock production is limited during the dry months. To address this challenge, methods such as silage and the fermented total mixed ration are used to preserve and enhance feed quantity and quality throughout the year. However, the characteristics of tropical crops and by-products, such as low water-soluble carbohydrates and high moisture content, make it difficult to produce high-quality silage. This chapter provides an overview of livestock raising, feed resources, silage production techniques, and their utilization in Southeast Asia.

Keywords Feed resource · Livestock raising · Silage · Southeast Asia

---

S. Bureenok (✉)
Rajamangala University of Technology Isan, Nakhon Ratchasima, Thailand
e-mail: smerjai.bu@rmuti.ac.th

Y. Yanti
Sebelas Maret University, Surakarta, Indonesia
e-mail: yuliyanti_fp@staff.uns.ac.id

V. Napasirth
National University of Laos, Vientiane, Laos

P. Napasirth
Udon Thani Rajabhat University, Udon Thani, Thailand
e-mail: pattaya.na@udru.ac.th

Y. Widyastuti
National Research and Innovation Agency, Cibinong, Indonesia
e-mail: yant001@brin.go.id

Y. Cai, K. Ataku (eds.), *Cultural History and Modern Production Technology of Silage*, https://doi.org/10.1007/978-981-96-5787-2_12

1 Silage Production and Livestock Raising in Thailand

Smerjai Bureenok

The population of livestock such as cattle, dairy cows, buffalo, and goats as reported by the Department of Livestock Development (DLD) in Thailand from 2015 to 2024 is shown in Fig. 12.1. According to the DLD, the government encourages the establishment of initiatives aimed at increasing cattle production for export. The cattle population in the country has been growing since 2015. In particular, the cattle population was approximately 9.39 million heads in 2022, representing a significant increase of 27.5% from 2021, and has remained stable until 2024. To increase meat production in the country, the number of buffalos has also been increased in recent years. The buffalo population has reached approximately 1.82 million heads, a significant increase of 16% from 2023. The DLD has also launched many projects to promote dairy farming. Since 1993, the school milk program has been promoted by the Thai Government to improve the health and welfare of the young generation. Recently, the dairy cow stock in Thailand was forecasted to amount to approximately 567,000 heads, though the number has fluctuated since 2015. The decrease in animal numbers and production is caused by the increasing prices of feed, labour, and operational costs. These increases have affected the profit margins for both the smallholder farmers and the small dairy cooperatives. Thus, the

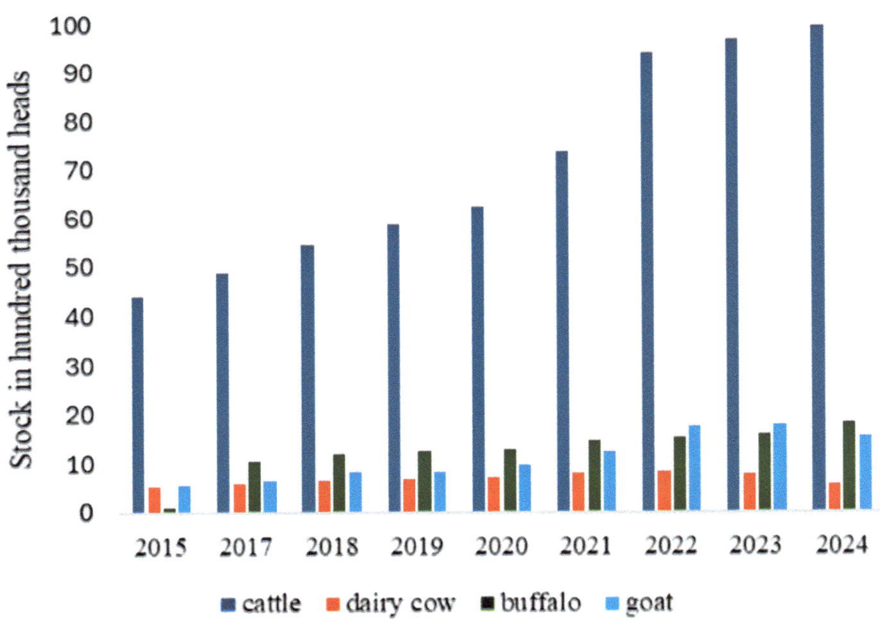

Fig. 12.1 Livestock stock in Thailand from 2015 to 2024

number of smallholder farmers has also decreased. Overall, the number of livestock increased from 2015 to 2022 and then remained stable until 2024.

In Thailand, a major limitation in raising beef and dairy cattle is insufficient feed, especially during the dry season (November to April). During this season, the quantity and quality of milk produced are low. Roughage is the main feed resource, and rice straw has traditionally been fed to ruminants in Thailand. However, dry rice straw is not a suitable feed for ruminants owing to its limited nutritive value (low crude protein and digestibility, high level of lignification and silicification). Developing high-quality roughage to support the feeding system throughout the year is important.

Generally, in Thailand, tropical forages such as para (*Brachiaria mutica*), ruzi (*Brachiaria ruziziensis*), guinea (*Panicum maximum*), signal (*Brachiaria decumbens*), atratum (*Paspalum atratum*), pangola (*Digitaria eriantha*), and Napier (*Pennisetum purpureum*) are used in animal feeding. Some legume forage crops such as *Stylosanthes guianensis*, *Stylosanthes hamata*, and *Leucaena leucocephala* are also available. By-products such as rice straw, corn stover, cassava leaves (*Manihot esculenta*), and sugarcane tops (*Saccharum officinarum*) are also available and widely used in this area after harvesting. However, the nutritional value of these by-products needs to be improved before feeding. These forage crops and by-products are available in the rainy season or after the main product harvesting. The large quantity of leftovers must be used within a short period to avoid quality loss. It will therefore be feasible to preserve their nutritional value and use them as a source of roughage to offer feed all year round.

Silage is the preservation of fresh forage by anaerobic fermentation, usually facilitated by *epiphytic* bacteria that convert soluble carbohydrates or sugars mainly to lactic acid and minor amounts of other volatile fatty acids. As a result, the pH is lowered, which inhibits or inactivates microbial growth and preserves the ensiled material. Generally, tropical forages contain low levels of water-soluble carbohydrates (WSC), low dry matter (DM), and high buffering capacity, especially in legume forages. Their coarse porosity, stemmy structures, and low number of epiphytic lactic acid bacteria (LAB) compared to other microorganisms make it more difficult to produce good silage. Acids, salts, fermentable carbohydrates, LAB cultures, and enzymes can all be used to control silage fermentation.

The application of a local additive like molasses as a substrate to promote bacterial growth is widely used due to its low cost and safety in handling. Some absorbents, such as rice bran, corn meal, or cassava pulp, may simultaneously increase both the soluble sugar concentrations and DM content of the ensiled product. Molasses (approximately, 73% DM and 25% WSC on a DM basis) (Bureenok et al. 2011) is widely added up to 10% w/w to provide fast fermentable carbohydrates for the ensilage of tropical forages. Bureenok et al. (2011) concluded that the addition of 5% molasses to Napier grass resulted in a 1.4 times higher intake of silage compared with the control. Furthermore, the reduced neutral detergent fibre (NDF) concentration of this silage is consistent with the increase in DM and organic matter digestibility for silages ensiled with molasses. However, Cherdthong et al. (2021) ensiled rice straw with 5% molasses, *Lactobacillus casei* TH14 plus 5% molasses, or *L. casei* TH14 with molasses and cellulase and found that the addition of *L. casei* TH14 and cellulase could inhibit the loss of crude protein during fermentation and decrease NDF content while increasing the digestibility of DM and organic matter in Thai-indigenous beef cattle.

Recently, many researchers have attempted to use LAB to accelerate the ensiling process. Moreover, the silage produced from starter culture technology is of higher quality, can be preserved for a long time, and may improve animal health and performance compared to naturally fermented silage (Liu et al. 2012). The successful fermentation of feed may also be affected by the high ambient temperatures (Li et al. 2019). Consequently, specific LAB might not be able to grow in hot climates (Weinberg et al. 2001). Thus, the inoculation of forage with heat-tolerable LAB before ensiling may be instrumental in achieving a well-preserved silage in tropical areas like Thailand. Many studies have aimed to screen and select LAB species suitable for use as silage additives in Thailand. The application of fermented juice of epiphytic lactic acid bacteria (FJLB), a local microbial starter, was recommended to improve the silage quality of tropical crops by increasing the LAB in the ensiling process (Bureenok et al. 2011, 2019). This method improved the digestibility of crude protein and increased cellulolytic bacteria population in cows. However, the FJLB contained many kinds of microbials that may induce the deterioration of silage after air exposure. Pitiwittayakul et al. (2021) selected the thermotolerant LAB strains isolated from the FJLB made from tropical crops and reported that four selected strains, *Lactiplantibacillus paraplantarum* St1 and St3, *Limosilactobacillus fermentum* N4, and *Pediococcus pentosaceus* R4, improved the fermentation quality and nutritive values of Stylo silage. They suggested that these strains could be used as a silage inoculant under tropical conditions. Khota et al. (2016) applied a local selected lactic acid bacterium, *L. casei* strain TH14, which could promote lactic acid production during silage fermentation. They suggested that when natural LAB are present on grass in sufficient quantity, commercial inoculants are unlikely to improve silage fermentation. Khota et al. (2017) reported that the addition of TH14 could improve silage quality but not nutrition value and in vitro DM digestibility, while it decreased methane production of sorghum silage. Therdtatha et al. (2024) reported that *L. plantarum* KUB-SP1-3 and *P. acidilactici* KUB-M6 isolated from Thai silage samples could improve the fermentation quality of signal grass silage compared with a commercial starter in a large-scale silage production (140-ton plastic bag silo).

Tropical crops and by-products contain high fibre content, which can interrupt the feed intake of animals. Plant cell wall-degrading enzymes have been suggested as silage additives to increase the amount of soluble carbohydrates available to LAB, expedite the pH decrease, and increase the digestibility of organic matter in the feed (Henderson 1993). Khota et al. (2016, 2017) reported that the cellulase enzyme can inhibit protein degradation and promote fibre degradation and concluded that it has great potential as an additive for tropical silage fermentation. However, the combination of cellulase and LAB did not improve fibre digestibility. This finding agrees with Bureenok et al. (2019), who reported that the combination of fibrolytic enzyme and FJLB did not improve nutrient digestion, suggesting that the fibre-degrading enzymes may have predigested the readily digestible fibre and left a slower and less degradable fraction. However, Kaewpila et al. (2020) reported that adding cellulase (0.2 g/kg fresh matter) to Napier grasses at the late maturity stage improved silage quality and in vitro DM digestibility of Napier grass silage. Khota et al. (2023) also found that *Acremonium* cellulase has the potential to be used as a silage additive for purple guinea grass by improving silage quality and increasing nutrient digestibility compared with the control. Further research is needed to investigate the long-term effects of fibrolytic enzymes on growth performance and ruminant production.

Silage storage containers are used to provide an environment for fresh silage to ferment and to ensure good preservation for longer period with minimum loss of nutrition value during ensiling and feeding. In Thailand, dairy production mainly relies on silage for feeding regimen among ruminants to improve milk production and quality, especially during the dry season. Nakamanee (2000) demonstrated different techniques of producing silage using various kinds of silos to dairy farmers and found that the farmers preferred bunker silos, plastic bucket technique, and black polythene bags, respectively. Most dairy farmers are considered to fall within the medium-sized group, with between 23 and 35 lactating cows. Some of these farmers lack the necessary machinery or equipment to produce silage. For this reason, smaller silos that can produce silage that is used up quickly might be better than larger silos. Plastic bags and reused feed bags are widely used for making corn silage for sale. The bags can hold an average of 25–30 kg of silage, which is suitable for transporting to the farmers, but prone to tear and damage from insects. Bunker silos, on the other hand, are the main types used for storing silage on larger farms. Recently, wrapping machines that are used to wrap bales on-site have been introduced in silage production (Fig. 12.2). Each bale has an average total weight of 50–110 kg, depending on the DM content of the material (Sasitorn Jorjong,

Fig. 12.2 Storage and production of silage in Thailand. (Photos by Smerjai Bureenok and Anan Petlum)

personal contact). The investment costs for specialized tools and equipment tend to be very high. However, DLD provides rental services for tools and equipment to farmers, making access to modern tools and equipment more feasible and investment costs more reasonable.

2 Silage Production and Livestock Raising in Indonesia

Yuli Yanti

Indonesia is one of the Southeast Asian countries and has the fourth-largest population in the world (Muzayanah et al. 2022). The livestock sector in the country contributes to meeting the animal protein needs of the people. Available evidence shows that despite the advancement in other sectors such as poultry, livestock development remains sub-optimal in meeting domestic needs, leading to the importation of red meat such as beef and buffalo meat. Indonesia has exotic livestock that can be developed including Bali and Madura cattle and Kacang goats. This type of livestock reportedly has low productivity. Consequently, since the 1980s, the government created a cross-breeding program with European livestock (*Bos taurus*) to increase productivity while retaining the ability to thrive in tropical environments (Agus and Widi 2018). Types of crossbred cattle include SimPO (Simental-Ongole-local cattle) and LimPO (Limousin-Ongole-local cattle).

In livestock farming in Indonesia, large ruminants are predominantly small scale with ownership of about two to four heads (Agus and Widi 2018; Fajar et al. 2023). Small-scale farms use simple technology, often do not require agricultural machinery, and even provide animal feed without considering the nutrient requirements (Fajar et al. 2023). This situation is because farmers raise livestock as a form of saving money and a side job, but not the main source of income. An overview of the smallholder farming system is presented in Fig. 12.3.

The forages often given are grasses, legumes, and agricultural waste. Examples of grasses include elephant (*Pennisetum purpureum*), odot (*Pennisetum purpureum cv. Mott*), pak cong, and field grass. Meanwhile, leguminous plants include turi

Fig. 12.3 Bali cattle (left), Ongole crossbreed cattle (centre), and Kacang goats (right) kept by small-scale farmers. (Photos by Yanti)

(*Sesbania grandiflora*), gamal (*Gliricidia sepium*), lamtoro (*Leucaena leucocephala*), and tarum (*Indigofera*) (Islamiyati et al. 2023). Furthermore, waste produced from the high agricultural yields can be used as feed. In this context, straw, a farming waste from rice, corn/tebon, soybean, and ground peanut, is widely used as animal feed due to its low cost and wide availability (Nugraha et al. 2023). However, rice straw and other agricultural waste are characterized by low nutritional content and require processing to increase its usefulness (Yanti and Yayota 2017). The feed that has recently become a favourite in Central Java and East Java is water spinach straw (*Ipomoea reptans*) used as a source of fibre with wide availability and long-lasting (Dahlan et al. 2013). Dried water spinach can be given directly to livestock (Purnamasari et al. 2021), or fermented, and prepared in pellet form (Adiwinarti et al. 2023).

Livestock concentrate feed is a combination of various feed ingredients. For this purpose, rice bran is widely available. Food industry waste such as tofu and soybean skin dregs can also be used as animal feed; they contain a high crude protein content of about 20–27% (Mulyasari et al. 2022). Furthermore, soybean meal waste, palm oil flour, corn waste, and onggok or waste from tapioca flour are used as animal feed. Feed ingredients such as pollard brand, soybean meal, and corn gluten meal are imported, posing a challenge to the livestock sector in the future.

Indonesia is a tropical country with two seasons, namely, rainy and dry. In the rainy season, forage grows very well, leading to high abundance. However, in the dry season, the available forage becomes limited, underscoring the need for a storage method. To address this problem, silage offers a suitable method of preserving feed ingredients and can help provide animal feed in the dry season (Yanti et al. 2021).

Generally, the use of silage has long been recognized by breeders in Indonesia. Agricultural institutions or departments and lecturers from campuses also often provide training to breeders on how to make silage correctly (Yanti et al. 2022) as shown in Fig. 12.4. Young breeders usually obtain information about silage production by searching on the Internet or from fellow breeders. However, the silage results are sometimes not optimal because there is no direct assistance from experts.

(a) (b)

Fig. 12.4 Silage-making training by the Department of Agriculture (left) and also by campus academics (right) for small-scale farmers in rural areas. (Photos by Yanti)

Silage-making in Indonesia has been implemented by both large-scale breeders and smallholder farms. Generally, silage is a feed preservation technology that can be accepted by farmers because it is easy to apply and provides benefits such as saving time and providing feed during times of drought (Lestari et al. 2020). It is given to livestock kept under intensive system of rearing. Livestock raised extensively and semi-intensively are rarely fed silage. On the island of Java, cattle and sheep farming is mostly kept entirely in pens. However, on other islands such as Sumatra, Kalimantan, Sulawesi, and Nusa Tenggara, livestock farming is mostly semi-intensive or extensive. In other words, the livestock are kept in pastures by consuming field grass growing wild or planted by previous breeders (Agus and Widi 2018).

The abundant rice straw on the island of Java is also used by farmers for animal feed by way of fermentation with the addition of urea or ammonia (Bata et al. 2020; Zain et al. 2019). To increase the usefulness, rice straw can be mixed with other feed to form a total mixed ratio (TMR) and then fermented (Yanti et al. 2022). The addition of molasses can increase lactic acid bacteria and ultimately reduce the pH value of silage (Baurhoo and Mustafa 2014).

According to previous reports, silage can be used after storage for at least 1 month (Yanti and Yayota 2019). Common silage forages made are elephant grass and corn stover (Bira et al. 2020; Fahmi et al. 2021). Grass can also be mixed with other forages and concentrates which are better known as complete feed to make silage. Elephant grass and indigofera in a ratio of 60:40 increased daily body weight gain in Bali cattle (Nurhayu et al. 2021), while a mixture of grass with sorghum straw and concentrate can be given to local Gorontalo goats (Sayuti et al. 2024). An increase in daily body weight gain also occurred in Bali cattle given a mixture of corn stover and peanut straw silage (Bahri et al. 2018). The use of vegetable waste can also increase feed conversion and efficiency as well as the body weight gain of local sheep (Muktiani et al. 2013). Additionally, coffee waste in Karo Regency was stored in the form of silage for goat feed (Patriani et al. 2019). Several qualities of silage from local feed ingredients are presented in Table 12.1.

Silage storage varies depending on the ability of the farmer to prepare the silo in the form of bunkers, plastic drums, and plastic bags (Mosebi et al. 2024). In general, each material has advantages and disadvantages. Bunker silos presented in Fig. 12.5 are easier to make but usually leak easily; hence, oxygen enters, and aerobic bacteria contaminate the edges of the silo, especially when not completely closed (Pereyra et al. 2011). Silo made from plastic bags presented in Fig. 12.6 has the advantages of being easy to make, movable, and cheaper. However, the disadvantage is that it is easily torn by grass twigs or eaten by birds or mice. This plastic bag silo is also only used once, giving rise to a new problem, which is plastic waste. Silos made from plastic drums (Fig. 12.7) are more durable, do not leak easily, are not eaten by mice, are easy to move, and can produce excellent silage (Campbell 2014), but the disadvantage is the high cost. In Indonesia, a plastic drum silo with a capacity of 160 L is priced at IDR 200,000 which is equivalent to 13 USD. This price is certainly expensive and unaffordable for small-scale farmers.

Silage feed has been widely fed to livestock such as cattle, goats, and sheep on Indonesian farms. This is to overcome the problem of poor quality of forage and

Table 12.1 pH values and nutrient content of silages prepared from some feed resources in Indonesia

Type of silage	pH	DM (%)	Crude protein (% DM)	Crude fibre (% DM)
Rice straw ensilaged with LAB[a]	4.6	94.2	8.8	25.7
Rice straw ensilaged with rumen fluid 15%[b]	4.6	69.21	7.69	25.54
Elephant grass and indigofera (50:50)[c]	4.5	88.7	16.74	–
Corn stover and ground nut stover (50:50)[d]	4.2	54.76	9.05	30.15
Limbah sayuran+onggok+rice bran[e]	4	56.42	16.27	–
Corn stover+corn flour (90:10)[f]	2.47	27.43	10.35	–

[a]Sarwono et al. (2022)
[b]Sio et al. (2022)
[c]Nurhayu et al. (2021)
[d]Bahri et al. (2018)
[e]Bakrie et al. (2014)
[f]Kurniawan et al. (2022)
DM dry matter

(a) (b)

Fig. 12.5 Bunker silos fermented rice straw covered with tarpaulin (left) and silos without covers (right) on smallholder farms for cattle feed. (Photos by Yanti)

scarcity during the dry season and also to save time cutting fresh grass daily. However, the problems often faced by farmers are failure to obtain good-quality silage, usually due to silo leaks, feed ingredients with extremely high or low water content, lack of WSC content, and the use of poor silage additives. In general, silage additives from *Lactobacillus* inoculants of rice plants provide TMR quality almost equivalent to commercial additives (Wahyudi et al. 2022). Based on the results, livestock in Indonesia is expected to optimize and produce good-quality silage and increase livestock productivity.

Fig. 12.6 Silage stored in plastic bag silos, both black and transparent. (Photos by Yanti)

Fig. 12.7 Process of chopping corn stover to make silage (left); bean and corn stover stored in plastic drum silos (center); silage in a silo drum that is ready to be given to livestock (right). (Photos by Yanti)

3 Silage Production and Livestock Raising in Loas

Viengsakoun Napasirth and Pattaya Napasirth

Livestock production, particularly cattle and buffalo, is a significant component of the agricultural sector in Lao PDR, contributing to both livelihoods and the national GDP. The livestock sector has witnessed rapid growth, particularly in cattle production. In 2023, there were 2407 livestock farms, marking a 6.9% increase from 2022. These farms housed 19.5 million animals, distributed across 713 cattle farms, 39 buffalo farms, 737 pig farms, 110 goat farms, 354 layer chicken farms, and 153 broiler chicken farms (Lao Statistics Bureau 2022). This surge in cattle farming, with a 319.41% increase in the number of farms since 2018, is likely driven by Laos' goal to export 50,000 cattle to China annually, coupled with increased domestic beef consumption due to population growth and rising per capita consumption (Napasirth and Napasirth 2018; Ministry of Agriculture and Forestry 202).

In 2022, the total buffalo population in Lao PDR reached 1.25 million head, with the central region housing the majority (48.80%). Savannakhet province held the largest buffalo population (332,000 heads, 26.65%), followed by Champasack province (162,000 heads, 13.00%). Smallholder farmers predominantly raised these animals, with 99.87% of the buffalo population under their care. Commercial buffalo farming remained limited, with 34 farms averaging 57 buffaloes each. Similarly, the cattle population in Laos totaled 2.43 million head in 2022, with the central region again housing the majority (50.95%). Savannakhet province led in cattle numbers (419,000 heads, 17.26%), followed by Xayabury province (285,000 heads, 11.74%).

Smallholder farmers played a significant role, raising 95.27% of the cattle population. Commercial cattle farming comprised 658 farms, averaging 174.41 per farm (Ministry of Agriculture and Forestry 202).

Most of the livestock production is undertaken by smallholder farmers, who typically raise indigenous breeds of cattle and buffalo. These ruminants are primarily raised for meat, although they have historically been used for the storage of wealth, for sale as meat, as a source of manure fertilizer, and for draught purposes as well (Xayalath et al. 2021; Nampanya et al. 2016). The average mature weight of Lao-native beef cattle is relatively low, with bulls reaching up to 350 kg and cows up to 250 kg (Stur et al. 2002). In addition to the indigenous breed, exotic breeds, such as the Red Sindhi and Red Brahman, have been introduced to improve meat production through crossbreeding. These crossbred cattle exhibit improved growth performance compared to the indigenous breed, with higher birth weights and faster growth rates (Keonouchanh et al. 2014). The primary buffalo breed in Laos is the swamp buffalo, characterized by its grey or pink colour. While there have been attempts to introduce Murrah buffalo from India for crossbreeding, these efforts have not been widely successful due to reproductive challenges (Napasirth and Napasirth 2018).

Husbandry practices for both cattle and buffalo are predominantly traditional and extensive. Smallholder farmers often allow their animals to graze freely on communal lands, fallow lands, and forests. However, this practice is increasingly constrained by factors such as the Land and Forest Allocation policy and the expansion of cash crop cultivation (Olmo 2020). Supplementary feeding is limited, and the nutritional quality of available feed resources, such as native grasses and rice straw, varies significantly depending on the season (Napasirth and Napasirth 2018). This reliance on natural grazing and low-quality feedstuffs contributes to the low productivity observed in the livestock sector.

This review aims to summarize and synthesize research findings related to silage production and utilization in Lao PDR, focusing on the use of various crops and by-products as silage materials, the impact of silage on animal performance, and the broader implications for sustainable livestock production in the country.

3.1 Silage Production from Diverse Crops and By-products

Research in Lao PDR has explored the potential of using a wide array of crops and agricultural by-products for silage production. These include traditional forage crops like taro (Kaensombath and Lindberg 2012a) and stylo (Kaensombath and Lindberg 2012b). Additionally, Cai et al. (2013) isolated LAB from various silage crops in Vientiane, Lao PDR. The results showed that seven strains of LAB were identified, primarily *Lactiplantibacillus plantarum* and *Weissella confusa*. These strains grew under a range of pH and temperature conditions in Lao PDR. Moreover, Napasirth et al. (2015) investigated the potential use of cassava residues (leaves, peel, and pulp) as silage for livestock feed in Lao PDR. The study found that these

Fig. 12.8 Preparing fermented total mixed ratio for Lao native cattle and buffaloes

residues had sufficient LAB populations for fermentation. The resulting silages, with or without LAB inoculants, were well-preserved, with low pH levels (below 4.0) after 60 days of ensiling. These findings suggest that cassava residues often discarded as waste can be effectively ensiled and potentially used as a valuable feed resource for livestock in Laos. The utilization of cassava pulp, a by-product of starch production, is particularly noteworthy as it addresses the challenge of waste disposal while providing a valuable feed resource (Napasirth 2021). Studies have investigated the fermentation characteristics and nutritional quality of silages made from these diverse materials, providing insights into their suitability for different livestock species (Fig. 12.8).

3.2 *Impact of Silage on Animal Performance*

The incorporation of silage into livestock diets has been shown to have positive effects on animal performance in Lao PDR, particularly when utilizing unconventional feed resources for ruminants. For example, Vongsamphanh et al. (2016) explored brewer's grain and cassava foliage as protein sources for Lao native beef cattle fed a basal diet of cassava pulp and urea. The study found that cattle fed a combination of brewer's grain and cassava foliage had similar live weight gain and feed conversion rates to those fed brewer's grain and Napier grass (Table 12.2). However, cattle fed solely cassava foliage had reduced feed intake and poor growth rates. This suggests that a combination of brewer's grain and cassava foliage can be an effective protein source for fattening cattle on a cassava pulp-urea diet. Napasirth et al. (2021) investigated the use of spontaneously fermented cassava pulp (SFCP) treated with *Lactiplantibacillus plantarum* in fermented total mixed rations (FTMR) for steers. The results showed improved average daily gain (ADG) and feed conversion ratio (FCR) after 60 days of feeding Lao-native beef cattle, indicating the potential of SFCP as a valuable energy source in cattle diets. Moreover, research has shown that spontaneously fermented cassava pulp can be an energy source for ruminants, with similar gas production kinetics to cassava chips (Napasirth 2021). Additionally, cassava peel is a potential feed ingredient for ruminants due to its high

Table 12.2 Effect of silage on livestock production in Lao PDR

References	Animal	Feed	Results
Napasirth et al. (2015)	–	Treating cassava leaves, peel, or pulp silage with *L. plantarum* and *L. rhamnosus*	*L. plantarum* and *L. rhamnosus* enhance cassava leaves, peel, or pulp silage fermentation quality
Keohavong and Onphachanh (2015)	Muscovy ducks	Taro silage combined with rice bran	Ducks fed taro silage had the lowest final body weight, the poorest feed conversion ratio (feed/gain), and the highest mortality rate
Kaensombath and Lindberg (2012b)	Landrace x Yorkshire crossbred and Lao native pig (Moo Lath)	Replacing crude protein from soybean 50% by using ensiled stylo	Growth performance and carcass traits did not negatively affect
Sivilai et al. (2018)	Native moo lath pigs	Replacing ensiled taro foliage with ensiled banana pseudo-stem and soybean meal	The biological value of protein in the ensiled taro foliage is superior to other treatment
Kandee et al. (2024)	Lao native pig gilts	Replacing soybean meal with cassava leaf silage (CLS) at 10%, 20%, and 30% in the diets	Increasing the proportion of CLS in the diet led to a linear decrease in ADG and FCR. While CLS did not significantly affect litter size or the number of live piglets born
Vongsamphanh et al. (2016)	Lao native beef cattle	Using brewer's grain and cassava foliage as protein sources	Combination of brewer's grain and cassava foliage can be an effective protein source for fattening cattle on a cassava pulp-urea diet
Napasirth et al. (2021)	Lao native beef cattle	Spontaneously fermented cassava pulp (SFCP) treated with *L. plantarum* in fermented total mixed rations	The results showed improved ADG and FCR after 60 days of feeding of Lao-native beef cattle

ADG average daily gain, *FCR* feed conversion ratio; *Lactiplantibacillus plantarum*; *Lactobacillus rhamnosus*

fibre content, serving as a roughage source. However, cassava peel has a lower nutritional value than cassava leaves and pulp, limiting its use in animal feed. Napasirth et al. (2015) found that treating cassava leaves, peel, or pulp silage with *Lactiplantibacillus plantarum* and *Lactobacillus rhamnosus* enhances fermentation quality, improving its suitability for animal consumption. On the other hand, research conducted on pigs revealed that replacing soybean protein with protein from ensiled taro leaves or stylo forage did not negatively affect growth performance or carcass traits (Kaensombath and Lindberg 2012a, b). Similarly, studies on goats demonstrated that cassava hay and Stylo 'CIAT 184' hay could serve as protein sources in diets based on gamba grass, leading to improved growth rates

(Xaypha 2005). Kandee et al. (2024) investigated the effects of replacing soybean meal with cassava leaf silage (CLS) at varying levels (10%, 20%, and 30%) in the diets of Lao native pig gilts during their first pregnancy. The study found that increasing the proportion of CLS in the diet led to a linear decrease in average daily gain (ADG) and an increase in feed conversion ratio (FCR). While CLS did not significantly affect litter size or the number of live piglets born, it did result in a linear decrease in litter weight and mean piglet weight at birth when included at 30% of the diet. The study concluded that CLS can be used as a protein source in the diets of Lao native pig gilts up to 20% without negatively impacting reproductive performance. Keohavong and Onphachanh (2015) investigated taro silage, combined with rice bran, as a feed for Muscovy ducks. Two feeding models were tested: mixed (taro silage and rice bran combined) and separated (taro silage and rice bran fed separately). Both models negatively impacted the ducks' growth performance compared to ducks fed duckweed or water hyacinth. Ducks fed taro silage had the lowest final body weight, the poorest feed conversion ratio (feed/gain), and the highest mortality rate. This suggests that taro silage, in the forms and combinations tested, may not be a suitable feed ingredient for Muscovy ducks. Kaensombath and Lindberg (2012b) found that replacing up to 50% of soybean crude protein with crude protein from ensiled stylo in the diets of growing Landrace x Yorkshire and Lao native pig (Moo Lath) in Lao PDR did not negatively affect growth performance or carcass traits (Fig. 12.9).

3.3 Conclusion

Research conducted in Lao PDR underscores the significant potential of silage production and utilization to address feed scarcity challenges and promote sustainable livestock production. The diverse range of crops and byproducts suitable for silage, coupled with the positive effects on animal performance, particularly in ruminants, highlights the adaptability and effectiveness of this approach. These findings highlight silage's potential to enhance the nutritional value of livestock diets and promote efficient growth, even when utilizing unconventional feed resources.

The research on silage production and utilization in Lao PDR has broader implications for sustainable livestock production in the country. By utilizing locally available crops and by-products, silage production reduces the reliance on imported feedstuffs, contributing to greater self-sufficiency in the livestock sector. Additionally, the positive impact of silage on animal performance further strengthens its role in enhancing productivity and profitability for smallholder farmers. Continued research and development in this area can further refine silage production techniques, optimize the utilization of local resources, and contribute to the long-term growth and resilience of the livestock sector in Lao PDR.

Fig. 12.9 Silage preparing and utilization in plastic drum and circular concrete tank at the Faculty of Agriculture, National University of Laos. (Photos by Napasirth)

4 Silage and TMR Prepared with Tropical Agricultural By-product

Yantyati Widyastuti

Livestock products, such as meat and milk, are essential food sources for humans as they provide animal protein. Protein is an essential nutrient with many body functions not limited to the supply of amino acids for the body's protein synthesis. Animal proteins are important parts of a healthy diet that complement plant foods and are especially important for health maintenance in vulnerable persons (Elmadfa and Meyer 2017). This is why the availability or supply of livestock products must be adequate to meet the demand, which is still increasing (Henchion et al. 2017).

Feed represents the primary cost driver, accounting for the highest production costs in livestock farming. Feeding has a strong influence on production. Quality feed in sufficient quantities for feeding ruminants must be available all year round to guarantee their good productivity. Recently, providing conventional feed like grass for ruminants has not been easy, because arable land to feed farmed animals is very inefficient. This inefficiency varies between areas and farming systems, depending on social and ecological conditions. In general, diets that include farmed animal products also contribute more to land degradation than diets that exclude them (Deckers 2016). Moreover, Makkar (2018) investigated various unconventional feed resources to address challenges in making the livestock sector more sustainable. Apart from providing ideal conditions for offering fresh forage to ruminants, there is no doubt that forage preserved through fermentation technology supports the provision of quality feed, such as silage and TMR. For this purpose, a wide range of agricultural by-products from various crops can be utilized (Wilkinson and Rinne 2017).

4.1 Silage and Ensiled Total Mixed Ration

Silage making, also known as ensilage, involves preserving the quality of forage in a fresh condition through natural anaerobic fermentation. The characteristics of raw materials for ensiling differ from those for solid-state fermentation processes. Typically, the raw material is fresh forage with high water content, and the fermentation process produces abundant active microorganisms (Dai et al. 2020). Additives are commonly used to effectively preserve forage. The primary additive, LAB, along with those naturally occurring in the forage, produces lactic acid and acts as preservatives. The forage and additives undergo fermentation over a specific period, progressing through several phases and undergoing various changes. An anaerobic environment, devoid of oxygen, is required to produce lactic acid through fermentation. Forage can be categorized as easy or difficult to ensile. The conditions that matter include non-structural carbohydrates, which are the main sugars that the LAB ferment to make lactic acid and other products. Buffering capacity refers to compounds that resist pH decline, and the moisture concentration of the forage affects the speed and extent of fermentation (Muck et al. 2020). TMR is a diet that consists of a mixture of feed ingredients such as forage, concentrate, and others, combined in a balanced manner. TMR is intended to be administered as a fresh, complete diet to ensure consistent intake and prevent fluctuations in rumen pH conditions. For long-term storage, it follows the preservation method used for silage and is often referred to as fermented TMR or TMR silage (Schmidt et al. 2017). Here, TMR refers to TMR silage. The use of TMR greatly improves efficiency in feed management.

4.2 Raw Materials for Silage and TMR

Raw materials for silage and TMR can include high-nutrient fresh forages or agricultural by-products categorized as waste. Agricultural by-products are the crop residues left after harvesting from farming. They usually include various straws and grain shells that are often discarded. Using agricultural by-products for feed has been a long-standing practice that continues to grow. Corn and rice are among the primary global food crops that produce abundant agricultural by-products. Corn is cultivated worldwide, spanning several parts of the world, while rice is predominantly grown in Asian countries (FAO 2023). Their by-products are corn stover and rice straw (Fig. 12.10), which are common forage sources for ruminants worldwide.

Fig. 12.10 Freshly harvested corn stover and rice straw from the field. (Photos by Widyastuti)

Table 12.3 Chemical composition of corn stover and rice straw

Item	Corn stover		Rice straw	
	(1)	(2)	(3)	(2)
DM (%)	21.91	37.10	31.50	38.77
CP (%)	9.01	6.02	3.89	5.92
EE (%)	nd	0.38	2.42	2.65
NDF (%)	64.54	73.76	63.01	62.66
ADF (%)	35.92	45.71	37.55	40.18
Ash (%)	9.62	9.70	18.21	15.25
GE (MJ/kg)	nd	17.00	nd	13.85

Data were cited from (1) Gao et al. (2019), (2) Liu et al. (2024), and (3) Du et al. (2024)
DM dry matter, *CP* crude protein, *EE* ether extract, *NDF* neutral detergent fiber, *ADF* acid detergent fiber, *GE* gross energy, *nd* not detected

Corn stover has potential as part of a ruminant's diet. The availability of corn stover can be calculated by determining the ratio of corn stover to corn grain, estimated to be 1.5. Separation of the whole corn stover revealed that the leaf blade has the highest protein content and the lowest NDF content (Li et al. 2014). Although parts of corn stover vary in nutritional value, it is not practical to fractionate it for feed preparation. Therefore, it is common to use whole corn stover for silage (Gao et al. 2019; Zhang et al. 2022). Corn stover and rice straw are characterized by high fibre content. Their chemical composition varies depending on various factors, such as plant types, varieties, planting location, soil type, and fertilizer (Table 12.3). For rice straw, the estimated quantity of rice straw production for every hectare of rice farm is the same as the weight of rice grain that can be harvested (Aquino et al. 2020). Rice straw is suitable for silage when it is fresh, retaining its nutritional value (Du et al. 2024; Sarwono et al. 2022).

Other potential agricultural by-products for use as feed include sugarcane bagasse (de Almeida et al. 2018; So et al. 2020; Iritani et al. 1996) and cassava peel (Napasirth et al. 2015; Kusmartono et al. 2022). Sugarcane bagasse is the by-product of extracting juice from sugarcane for the sugar industry, while cassava peels are remnants of the process of making tapioca flour. These two materials are secondary products not directly harvested from farmlands. It is more appropriate to utilize agro-industry by-products in animal feeds. These by-products are well suited for sustainable technology and animal farming management (Yang et al. 2021). Retail food waste, including discarded fruits, vegetables, bread, and bakery products from grocery stores, can also be utilized to produce high-quality silage. However, the collected samples must be processed immediately for effective silage production (Rodriguez et al. 2024).

TMR silage can be prepared by mixing fresh raw materials and fermented feed or other silage materials. Kim et al. (2012) prepared a fermented mixture of rice bran and wheat bran (1:1 ratio) supplemented with 0.1% *Lactobacillus acidophilus* (KCTC3140) and 0.1% *Saccharomyces cerevisiae* (KACC30068). This mixture, referred to as fermented feed, was then mixed with other materials, including rice straw, to prepare TMR. Yanti et al. (2019) utilized rice straw, brewer's grain, tofu waste, steam-flaked corn, a mineral mixture, and corn stover silage in their study. The corn stover was initially ensiled for about 2 months and then combined with other raw materials to prepare TMR. The TMR was treated with either fermented juice from epiphytic LAB or a commercial additive before fermentation and was allowed to ferment for a minimum of 2 months. Song et al. (2023) also utilized corn silage for fresh TMR. The TMR silage was prepared with the addition of a combination of LAB and fibrolytic enzymes for comparison in dairy cows. An advantage in TMR preparation is achieved by combining raw materials with different characteristics before ensiling, which enhances the preservation process. Mixing dry with wet raw materials may reduce the risk of effluent production and prevent undesirable fermentation. TMR silage typically has a markedly higher aerobic stability than fresh TMR. Ensiling TMR is generally more stable when exposed to air compared to fresh TMR. During the ensiling of TMR, proteolysis occurs, resulting in an increased breakdown of proteins in the rumen when the silage is fed to animals. Ingredients with high levels of soluble sugars and homofermentative LAB can improve the fermentation process. On the other hand, using ingredients that promote heterofermentation can enhance the stability of TMR silage when exposed to air (Bueno et al. 2020).

4.3 Additives for Silage and TMR

Silage additives are widely utilized globally, and significant advancements have been made in their use (Wilkinson and Rinne 2017). Muck et al. (2018) stated that there are six primary categories of additives, with a focus on microbial additives. These are (1) homofermentative LAB, which rapidly decreases pH and increases lactic acid relative to the other fermentation products; (2) obligate heterofermentative LAB, with *Lactobacillus buchneri* as the dominant species; (3) combination inoculants containing obligate heterofermentative LAB plus homofermentative LAB, aiming to provide the aerobic stability benefit of *Lb. buchneri* with the silage fermentation efficiency and animal productivity benefits of homofermentative LAB; (4) novel potential non-LAB species such as *Streptococcus bovis* within a homofermentative LAB and some yeast on inhibiting moulds and other detrimental silage microorganisms; (5) chemicals, such as formic acid which causes direct acidification, suppressing *Clostridia* and other undesired bacteria and improving protein preservation during ensiling, and sorbic, benzoic, propionic, and acetic acids (HAC), which improves silage aerobic stability at feed out through direct inhibition of yeast and moulds; and (6) enzymes such as cellulase and hemicellulase are added to forage primarily to break down plant cell walls during ensiling to improve silage fermentation by providing sugars for the LAB and enhance the nutritional value of the silage by increasing the digestibility of cell walls as well as ferulic acid esterase (FAE) manufactured by *Lb. buchneri* which enhances the efficiency of silage digestion. Andrada et al. (2023) studied the fermentation and nutritional parameters of corn stover silage using FAE from strains of *Lactobacillus johnsonii, L. plantarum, L. fermentum,* and *L. brevis*. The results in inoculated silages showed no correlation with FAE activity, leading to varying effects on dry matter (DM) loss, increased organic matter (OM), and digestibility of corn stover. Optimal utilization of low-quality straw as agricultural by-products can be enhanced by applying microbial inoculants (*L. plantarum*, *Lb. buchneri*, *Pediococcus pentosaceus*, *Aspergillus niger*, *Bacillus subtilis*, and *Candida utilis*) and enzyme-based silage cocktails (xylanase, β-mannanase, and glucanase). The results indicated a significant improvement in the fermentation and nutritional quality of corn stover and rice straw silages (Liu et al. 2024). Furthermore, LAB plays a crucial role in modulating microbial communities in various silages and during ensilage. LAB has diverse functions, including high antioxidant potential, bacteriocin production, and pesticide degradation (Guo et al. 2023).

4.4 Feeding Value of Silages and TMR for Ruminants

Song et al. (2023) investigated the impact of feeding FTMR containing corn stover on rumen fermentation parameters and microbiota in dairy cows. Twelve dairy cows were randomly assigned to two treatment groups ($n = 6$ per group) and were fed either fresh TMR or FTMR for a period of 12 weeks. The results showed that the FTMR group had significantly higher bacterial and fungal diversity compared to the fresh TMR group. The dominant bacterial genera in the bacteria, fungi, protozoa, and archaea showed significant differences between TMR and FTMR. The results imply that FTMR can modulate the rumen microbiome and improve ruminal fermentation. Sakai et al. (2015) studied the impact of TMR silage, which included brewer's grain and rice straw, as a replacement for conventional concentrate, on feed intake and milk production in middle-to-late lactation Murrah buffaloes. The TMR silage had higher NDF and ADF contents and digestibility compared to the concentrate ($P < 0.05$), with similar CP and total digestible nutrient (TDN) contents as the concentrate. The buffaloes were divided into three groups: a control group fed concentrate at 0.6% of body weight (BW) and T1 and T2 treatment groups fed TMR silage at 0.6% and 1.2% of BW on a DM basis, respectively, with rice straw ad libitum. The daily intakes of DM, CP, TDN, and BW change were found to be higher in T2 compared to the control and T1 groups ($P < 0.05$). The milk composition did not differ among the treatments, but the milk yield was higher in T2 ($P < 0.05$). There were no significant differences in milk yield per dry matter intake (DMI) and milk yield per TDN intake among the treatments. The increase in BW and milk yield in buffaloes might have been due to the high TDN intake from supplementary TMR silage. Zhang et al. (2022) evaluated the feeding of corn stover after fresh (CF) or ripe corn (CR) on digestibility and rumen fermentation in growing beef cattle.

The CR diet was formulated by replacing 50% of the forage in CF silage with CR silage on a DM basis. Feeding the CR diet resulted in increased feed intake, but a reduction in fibre digestibility, coupled with a decrease in the abundance of ruminal fibre degraders, as well as lower growth performance in beef cattle compared to the CF diet. It was concluded that extending the harvest time of corn stover silage by 20 days increases the fibre and DM content of the forage, thereby promoting feed intake while decreasing fibre degradation, without affecting rumen fermentation or growth performance.

Majeed et al. (2022) evaluated replacing wheat straw with corn stover on growth performance, behaviour characteristics, blood metabolites, and nutrient digestibility in Beetal bucks. The dietary treatments included conventional feed (25% wheat straw and 75% concentrate), corn stover 50 (50% of wheat straw (12.5%) was replaced with corn stover), and corn stover 100 (wheat straw was completely replaced with corn stover). Replacement of wheat straw with corn stover increased DMI and average daily gain significantly ($P \leq 0.05$) by 10% and 26%, respectively. Rumen pH and faecal score, however, remained unaffected but resulted in increased feeding and rumination time, improved lying time and length, and decreased bouts. Blood glucose, urea, bilirubin, and calcium levels remained unaffected; however, the blood phosphorous level was lower. Bucks fed corn stover-based ration showed an increase in organic matter, CP, NDF, and ADF digestibility. It was concluded that the replacement of wheat straw with corn stover as fibre residue in the diet of bucks resulted in an increased DMI and higher average daily gain, improved behavioural characteristics, and higher nutrient digestibility.

The in vivo studies show improvement in ruminal fermentation and intake of TMR silage-containing agricultural by-products. Additionally, in vitro technique is often used to simplify the in vivo approach. Cherdthong et al. (2020) used an in vitro gas production technique to assess ensiled rice straw, followed by a $2 \times 2 \times 2$ factorial arrangement in a completely randomized design with three replications. Rice straw was treated with *L. casei* TH14 (T) at 0 and 105 colony-forming units (cfu)/kg fresh matter (FM), cellulase (C) at 0 and 10,000 units/kg FM, and molasses (M) at 0 and 50 g/kg FM. The digestibility of rice straw increased significantly after being treated with T × C ($P < 0.01$), T × M ($P < 0.05$), and C × M ($P < 0.05$) after 48 hours of incubation. Ensiled rice straw of T × C × M could improve gas kinetics and fibre digestion and contribute to positive ruminal characteristics. Wang et al. (2017) evaluated the in vitro digestibility of corn stover silage mixed with four levels of FM (0, 10, 20, and 30%) of common vetch (*Vicia sativa* L.) and alfalfa (*Medicago sativa* L.), respectively. The results showed that corn stover silage containing 10% common vetch performed best, as indicated by a significantly ($P < 0.05$) higher in vitro DM and NDF digestibility. These findings show the benefit of incorporating legumes into the ensiling process as an effective strategy to improve the nutritional quality of corn stover as an agricultural by-product.

4.5 Future Directions

Silage and TMR are practical and cost-effective options for feeding ruminants, particularly in regions that generate significant agricultural by-products. This would be a smart way to deal with leftover harvest residue that could otherwise harm the environment. To maximize successful silage and TMR production, it is more effective to disseminate silage and TMR technology to the farmers in groups such as a cooperative, rather than individually. It is entirely possible to establish a TMR centre similar to the existing ones in some developed countries. In addition, groups can also be provided with all necessary materials, such as additives and equipment. Microbial additives require extensive study to find ones suitable for silage and TMR prepared with agricultural by-products. It would be even more beneficial if the local government provided financial support to produce silage and TMR to meet the needs of livestock farmers and facilitate sustainable livestock development by providing modern facilities for the production of silage and TMR. Therefore, there is an urgent need for investment in silage and TMR production by the local government.

References

Adiwinarti R, Rianto E, Purbowati E, Restitrisnani V, Purnomoadi A. Comparative feed management system in sheep fed different physical forms of ration containing Ipomoea aquatica on the performance, rumen characteristics, and chewing activity. J Adv Vet Anim Res. 2023;10(4):677–84.

Agus A, Widi TSM. Current situation and future prospects for beef cattle production in Indonesia – a review. Asian-Austral J Anim Sci. 2018;31:976–83.

Andrada E, Marquez A, Chagra Dib EP, Gauffin-Cano P, Medina RB. Corn stover silage inoculated with ferulic acid esterase producing *L. johnsonii*, *L. plantarum*, *L. fermentum*, and *L. brevis* strains: fermentative and nutritional parameters. Fermentation. 2023;9:331.

Aquino D, Barrio AD, Trach NX, Hai NT, Khang DN, Toan NT, et al. Rice straw-based fodder for ruminants. In: Gummert M, Hung N, Chivenge P, Douthwaite B, editors. Sustainable rice straw management. Cham: Springer; 2020. p. 111–30.

Bahri S, Sirajuddin SN, Hasan S, Natsir A. Quality of silage made from a combination of corn straw and peanut straw and its effects on Bali cattle performance. Pak J Biotechnol. 2018;15:671–5.

Bakrie B, Sastro Y, Bahar S, Sente U and Andayani D. Effectiveness of cassava pomace or cassava flour as additive in the processing of vegetable waste silage. Proceeding of the 16th AAAP animal science congress Vol. II. 2014. pp. 1658–1661.

Bata M, Rahayu S, Rimbawanto EA. Nutrient digestibility, intake rate, and performance of Indonesian native cattle breeds fed rice straw ammoniation and concentrate. IOP Conf Ser Earth Environ Sci. 2020;746:012006.

Baurhoo B, Mustafa A. Short communication: effects of molasses supplementation on performance of lactating cows fed high-alfalfa silage diets. J Dairy Sci. 2014;97(2):1072–6.
Bira GF, Tshuk PK, Seran T. The effect of using different types of forage for making complete silage to nutrient content produced. J Trop Anim Sci Technol. 2020;2:43–51.
Bueno AVI, Lazzari G, Jobim CC, Daniel JLP. Ensiling total mixed ration for ruminants: a review. Agronomy. 2020;10:879.
Bureenok S, Suksombat W, Kawamoto Y. Effects of the fermented juice of epiphytic lactic acid bacteria (FJLB) and molasses on digestibility and rumen fermentation characteristics of ruzigrass (*Brachiaria ruziziensis*) silages. Livest Sci. 2011;138:266–71.
Bureenok S, Langsoumechai S, Pitiwittayakul N, Yuangklang C, Vasupen K, Saenmahayak B, et al. Effects of fibrolytic enzymes and lactic acid bacteria on fermentation quality and in vitro digestibility of Napier grass silage. Ital J Anim Sci. 2019;18:1438–44.
Cai Y, Pang H, Napasirth V. Characterization and identification of lactic acid bacteria isolated from silage prepared in Lao PDR. In Michalk D.L., Millar G.D., Badgery W.B., and Broadfoot K.M. (Eds), Advances in Forage Conservation to Improve Quality. The Proceeding of the XXII International Grassland Congress. 2013. pp. 770-771. https://uknowledge.uky.edu/igc/22/1-11/
Campbell B. Small-scale silage production: a resource for smallholder farmers. An Echo Release. 2014.
Cherdthong A, Suntara C, Khota W. *Lactobacillus casei* TH14 and additives could modulate the quality, gas kinetics, and the in vitro digestibility of ensilaged rice straw. J Anim Physiol Anim Nutr. 2020;104:1–14.
Cherdthong A, Suntara C, Khota W, Wanapat M. Feed utilization and rumen fermentation characteristics of Thai-indigenous beef cattle fed ensiled rice straw with *Lactobacillus casei* TH14, molasses, and cellulase enzymes. Livest Sci. 2021;245:104405.
Dahlan M, Wardoyo and Prasetyo H, Suplay produksi bahan kering jerami kangkung sebagai bahan pakan ternak ruminansia di Kabupaten Lamongan (Studi Musim Tanam MK II Tahun 2012). Jurnal Ternak. 2013;4(2):11–21.
Dai Z, Cui L, Li J, Wang B, Guo L, Wu Z, et al. Fermentation techniques in feed production. In: Animal agriculture, sustainability, challenges, and innovations. London: Academic Press; 2020. p. 407–29.
de Almeida GAP, de Andrade Ferreira M, de Lima Silva J, Chagas JCC, Véras ASC, de Barros LJA, et al. Sugarcane bagasse as exclusive roughage for dairy cows in smallholder livestock systems. Anim Biosci. 2018;31(3):379–85.
Deckers J. Animal (de)liberation: should the consumption of animal products be banned? London: Ubiquity Press; 2016. p. 13–50.
Du Z, Nakagawa A, Fang J, Ridwan R, Astuti WD, Sarwono KA, et al. Cleaner anaerobic fermentation and greenhouse gas reduction of crop straw. Microbiol Spectr. 2024;12:7.
Elmadfa I, Meyer AL. Animal proteins as important contributors to a healthy human diet. Annu Rev Anim Biosci. 2017;5(1):111–31.
Fahmi M, Utomo R, Suhartanto B, Astuti A, Umami N. Chemical quality and digestibility value in silage of Pennisetum purpuphoides and Pennisetum purpureum gamma with different levels of molasses supplementation. Key Eng Mater. 2021;884:204–11.
Fajar MM, Riyanto J, Dewanti R, Cahyadi M, Wati AK, Pawestri W, et al. The characteristic of buffalo production (Bubalus bubalis) at the smallholder farmers in Sukoharjo regency, Central Java, Indonesia. IOP Conf Ser Earth Environ Sci. 2023;1:1246.
FAO. World Food and Agriculture—Statistical Yearbook. Rome: FAO; 2023.
Gao J, Wang P, Zhou C, Li P, Tang H, Zhang J, et al. Chemical composition and in vitro digestibility of corn stover during field exposure and the fermentation characteristics of silage prepared with microbial additives. Anim Biosci. 2019;32(12):1854–63.
Guo X, Xu D, Li F, Bai J, Su R. Current approaches on the roles of lactic acid bacteria in crop silage. Microbial Biotechnol. 2023;16:67–87.
Henchion M, Hayes M, Mullen AM, Fenelon M, Tiwari B. Future protein supply and demand: strategies and factors influencing a sustainable equilibrium. Food Secur. 2017;6:53.

Henderson N. Silage additives. Anim Feed Sci Technol. 1993;45:35–56.

Iritani S, Mitsuhashi M, Chaen H, Miyake T. Fermented bagasse feed, and its preparation and uses. 1996; US005547692A.

Islamiyati R, Hasan S, Rusdy M, Nompo S, Nohong B, Lahay N, et al. Teknologi pakan komplit dan pemanfaatan leguminosa sebagai konsentrat hijau di Kecamatan Bontonompo Kabupaten Gowa. Jdistira. 2023;3:32–8.

Kaensombath L, Lindberg BE. Effect of defoliation interval on biomass yield and chemical composition of CIAT Stylo 184 (*Stylosanthes guianensis* (Aubl.) Sw. var Guianensis). Grassl Sci. 2012a;58:221–6.

Kaensombath L, Lindberg BE. Effect of harvesting interval on biomass yield and chemical composition of taro (*Colocasia esculenta* (L.) Schott) for feeding pigs in Laos. Field Crop Res. 2012b;128:71–5.

Kaewpila C, Khota W, Gunun P, Kesorn P, Cherdthong A. Strategic addition of different additives to improve silage fermentation, aerobic stability, and in vitro digestibility of Napier grasses at late maturity stage. Agriculture. 2020;10:262.

Kandee T, Chanthakhoun V, Polyorach S. Effects of cassava leaves silage as a protein source in dietary on growth and reproductive performance of Lao native pig gilts. IJAST. 2024;20(3):1075–82.

Keohavong B, Onphachanh X. Growth Performance of Muscovy Ducks fed Duckweed, Taro Silage, and Water Hyacinth. Souphanouvong Journal. 2015; 1: 26-30.

Keonouchanh S, Phengsavanh P, Somchanh K. Complied for study and research on Lao local livestock breeds. Lao Livestock Development and Research Center, National Agriculture and Forestry Research Institute (Report NAFRI Rep. 2014) 2014. pp. 30–43.

Khota W, Pholsen S, Higgs D, Cai Y. Natural lactic acid bacteria population of tropical grasses and their fermentation factor analysis of silage prepared with cellulase and inoculant. J Dairy Sci. 2016;99:9768–81.

Khota W, Pholsen S, Higgs D, Cai Y. Fermentation quality and in vitro methane production of sorghum silage prepared with cellulase and lactic acid bacteria. Asian Australas J Anim Sci. 2017;30:1568–74.

Khota W, Pholsen S, Subepang S, Thip-uten S, Kaewpila C. Effects of purple Guinea grass silage prepared with lactic acid bacteria and fibrolytic enzyme on feed intake, digestibility, and nitrogen balance in post-weaning Thai native beef cattle. J Res Agric Innov. 2023;1:1–11.

Kim SH, Alam MJ, Gu MJ, Park KW, Jeon CO, Ha JK, et al. Effect of total mixed ration with fermented feed on ruminal in vitro fermentation, growth performance and blood characteristics of Hanwoo steers. Anim Biosci. 2012;25(2):213–23.

Kurniawan W, Lestari NA, Isnaeni PD, Sandiah N. Different composition of Indigofera zolingeriana and corn stover mix on silage quality and fermentation characteristic. Proceedings of the international conference on improving tropical animal production for food security (ITAPS 2021). Adv Biol Sci Res. 2022;20:335–9.

Kusmartono, Retnaningrum S, Mashudi, Harper KJ, Poppi DP. Improving live weight gain of crossbred *Limousin bulls* with cassava peel silage. Animal. 2022;16:100524.

Lao Statistics Bureau. Report on International Trade Statistics of Goods of Laos (in Lao language). 2022.

Lestari VS, Rahardja DP, Saleh IM. Perception of cattle farmers toward silage as animal feed. The 5th international seminar on agribusiness. IOP Conf Ser Earth Environ Sci. 2020;518:012070.

Li H, Xu L, Liu W, Fang M, Wang N. Assessment of the nutritive value of whole corn stover and its morphological fractions. Anim Biosci. 2014;27(2):194–200.

Li D, Ni K, Zhang Y, Lin Y, Yang F. Fermentation characteristics, chemical composition, and microbial community of tropical forage silage under different temperatures. Asian Australas J Anim Sci. 2019;32:665–74.

Liu Q, Chen M, Zhang J, Shi S, Cai Y. Characteristics of isolated lactic acid bacteria and their effectiveness to improve stylo (*Stylosanthes guianensis* Sw.) silage quality at various temperatures. Anim Sci J. 2012;83:128–35.

Liu E, Wang S, Wang S, Khan NA, Zhou X, Tang S, et al. Bacterial inoculants and enzymes-based silage cocktails boost the ensiling quality of biomasses from reed, corn, and rice straw. Chem Biol Technol Agric. 2024;11:29.

Majeed F, Ahmed S, Rashid MA, Shahid MQ, Mohsin I. Effects of replacement of wheat straw with corn stover based TMR on growth performance, behavioral characteristics, selected blood metabolites, and nutrient digestibility in Beetal bucks. Trop Anim Health Prod. 2022;54:115.

Makkar HPS. Feed demand landscape and implications of food-not-feed strategy for food security and climate change. Animal. 2018;12(8):1744–54.

Ministry of Agriculture and Forestry. Agricultural statistics yearbook 2022. Department of Statistics. Lao People's Democratic Republic. 2023. URL: https://dlf.mis-maf.gov.la/pdf/Agricultural%20statistic%20yearbook%202022.pdf.

Mosebi PE, Madakadze IC, Ntakatsane MP, Ratsele R. Fermentation characteristics of maize–forage legume mixtures ensiled in small-scale silos. Afr J Range Forage Sci. 2024;41:133–41.

Muck RE, Nadeau EMG, McAllister TA, Contreras-Govea FE, Santos MC, Kung L Jr. Recent advances and future uses of silage additives. J Dairy Sci. 2018;101:3980–4000.

Muck RE, Kung L, Collins M. Silage production. In: Forages, the science of grassland agriculture. Chichester: Wiley; 2020. p. 767–87.

Muktiani A, Achmadi J, Tampoebolon BIM, Setyorini R. The use of vegetable waste silage supplemented with mineral and alginate as feeding for sheep. J Ilmu dan Teknologi Peternakan. 2013;2:114–51.

Mulyasari, Subaryono, Yosmaniar. Utilization of tofu waste from Indonesian small scale industry as fish feed through a fermentation process. IOP Conf Ser Earth Environ Sci. 2022;978:012042.

Muzayanah IFU, Lean HH, Hartono D, Indraswari KD, Partama R. Population density and energy consumption: a study in Indonesian provinces. Heliyon. 2022;8:1–9.

Nakamanee G. Successful smallholder silage production: a case study from Northeast Thailand. FAO plant production and protection papers. 2000. pp. 15–16.

Nampanya S, Khounsy S, Young JR, Napasirth V, Bush RD, Windsor PA. Smallholder large ruminant health and production in Lao PDR: challenges and opportunities for improving domestic and regional beef supply. Anim Prod Sci. 2016;57:1001.

Napasirth P. Using spontaneously fermented cassava pulp (SFCP) neglected in an open-pit as the energy source of fermented total mixed ration (FTMR) on the in vitro gas production technique. Int J Agric Technol. 2021;17(1):227–36.

Napasirth P, Napasirth V. Current situation and future prospects for beef production in Lao People's Democratic Republic – a review. Asian Australas J Anim Sci. 2018;31(7):961–7.

Napasirth V, Napasirth P, Sulinthone T, Phommachanh K, Cai Y. Microbial population, chemical composition, and silage fermentation of cassava residues. Anim Sci J. 2015;86(10):842–8.

Napasirth P, Wachirapakorn C, Napasirth V. The effects of *Lactobacillus plantarum* inoculation of spontaneously fermented cassava pulp (SFCP) on the quality, feed intake, and growth of Lao-native beef steers. IJAT. 2021;17(5):1847–56.

Nugraha RI, Shiddieqy MI, Laconi EB, Jayanegara A. Estimation of rice straw adequacy as ruminants feed in Java based on yield area conversion. IOP Conf Ser Earth Environ Sci. 2023;1187:012004.

Nurhayu A, Saenab A, Ella A, Ishak ABL, Qomariyah N. The effects of elephant grass silage combined with Indigofera sp. on the performance of Bali cattle. Anim Health Prod. 2021;9:229–35.

Olmo L. Investigating smallholder cattle and buffalo reproductive health and management in Lao PDR to enhance reproductive efficiency and improve livelihoods. Doctoral thesis, The University of Sydney. 2020. https://laofab.org/document/view/4212

Patriani P, Ginting N, Hasanah U, Mirwandhono RE. Application of silage waste fruit leather technology in Suka village, Kecamatan Tigapanah, Karo district. Abdimas Talenta. 2019;4:192–9.

Pereyra MLG, Chiacchiera SM, Rosa CAR, Sager R, Dalcero AM, Cavaglieri L. Comparative analysis of the mycobiota and mycotoxins contaminating corn trench silos and silo bags. J Sci Food Agr. 2011;91:1474–81.

Pitiwittayakul N, Bureenok S, Schonewille JT. Selective thermotolerant lactic acid bacteria isolated from fermented juice of epiphytic lactic acid bacteria and their effects on fermentation quality of stylo silages. Front Microbiol. 2021;12:673946.

Purnamasari I, Sari IW, Rahayu S, Yamin M. The effect of grass substitution with dry kale and sprout waste on Garut sheep's performance. J Peternak Indones. 2021;23:25–32.

Rodriguez VG, Vandestroet L, Abeysekara VC, Ominski K, Bumunang EW, McAllister T, et al. Optimizing silage strategies for sustainable livestock feed: preserving retail food waste. Agriculture. 2024;14:122.

Sakai T, Devkota NR, Oishi K, Hirooka H, Kumagai H. Evaluation of total mixed ration silage with brewers grains for dairy buffalo in Tarai. Nepal Anim Sci J. 2015;86:884–90.

Sarwono KA, Rohmatussolihat, Watman M, Ratnakomala S, Astuti WD, Fidriyanto R, et al. Characteristics of fresh rice straw silage quality prepared with the addition of lactic acid bacteria and crude cellulase. AIMS Agric Food. 2022;7(3):481–99.

Sayuti M, Syahrudin, Sahara L, Rachman AB, Metaragakusuma AP, Febrisiantosa A. Evaluating complete silage for goat feeding in Gorontalo, Indonesia. Curr Res Biosci Biotechnol. 2024;5:17–21.

Schmidt P, Restelatto R, Zopollatto M. Ensiling total mixed rations – an innovative procedure. In Proceedings of the V international symposium on forage quality and conservation, Piracicaba; Brazil. Piracicaba, Brazil. 2017. pp. 7–20.

Sio S, Bira GF, Batu MS, Pardosi L, Mau RJ, Klau MO, Hoar J. Organoleptic quality and nutrition of rice straw silage utilizing local microorganisms (mol) of cattle rumen fluid at different inoculum levels. J Adv Vet Res. 2022;12(1):36–41.

Sivilai B, Hang DT, Linh NQ, Preston TR. Effect of replacing ensiled taro foliage (*Colocasia esculenta*) with ensiled banana pseudo-stem (*Musa* spp) and soybean meal on intake, digestibility, and nitrogen retention in moo lath pigs. Livest Res Rural Dev. 2018:30.

So S, Cherdthong A, Wanapat M. Improving sugarcane bagasse quality as ruminant feed with *lactobacillus*, cellulase, and molasses. J Anim Sci Technol. 2020;62(5):648–58.

Song J, Ma Y, Zhang H, Wang L, Zhang Y, Zhang G. Fermented total mixed ration alters rumen fermentation parameters and microbiota in dairy cows. Animals. 2023;13:1062.

Stur W, Gray D, Bastin G. Review of the livestock sector in the Lao People's Democratic Republic. Nairobi: International Livestock Research Institute; 2002.

Therdtatha P, La-Ongkham O, Nakphaichit M, Mapato C, Rungruang S, Nakayama J, Nitisinprasert S. Effect of lactic acid bacterial starter KUB-G2 on grass silage quality and its microbial community performed using 140-ton plastic bag silos: a large-scale study. Qual Assur Saf Crop Foods. 2024;16(SP1):1–13.

Vongsamphanh P, Preston TR, Napasirth V, Napasirth P. Brewers' grain and cassava foliage (*Manihot esculenta Cranz*) as protein sources for local "yellow" cattle fed cassava pulp-urea as basal diet. Livest Res Rural Dev. 2016;28(11)

Wahyudi A, Hendraningsih L, Sutawi S, Siskawardani DD. Application of lactobacillus inoculant from various rice paddy (*Oryza sativa*) total mixed ration silage microbial composition. Biodiversitas. 2022;23:3387–94.

Wang S, Yuan X, Dong Z, Li J, Shao T. Effect of ensiling corn Stover with legume herbages in different proportions on fermentation characteristics, nutritive quality, and in vitro digestibility on the Tibetan Plateau. Grassl Sci. 2017;63:236–44.

Weinberg ZG, Szakacs G, Ashbell G, Hen Y. The effect of temperature on the ensiling process of corn and wheat. J Appl Microbiol. 2001;90:561–6.

Wilkinson JM, Rinne M. Highlights of progress in silage conservation and future perspectives. Grass Forage Sci. 2017;73:1–13.

Xayalath S, Mujitaba MA, Ortega ADSV, Rátky J. A review on the trend of livestock breeds in Laos. Acta Agraria Debreceniensis. 2021;1:227–37.

Xaypha S. Goat production in smallholder farming systems in lowland Lao PDR and an evaluation of different forages for growing goats. A thesis submitted in fulfillment of the requirements for the degree of degree. Swedish University. 2005.

Yang K, Qing Y, Yu Q, Tang X, Chen G, Fang R, et al. By-product feeds: current understanding and future perspectives. Agriculture. 2021;11:207.

Yanti Y, Yayota M. Agricultural by-products as feed for ruminants in tropical area: nutritive value and mitigating methane emission. Rev Agric Sci. 2017;5:65–76.

Yanti Y, Yayota M. Effect of epiphytic lactic acid bacteria fermented juice inclusion on quality of total mixed ration silage of agricultural and food by-products. Indian J Anim Sci. 2019;89:1002–8.

Yanti Y, Kawai S, Yayota M. Effect of total mixed ration silage containing agricultural by-products with the fermented juice of epiphytic lactic acid bacteria on rumen fermentation and nitrogen balance in ewes. Trop Anim Health Prod. 2019;51:1141.

Yanti Y, Riyanto J, Dewanti R, Cahyadi M, Wati AK, Pawestri W. The fermentation quality of complete feed with FJLB silage additive from tropical grass. IOP Conf Ser Earth Environ Sci. 2021;824:012060.

Yanti Y, Setyawati A, Sumani S, Ariyanto DP, Komariah K. Pelatihan pembuatan silase pakan komplit dengan aditif FJLB di kelompok ternak Putra Rahayu dan Ngudi Rejeki Wonogiri. Bubungan Tinggi. 2022;4:886–94.

Zain M, Ningrat RWS, Erpomen, Putri EM, Makmur M. The effects of leguminous supplementation on ammoniated rice straw based completed feed on nutrient digestibility on in vitro microbial protein synthesis. IOP Conf Ser Earth Environ Sci. 2019;287:012018.

Zhang M, Wang R, Wu T, Yang Y, He Z, Ma Z, et al. Comparisons of corn Stover silages after fresh- or ripe-corn harvested: effects on digestibility and rumen fermentation in growing beef cattle. Animals. 2022;12:1248.

Chapter 13
Silage Production and Utilization in Europe

Marketta Rinne

Abstract Grassland-based livestock production systems in Europe have evolved based on local conditions which vary significantly across the continent. To ensure year-round feeding in all European regions, preserving forage is essential, with ensiling being the predominant method used for grass preservation. In recent years, maize cultivation has experienced rapid growth, and maize silage has become an integral component of most dairy cow diets in Europe. Advances in plant breeding and climate change have contributed to the expansion of maize cultivation towards northern latitudes. In addition to forages, ensiling is also utilized to stabilize moist cereal grains and various food industry by-products. Forage plants are efficient at utilizing solar radiation and nutrients, leading to the incorporation of ensiled maize and grass in biogas production. In some areas, this shift has created competition for biomass between energy production and the food supply chain. The science and technology behind silage-making continue to evolve, driven by the increasing demands for sustainable intensification, which includes reducing the environmental impact of livestock production, addressing economic pressures, and ensuring food safety and security. Additionally, farm sizes are steadily increasing across Europe, enabling greater use of advanced technology and investments in silage production.

Keywords Europe · Food security · Ruminant · Silage · Sustainable intensification

1 Overview of Silage Production in Europe

European agricultural policy has a strong focus on robust and resilient food systems while reducing the impact of agriculture on climate change and environmental degradation and addressing biodiversity decline (European Commission 2020, 2021). The grassland-based livestock production systems have evolved based on local

M. Rinne (✉)
Natural Resources Institute Finland, Jokioinen, Finland
e-mail: marketta.rinne@luke.fi

Y. Cai, K. Ataku (eds.), *Cultural History and Modern Production Technology of Silage*, https://doi.org/10.1007/978-981-96-5787-2_13

conditions that vary widely within Europe. They have traditionally a strong role in various regions ranging from highly intensive milk production of dairy cows to free foraging of semidomesticated reindeer in the North of Europe and numerous variable production systems in between. Provision of food from grassland-based production systems plays an important role in food security as low-cost human inedible resource (grassland) is converted by ruminants into high-quality foods such as dairy and meat products. In many cases this is possible without directly competing with crop production (Van Kernebeek et al. 2016).

Agricultural land covers on average 39.1% of the land area in the European Union in 2018 (Eurostat 2024). The range is large from 63.7% in Denmark to 6.9% in Finland. A more detailed description of land area use is provided in Fig. 13.1. None of the European regions can provide year-around fresh forage to animals so that preservation of forage is required in all production systems. Traditionally, grass was dried to provide hay for winter feeding, but supported by the theoretical development of ensiling highlighted by the Nobel prize in Chemistry in 1945 by A.I. Virtanen (Nobel Prize 2024), ensiling has almost completely overridden hay as a grass preservation method (Wilkinson and Rinne 2018). There are however specialized production systems where ensiled forages are not allowed, e.g. the production of Parmigiano Reggio cheese in Northern Italy (Mantovi et al. 2015).

There are no direct statistics within European Union to report the amount of silage produced as products of many land use categories can be preserved as silage. In the early 2000s, Wilkinson and Toivonen (2003) provided an estimation of forage production in most European countries. At that time, on dry matter (DM) basis, hay represented 34% of forage preservation, while ensiled grass, maize, and other crops

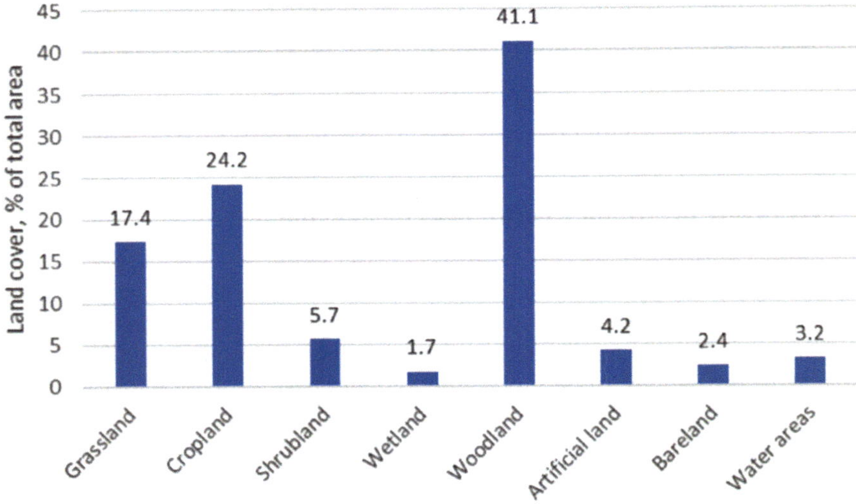

Fig. 13.1 Land cover within the European Union in 2018. (Source: Eurostat, DOI:10.2908/lan_lcv_ovw)

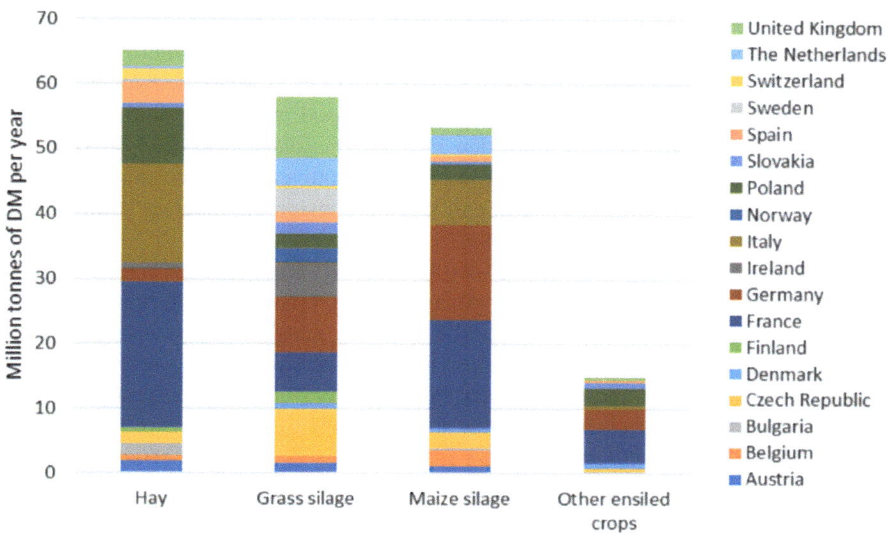

Fig. 13.2 Production of hay and different types of silage in European countries. (Source: Wilkinson and Toivonen 2003)

covered 30%, 28%, and 8%, respectively (Fig. 13.2). Based on recent trends, it can be assumed that the proportion of hay has decreased and that of maize increased since these estimates were compiled. It is also noteworthy that these figures are presented on DM basis. The different types of forages have distinctly different average DM concentrations so that the results would look very different if presented on fresh matter basis. Grass hay has a DM concentration of ca. 860 g/kg and maize silage 300–350 g/kg. For grass silage, the variation is the largest depending on the target (and success) of field wilting prior to ensiling and can be around 200 g/kg in direct-cut silages while up to 600–700 in highly wilted haylage. However, most of the silage prepared for intensive cattle production is preserved in bunker or clamp silos with target DM concentration around 300–400 g/kg.

2 European Grassland Characteristics

The environmental conditions vary widely within Europe, which results in distinct types of grassland production, preservation, and utilization in the different regions. In the 18th Symposium of the European Grassland federation (EGF 2015), one of the themes was grassland production around Europe, and several reviews were produced to describe them for interested readers. A brief overview related to grassland production is given here.

Grasslands have the opportunity to provide several ecosystem services (Richter et al. 2024). Grazing of pastures provides biodiversity and cost benefits and allows animals to present their natural behaviour so that it should be promoted over indoor feeding whenever possible but is not feasible year-around anywhere in Europe. Grazing may also be limited by lack of suitable pastures close to the livestock units, particularly when automatic milking systems are used, although solutions exist to overcome barriers for grazing (van den Pol-van Dasselaar et al. 2020). Although harvesting grass as silage does not promote, e.g. biodiversity to the same extent as grazing, perennial multispecies leys can still be considered superior compared to annual monocultures of arable cropping.

The most popular grass species grown for silage in Europe is perennial ryegrass (*Lolium perenne* L.), which has many positive characteristics such as high yield, fast regrowth ability, and good feed values. Regarding silage making, relatively high water-soluble carbohydrate concentration facilitates adequate lactic acid production contributing to low pH and good preservation quality of ensiled material. The winter hardiness of ryegrass is however low so that in the northern parts of Europe species such as timothy (*Phleum pratense* L.) and meadow fescue (*Festuca pratensis* Huds.) are used instead of ryegrass (Virkajärvi et al. 2015). Multi-species swards are currently being emphasized as they may be more resilient in the changing climate linked with higher pressure of abiotic and biotic stressors.

The use of forage legumes dropped drastically when mineral N fertilizers became widely available during the latter part of the 1900s. Numerous research projects and advisory attempts have taken place to advocate wider use of them, but the popularity is increasing rather slowly, although there is evidence of benefits in both agronomic and animal performance (Egan et al. 2018). In central Europe, lucerne (*Medicago sativa* L.) and white clover (*Trifolium repens* L.) are the most common forage legumes, while in Northern Europe, red clover (*Trifolium pratense* L.) is most persistent (Virkajärvi et al. 2015).

Legumes may be more difficult to ensile than other types of crops due to their inherently high moisture content and buffering capacity and low water-soluble carbohydrate concentration. Thus, emphasis should be put on sufficient wilting and other management factors including use of additives, to ensure good preservation quality of them (Clavin et al. 2017; Rinne et al. 2023).

3 Annual Forage Crops are Commonly Ensiled

The popularity of maize (*Zea mays* L.) cultivation in Europe has increased rapidly, and maize silage is currently an integral part of most European dairy cow diets. Also globally, whole crop maize silage is the most important forage type in ruminant livestock feeding (Wilkinson and Rinne 2018). Developments in plant breeding and production technology (e.g. plastic mulch at seeding) as well as climate warming have supported the spread of forage maize cultivation towards northern latitudes. An evaluation of the performance of forage maize in the very north of Europe, in Finland, concluded that annual variability in the maturity achieved by forage maize was still very large and in general unsatisfactory in the main milk producing areas (around 63°N latitude) (Lehtilä 2024). A study in Finland also evaluated the preservation quality of maize at lower than optimal stage of development when ensiled, but no major challenges were observed (Rinne et al. 2024).

An alternative to maize is harvesting small grain cereals (most commonly wheat and barley) as whole crop cereal silage (Dewhurst 2013; Rustas 2009). Some of the benefits of forage maize can be achieved (single harvest per growing season, annual planting facilitating efficient slurry use) with less risk of crop failure due to annually fluctuating weather conditions, but the high yield and feed value of maize may not be reached. Grain legumes pea, faba bean, and lupins can also be harvested as whole crop silages instead of collecting just the seed yield with positive ensiling and animal production performance demonstrated (Dewhurst 2013; Kleinmans et al. 2019; Palmio et al. 2022).

There is increasing interest also towards novel plants to be introduced in areas within Europe where they have not traditionally been used. The changing climate motivates this search to identify alternatives with better tolerance to varying climatic conditions. Tropical plants introduced to Europe include, for example, sorghum (*Sorghum bicolor*) (Getachew et al. 2016), while a traditional European grass species tall fescue (*Festuca arundinacea*) has gained renewed interest due to its good drought tolerance. In addition, breeding of the currently common forage plants aims to produce new varieties with increased tolerance to biotic and abiotic stressors such as drought, waterlogging, diseases, and pests.

4 Ensiled Grains and By-products

Crimping and ensiling of cereal grains and other seeds has gained popularity during last decades particularly in the more humid areas of western and northern Europe. In these areas, the moisture content of the grains at harvest is so high that preservation requires artificial drying. An alternative method for drying is crimping, e.g. crushing slightly the moist grains, and preserving them air-tight, resulting in ensiled grains. The preservation method relies on the same principles as forage preservation, i.e. lactic acid fermentation lowers the pH in anaerobic conditions. Different types of silage additives can be used to improve the fermentation quality and aerobic stability of the crimped grains (Franco et al. 2022). As a preservation method, ensiling of grains is cost and energy effective compared to drying as investments of drying buildings and their running costs are not needed (Jokiniemi et al. 2014). There are indications that fermentation during preservation may improve the digestibility of maize starch (Allen and Ying 2021), and in case of faba beans, harmful substances (vicin and convisin) were degraded during the fermentation phase (Rinne et al. 2020a). In vivo feeding trials have also demonstrated superior (Huuskonen et al. 2020) or equal (Rinne et al. 2022) performance of cattle when fed crimped ensiled grains compared to dry ones.

Several moist by-products from food industry can be ensiled to be used as livestock feeds. Stenmarck et al. (2016) estimated that about 88 million tons of food waste is generated in the entire food chain of Europe, which is about 20% of the total food produced. In fruit and vegetable industry, the losses can be up to 50% (Bakshi et al. 2016). Some by-products such as sugar beet pulp are commonly preserved by ensiling, but, for example, vegetable residues have very low DM, spoil rapidly, and are typically produced as rather small batches which makes it difficult to efficiently recycle them back into the food chain.

There are however examples of successful use of ensiled vegetable by-products (Arco-Pérez et al. 2017). Use of proper ensiling management techniques including the use of additives can aid in controlling the fermentation and improve the aerobic stability of the material during feed-out phase (Orosz and Davies 2015).

5 Livestock Utilization of Silage in Europe

The variability of silage utilization in various European livestock production systems is large, and describing them in detail is beyond the scope of this review. The numbers of most important livestock animal species consuming silage in the European Union in 2023 (Eurostat 2024) are as follows:

Bovines 73.7 million
Sheep 57.5 million
Goats 10.7 million

The genetic potential for milk production of the current dairy cows used in intensive milk production is so high that supplementing forages with energy and protein-rich concentrate feeds is currently done in most European dairy systems. Alleviating feed-food competition and utilizing as much as possible of non-human-edible feeds for livestock are on the other hand important targets. Biologically, ruminants are rather flexible in using different types of diets, and the forage-to-concentrate ratio can be manipulated to a large extent. Figure 13.3 shows results from dairy cow experiments, where forage-to-concentrate ratio ranges from 0 to 0.8, and although there is a trend of increasing milk production with higher concentrate proportion, the increase is relatively modest and tends to level of at higher levels of concentrate use. The economic outcome depends on the prices of forages vs. concentrate feeds and should be evaluated using real feed costs and accurate estimates of the milk production responses to diet changes.

On top of quantity of silage in the ration, silage quality plays a crucial role in successful livestock production. Silage quality is defined by two major factors: the initial nutritional quality of the raw material at harvest and the extent and type of modifications during the preservation and feed-out phases. The science and technology of silage making continue to develop the main drivers being increasing requirements for sustainable intensification, i.e. reduced environmental impact of livestock production, economic pressures, and needs to ensure food safety and security. The farm sizes continue to increase all over Europe, which facilitates use of increasing technology and investments. It is imperative that the opportunities provided by

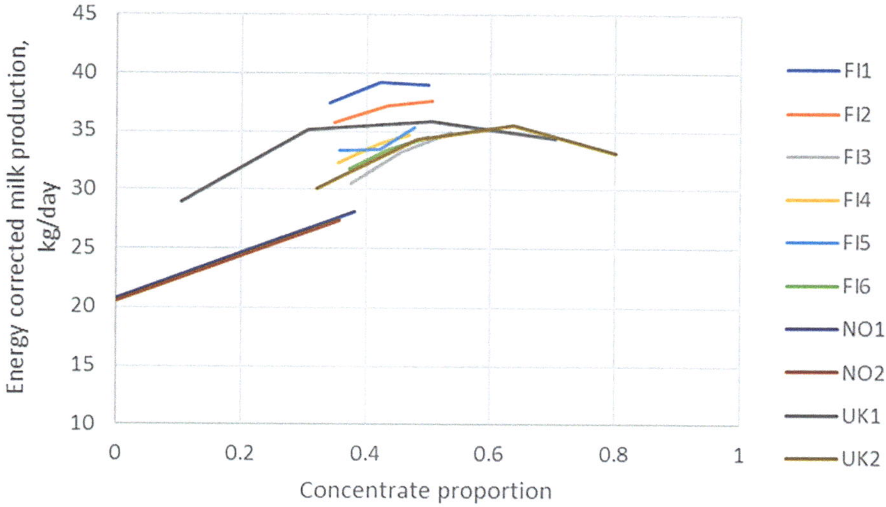

Fig. 13.3 Milk production responses to increasing concentrate proportion (on dry matter basis) on different grass silage-based diets according to Sairanen et al. (2022; FI1–FI6), Steinshamn and Thuen (2008; NO1 and NO2) and Ferris et al. (2001; UK1 and UK2)

developing technologies are used also in the silage industry (Vyas and Adesogan 2023).

6 Novel Uses of Silage

Within EU, there is a target to replace fossil fuels by bioenergy (EU Renewable Energy Directive EU/2018/2001 and its amendment in 2023 EU/2023/2413). Forage plants are efficient utilizers of solar radiation and nutrients so that there is a lot of scope to increase the production of green biomass in many areas. In some European countries (e.g. France, Germany), there is currently a direct competition of agricultural land between renewable energy production (whole crop maize and grass for biogas production in anaerobic fermenters, solar panels installed on the fields). Biogas production can also be used to valorize surplus silage and utilize batches with quality deviations that make them unsuitable as animal feeds.

Green biomass could potentially provide also higher-added-value products than energy, when processed in so-called green biorefineries. Novel sustainable products such as paper and packaging, nanofibers, animal bedding, novel protein feeds, extracted proteins, biochemicals, nutraceuticals, bioactive compounds, and biogas could be produced from grass (Gaffey et al. 2023). This approach widens the scope of grass use from the traditional ruminant livestock-related feed use to a variety of innovative applications. Simultaneously the benefits of grasslands and the ecosystem services they provide are maintained and potentially even expanded.

Although most biorefinery approaches are based on fresh grass (Jørgensen et al. 2022; Gaffey et al. 2023), using ensiled feedstock would provide important advantages by providing a year-around feedstock and improved stability also for the products (Rinne 2024). The trade-off, however, is the modification of compounds during biomass fermentation, which may, depending on desired end-products, be unfavourable. There are only a few studies directly comparing fresh and ensiled green biomass in terms of biorefinery output. Ayanfe et al. (2023) demonstrated that ensiling improved liquid separation efficiency compared to fresh grass, and results by Rinne et al. (2020b) suggest that the ensiling process can be optimized as a pre-treatment for a biorefinery, e.g. by fibrolytic enzyme application. At the moment, the development of the green biorefinery industry is still in its infancy, and future will show how important user of silage it may eventually be.

References

Allen MS, Ying Y. Effects of corn grain endosperm type and conservation method on site of digestion, ruminal digestion kinetics, and flow of nitrogen fractions to the duodenum in lactating dairy cows. J Dairy Sci. 2021;104:7617–29.

Arco-Pérez A, Ramos-Morales E, Yáñez-Ruiz DR, Abecia L, Martín-García AI. Nutritive evaluation and milk quality of including of tomato or olive by-products silages with sunflower oil in the diet of dairy goats. Anim Feed Sci Technol. 2017;232:57–70.

Ayanfe N, Franco M, Stefański T, Pap N, Rinne M. The effects of grass biomass preservation methods, organic acid treatment and press type on the separation efficiency in the green biorefinery. Bioresour Technol Rep. 2023;21:101356.

Bakshi MPS, Wadhwa M, Makkar HPS. Waste to worth: vegetable wastes as animal feed. CAB Rev. 2016;11:1–26.

Clavin D, Crosson P, Grant J, O'Kiely P. Red clover for silage: management impacts on herbage yield, nutritive value, ensilability and persistence, and relativity to perennial ryegrass. Grass Forage Sci. 2017;72:414–31.

Dewhurst R. Milk production from silage: comparison of grass, legume and maize silages and their mixtures. Agric Food Sci. 2013;22:57–69.

Egan M, Galvin N, Hennessy D. Incorporating white clover (*Trifolium repens* L.) into perennial ryegrass (*Lolium perenne* L.) swards receiving varying levels of nitrogen fertiliser: effects on milk and herbage production. J Dairy Sci. 2018;101:3412–27.

EGF. Grassland and forages in high output dairy farming systems: proceedings of the 18th symposium of the European Grassland Federation, Wageningen, The Netherlands. Grassl Sci Eur. 2015;20:51–66.

European Commission. A farm to fork strategy for a fair, healthy and environmentally-friendly food system. Brussels: European Commission; 2020.

European Commission. Directorate-General for Environment, EU biodiversity strategy for 2030—bringing nature back into our lives. Brussels: European Commission; 2021.

Eurostat. Livestock. Available: https://ec.europa.eu/eurostat/web/main/data/database. 2024.

Ferris CP, Gordon FJ, Patterson DC, Kilpatrick DJ, Mayne CS, McCoy MA. The response of dairy cows of high genetic merit to increasing proportion of concentrate in the diet with a high and medium feed value silage. J Agric Sci. 2001;136:319–29.

Franco M, Tapio I, Rinne M. Preservation characteristics and bacterial communities of crimped ensiled barley grain modulated by moisture content and additive application. Front Microbiol. 2022;13:1092062.

Gaffey J, Rajauria G, McMahon H, Ravindran R, Dominguez C, Ambye-Jensen M, et al. Green biorefinery systems for the production of climate-smart sustainable products from grasses, legumes and green crop residues. Biotechnol Adv. 2023;66:108168.

Getachew G, Putnam DH, De Ben CM, De Peters EJ. Potential of sorghum as an alternative to corn forage. Am J Plant Sci. 2016;07:1106–21.

Huuskonen A, Rinne M, Manni K. Effects of different barley grain preservation techniques on intake, growth and carcass traits of finishing dairy bulls fed grass silage-based rations. J Agric Sci. 2020;158:748–55.

Jokiniemi T, Jaakkola S, Turunen M, Ahokas J. Energy consumption in different grain preservation methods. Agron Res. 2014;12:81–94.

Jørgensen U, Jensen SK, Ambye-Jensen M. Coupling the benefits of grassland crops and green biorefining to produce protein, materials and services for the green transition. Grass Forage Sci. 2022;77:295–306.

Kleinmans J, McDougall S, Holmes A. Yield and quality of whole crop faba beans harvested and ensiled at different maturities. Agron N Z. 2019;49:1–8.

Lehtilä A. Cultivation and environmental impacts of forage maize in boreal conditions. Doctoral thesis. Dissertationes Universitatis Helsingiensis. 2024; 202.

Mantovi P, Dal Prà A, Pacchioli MT, Ligabue M. Forage production and use in the dairy farming systems of northern Italy. In: Grassland and forages in high output dairy farming systems: proceedings of the 18th symposium of the European Grassland Federation, Wageningen, The Netherlands. Grassl Sci Eur. 2015;20:51–66.

Nobel Prize. The Nobel Prize in Chemistry 1945. Nobel Prize Outreach AB. 2024.

Orosz Sz, Davies DR. Short and long term storage of wet by-products fed by ruminants. In: Daniel JLP, Morais G, Junges D, Nussio LG, editors. Proceedings of the XVII international silage conference. Piracicaba, Brazil. 2015. pp. 200–242.

Palmio A, Sairanen A, Kuoppala K, Rinne M. Milk production potential of whole crop faba bean silage compared with grass silage and rapeseed meal. Livest Sci. 2022;259:104881.

Richter FJ, Suter M, Lüscher A, Buchmann N, Benni N, El Feola Conz R, et al. Effects of management practices on the ecosystem-service multifunctionality of temperate grasslands. Nat Commun. 2024;15:3829.

Rinne M. Novel uses of ensiled biomasses as feedstocks for green biorefineries. JASB. 2024;15:36.

Rinne M, Leppä MM, Kuoppala K, Koivunen E, Kahala M, Jalava T, Salminen JP, Manni K. Fermentation quality of ensiled crimped faba beans using different additives with special attention to changes in bioactive compounds. Anim Feed Sci Technol. 2020a;265:114497.

Rinne M, Winquist E, Pihlajaniemi V, Niemi P, Seppälä A, Siika-aho M. Fibrolytic enzyme treatment prior to ensiling increased press-juice and crude protein yield from grass silage. Bioresour Technol. 2020b;299:122572.

Rinne M, Franco M, Tapio I, Stefański T, Bayat A, Mäntysaari P. Effects of grass silage additive type and barley grain preservation method on rumen fermentation, microbial community and milk production of dairy cows. Agriculture. 2022;12:266.

Rinne M, Franco M, Manni K, Huuskonen A. Evaluating the effects of wilting, mixing with timothy and silage additive application on red clover silage quality. Agric Food Sci. 2023;32:207–18.

Rinne M, Franco M, Kuoppala K, Jalava T, Stefański T, Kokkonen T. Effects of harvest date and additives on maize silage quality under Boreal conditions. Agric Food Sci. 2024;33:127–38.

Rustas B-O. Whole-crop cereals for growing cattle: effects of maturity stage and chopping on intake and utilisation. Doctoral thesis. Acta Universitatis Agriculturae Sueciae 74. Swedish University of Agricultural Sciences. 2009.

Sairanen A, Juutinen E, Rinne M. The effect of harvesting strategy of grass silage and concentrate level on feed intake, digestibility and milk production of dairy cows. Agric Food Sci. 2022;31:160–74.

Steinshamn H, Thuen E. White or red clover-grass silage in organic dairy milk production: Grassland productivity and milk production responses with different levels of concentrate. Livest Sci. 2008;119:202–15.

van den Pol-van Dasselaar A, Hennessy D, Isselstein J. Grazing of dairy cows in Europe-an in-depth analysis based on the perception of grassland experts. Sustain For. 2020;12:1098.

Van Kernebeek HRJ, Oosting SJ, Van Ittersum MK, Bikker P, De Boer IJM. Saving land to feed a growing population: consequences for consumption of crop and livestock products. Int J Life Cycle Assess. 2016;21:677–87.

Virkajärvi P, Rinne M, Mononen J, Niskanen O, Järvenranta K, Sairanen A. Dairy production systems in Finland. In: Grassland and forages in high output dairy farming systems. Proc Grassl Sci Eur. 2015;20:51–66.

Vyas D, Adesogan AT. Smart silage of the future. In: International silage conference. 2023. pp. 89–94.

Wilkinson JM, Rinne M. Highlights of progress in silage conservation and future perspectives. Grass Forage Sci. 2018;73:40–52.

Wilkinson JM, Toivonen ML. World silage. Painshall, Lincoln: Chalcombe Publications; 2003.

Chapter 14
Silage Production and Livestock Feeding in China

Peng Wang and Yang Cao

Abstract The development of grass and animal husbandry in China has made silage increasingly important as a key feed resource for ruminant production. Silage or fermented total mixed ration provides a nutritionally balanced and stable diet for Chinese herbivorous livestock. The fermentation process can utilize unconventional feed resources, such as industrial by-products, energy crops, and certain waste materials from agricultural and industrial production. This chapter mainly focuses on the importance and practical application of silage in China's animal husbandry sector and briefly introduces recent research progress in fermented total mixed ration in China.

Keywords China · Fermented total mixed ration · Livestock raising · Silage production

With the continuous development of China's economy and the improvement of people's living standards, the development and change of the animal husbandry industry, as an important part of the national economy, have been widely concerned by all sectors of society. As a big country of animal husbandry, China's ruminant stock ranks among the top in the world, and there is a great demand for feed in the market. Although China has jumped to be the world's largest feed producer since 2012, with a total feed output of up to 194 million tones, accounting for 36% of the total grain output, it is still unable to meet the market demand. In recent years, silage, as a nutritious and easy-to-store high-quality feed, has been widely used and recognized in the livestock industry. Silage industry gradually plays an increasingly important role in herbivore farming. In the current breeding practice of dairy cattle, beef cattle, and sheep, silage occupies about 65%, 26%, and 11% of roughage

P. Wang (✉)
Jilin University, Changchun, China
e-mail: pengwang@jlu.edu.cn

Y. Cao
Heilongjiang Bayi Agricultural University, Daqing, China

Y. Cai, K. Ataku (eds.), *Cultural History and Modern Production Technology of Silage*, https://doi.org/10.1007/978-981-96-5787-2_14

Fig. 14.1 Application of silage in animal production. (Photos by Wang)

(Fig. 14.1). With the structural adjustment of the agricultural industry and the promotion of modern grassland animal husbandry in the south, the development of the silage industry will be taken to the next level.

1 Development of Silage Production in China

Peng Wang

Silage is plant feed containing a large amount of moisture that is sealed and fermented as animal feed (Mahmoud et al. 1979). In traditional farming methods, most of them are mainly hay feeding, which leads to a drastic decrease in the nutrient content and palatability of the forage. However, silage made from silage processing, which is more resistant to storage than fresh fodder and its nutritional content is stronger than dry fodder, can effectively solve the problem of forage shortage in the autumn and winter seasons (Nishino et al. 2004). In addition, the storage of silage takes up little space and there is no fire problem. What's more, crushing rice straw, then silage, ammonization, kneading, and micro-storage, and then used for breeding, not only can save feed costs, but also let rice straw in animal manure over the belly to return to the field, thus promoting the benign cycle of agriculture, which is a high-efficiency method of utilization (Helander et al. 2015).

There are two types of silage production processes in general, the integrated harvesting and processing of field harvesting, chopping, and throwing for storage and the segmented harvesting and processing of field harvesting, transport, and chopping for storage. In the high degree of intensification and flat terrain area, the integrated harvesting and processing is the main one, and in the low degree of intensification, mountainous or discontinuous area, the sectional harvesting and processing is the main one. The harvesting of yellow silage straw is similar to the harvesting of green forage, with one type of harvesting the straw at the same time as harvesting the seeds and another type of harvesting the straw after harvesting the seeds. Due to the significant differences in climate in China from south to north and from east to west, silage is processed from April in spring to November in autumn and winter.

Table 14.1 Inventory of beef cattle and dairy cow, livestock product production, and silage planting in China, 2014–2023

	2014	2015	2016	2017	2018	2019	2020	2021	2022	2023
Inventory (millions)										
Total cattle	90.1	90.6	88.3	90.4	89.2	91.4	95.6	98.2	102.2	105.1
Beef cattle	70.4	73.7	74.4	66.2	66.2	70.0	76.9	80.0	84.5	–
Cow	8.4	8.5	7.6	7.7	5.0	4.7	5.2	6.2	6.4	6.6
Slaughter (millions) total cattle										
	42.0	42.1	42.7	43.4	44.0	45.3	45.7	47.1	48.4	50.2
Yield (millions of tons)										
Beef	6.16	6.17	6.17	6.35	6.44	6.67	6.73	6.98	7.18	7.53
Mutton	4.28	4.40	4.60	4.71	4.75	4.88	4.92	5.14	5.25	5.31
Milk	31.6	31.8	30.6	30.4	30.8	32.0	34.4	36.8	39.3	42.0
Silage crop area (1000 ha)										
	1584	1634	1814	1874	1971	2115	2199	2575	2631	–

Source: National Bureau of Statistic of China

In recent years, China's silage industry has continued to expand, with a market size of more than tens of billions of RMB and a trend of steady growth. In recent years, China's beef and sheep inventory has overall grown; as of 2023 China's dairy cattle inventory is 6.6 million, and beef cattle inventory is 105.1 million, assuming that the usual standards of Western developed countries to feed a cow needs 3 acres of silage per year and beef cattle need to be 2 acres, then with the silage market space of up to 229 million mu (Table 14.1). It can be seen that China's silage industry has a broad market prospect. The market is rich in silage, mainly divided into three categories: crop straw silage, annual forage silage, and perennial forage silage. The National Bureau of Statistics data show that China's silage forage area in 2010–2023 showed an overall upward trend, rising from 34.98 million hectares in 2010 to 613.91 million hectares in 2023. China's perennial silage production is more than 28,000 million tones, of which crop straw silage production accounts for a larger proportion, followed by annual forage silage and perennial forage silage. This is mainly due to the promotion of the following aspects.

1.1 Analysis of Industry Development Drivers

1.1.1 Technical Progress

In 2013, the Ministry of Agriculture launched the silage special project 'Research and demonstration of forage silage technology and supporting facilities and equipment'. The design of new silage cellar, the research and development of high-efficiency silage equipment, and the improvement of silage fermentation technology all provide strong support for the development of silage industry. Research work has been carried out on the physical, chemical, and biological characteristics of raw materials that affect the quality of forage silage; formed the processing technology of additives and forage with the deterioration of nutrients in forage silage; integrated the technology of screening, identification, cultivation, and preparation of bacterial agents of lactic acid bacteria (LAB) that improve the quality of silage; standardized the technical procedures of forage silage; put forward the indicators of the evaluation of the quality of silage products with the grade value; and established an animal forage-feed nutrient balanced feeding technology.

1.1.2 Policy Support Management Macro and Micro in Place

The government's support for the livestock industry has been increasing, and a series of supportive policies have been introduced to promote the healthy development of the silage industry. From 2004 to 2015, the central government's No. 1 document, which focused on the 'three rural issues' (agriculture, rural areas, and farmers) for 12 consecutive years, proposed the idea of agricultural structural adjustment each time, promoting the development of the silage industry from a macro level. The Ministry of Agriculture issued the 'National grain-saving animal husbandry development plan (2011–2020)', 'Ministry of Agriculture on the promotion of grass-fed animal husbandry to accelerate the development of the guiding opinions', etc. which are from the top design, including silage industry and the construction of the modern forage industry system to provide a macro-management decisions. Provincial, municipal, and county management departments at all levels, in view of regional characteristics, focusing on fine, fine, strict, and practical, have introduced corresponding management and support policies to promote the development of silage industry, consolidate the forage base, enhance technical services, and safeguard the market in an orderly manner.

1.1.3 Demand for Market Expansion

Straw yellow silage has been promoted in large quantities since the 1950s, and whole-plant maize silage has been promoted and demonstrated since the 1980s. With the expansion of the scale of animal husbandry and the improvement of breeding technology, the demand for high-quality and high-nutritional feed has been growing, and silage has been favoured by the market with its unique advantages. At present, a circular industrial chain has been formed with forage silage seed enterprises, silage cultivation trusteeship intermediaries, harvesting and processing machinery service cooperatives, silage production enterprises, silage consumption enterprises, and farming manure treatment enterprises, and there are also related consumables, additives, machinery, information, and other industries.

1.2 Application and Importance of Silage in Animal Husbandry

The application of silage in animal husbandry is mainly reflected in the following aspects.

1.2.1 Alleviate the Pressure of Grassland

In the area of grassland resource tension, through the production and application of silage, it can effectively alleviate the pressure of grassland and reduce the burden of grassland ecological environment.

1.2.2 Improve the Quality of Forage

Silage can preserve most of the nutrients of raw materials, especially vitamins and minerals, and after fermentation, the palatability and digestibility of feed are improved, thus improving the quality and nutritional value of forage.

1.2.3 Convenient Forage Preservation

Silage is easy to store and transport, which can preserve the nutritional value of forage for a long time and provide a stable supply of forage for the animal husbandry industry.

1.2.4 Promote the Development of Animal Husbandry

The application of silage makes the animal husbandry industry able to develop more stably and improves the production efficiency and economic benefits of the animal husbandry industry.

1.3 Conclusion

In conclusion, the scale of China's silage industry continues to expand, the market size is growing steadily, and the application and importance of silage in animal husbandry are becoming more and more prominent. In modern animal husbandry, silage is an important source of diet, which can effectively maintain the original nutrient composition of grass, and also breaks through the seasonal limitation of grass feeding under natural conditions, which plays an important role in the current development of animal husbandry. In the future, with the continuous progress of technology and the continuous support of policy, the silage industry will continue to maintain a rapid development trend, providing strong support for the sustainable development of animal husbandry. Therefore, we should increase the scientific research on silage and explore the high-quality silage production process, so as to promote the healthy, sustainable, and green development of China's livestock industry. We have reason to believe that the silage industry will usher in a broader development space and a better development prospect under the background of the sustainable development of China's livestock industry.

2 Fermented Total Mixed Ration Production and Livestock Raising in China

Yang Cao

Total mixed ration (TMR) is a nutritionally balanced daily feed tailored to the nutritional needs of ruminants by mixing concentrate feed, roughage, and feed additives in the required proportions. TMR provides balanced nutrition for ruminants, enhances their growth performance, and improves feed conversion efficiency. However, due to its moisture content, TMR is difficult to transport over long distances and requires specialized equipment for production, which limits its application in small-scale farming. By thoroughly mixing TMR with fermentation agents, compressing the mixture, and sealing it with plastic stretch film to create an anaerobic environment, fermented total mixed ration (FTMR) can be produced. Anaerobic fermentation of TMR can increase the energy and protein content of the ration, enhance the probiotics content, inhibit the growth of harmful microorganisms in FTMR, and improve the feed's palatability. This results in a nutritionally balanced, quality-stable feed that can be stored and transported over long periods. When applied in the farming of ruminants such as cattle and sheep, FTMR can improve animal production performance, promote healthy growth, and enhance milk yield and meat quality. Compared to traditional TMR preparation, the FTMR preparation process involves a higher degree of mechanization, saving substantial labour costs and effectively increasing feeding efficiency.

Non-conventional feed refers to less commonly used or unusual feeds, including industrial by-products, energy crops, and certain waste materials from agricultural and industrial production. These feeds are rich in nutrients and, with proper processing, can be used in animal production, reducing production costs and conserving feed resources (Cao et al. 2016). In China, annual production of non-conventional feeds exceeds 1.087 billion tons, with straw feed accounting for a significant portion at approximately 670 million tons and various by-products like brewers' grains reaching over 30 million tons. However, the utilization rate of non-conventional feeds in feed is relatively low, and improper handling of these feeds can lead to environmental pollution. Therefore, rational development and utilization of non-conventional resources to replace part of the grain used for animal feeding can help alleviate the current resource shortage and promote livestock development. Currently, industrial by-products and energy crops are commonly used in FTMR (Qiu et al. 2019).

Industrial by-products and some waste products from food production (such as brewers' grains and apple pomace) can be used to prepare FTMR for feeding beef cattle. Adding industrial by-products like soy sauce lees, rice husks, persimmon peels, pumpkin seed by-products, and mushroom residues to FTMR can effectively improve its quality, enhance palatability, and boost the growth performance of ruminants (Long et al. 2023; Yang et al. 2021). However, the appropriate addition levels vary for different materials, with soy sauce lees and mushroom residues being suitable at 30%, persimmon peels and pumpkin seed by-products at 20%, and rice husks at 5%. Rational use of industrial by-products in FTMR not only improves fermentation quality and ruminant production performance but also conserves feed resources and increases farming efficiency.

Energy crops such as corn and sugar beets, typically used for fuel production, can also be ingredients in fermentation mixed rations (Cao et al. 2011). Energy crops like oilseed rape straw, corn stover, and wet corn can replace roughage in ruminant diets. The replacement amount depends on the nutritional value of the specific energy crop. For example, the best fermentation results are achieved when adding 7.5% oilseed rape straw. Using 13% corn stover and 5% wet corn as substitutes for alfalfa in feed results in optimal fermentation, maintaining effective nutrient content and improving nitrogen balance. When using only straw to replace roughage, a replacement level of 40% yields high fermentation quality and aerobic stability, maximizing straw utilization. Proper use of energy crops in FTMR can save high-quality roughage resources while ensuring good fermentation quality.

In China, increasing beef and mutton consumption and the consequent growing demand for cattle or sheep have become major issues, necessitating to maximize beef or mutton production efficiency. Current research in beef cattle or sheep breeding has focused on solving these issues. However, in most of China, severe grassland degradation has led to a decrease in grasses such as Chinese wild rye (*Leymus chinensis* (Trin.) Tzvel.), an important forage grass for grazing animals, leading to shortages from winter to early spring when snow covers the ground; for this reason, the fodder provided by the grassland was not enough for the local need of the livestock industry. It is therefore essential to find an alternative forage resource to make up for the lack of forage grass. Corn accounts for 30–40% of all cultivated areas in China, and corn stover is the main crop residue in most of China with a yield of approximately 6.40×10^8 t year^{-1}. Dry corn stover is an abundant by-product of corn grain harvesting, although in China, some is used as roughage for ruminants by ensiling, ammoniating, or drying; it is also used as fuel for cooking or heating by baling or stacking. Due to economic and environmental concerns, there is an increasing demand for the efficient use of crop residues, including corn stover. Increased production costs in particular have forced commercial cattle and sheep producers to minimize costs wherever possible. Especially in villages where most of the land plant is corn, there is almost no grass for grazing, or rest grazing and yard feeding were carried out according to the national policy on protecting ecological environment of grassland and inverting cultivation to forest or grassland and where villagers use a dry corn stover as forage feeding ruminants. Dry corn stover is high in fibre and challenging for animals to digest. When fed in its dry, hard state, it has the drawback of poor intake by ruminants due to its low moisture content and large particle size, which makes it difficult for them to chew. Thus, the effective utilization of dry corn stover is an important issue in these regions. When fresh TMR was prepared with corn stover to feed ruminants, because of its hard and poor palatability, some of corn stover was refused. Cao et al. (2011) reported that ensiling could increase the digestibility of the TMR for ruminants and that the unpalatable by-products or crop residues could be incorporated into TMR and that adjusting the moisture content by adding water and lactic acid bacteria (LAB) to ensile the fermented TMR with dry corn stover would improve preference and digestibility of the corn stover in the TMR. Although in general fermented TMR is usually prepared using materials such as food by-products containing high moisture, but except for the corn stover, there are almost no other food by-products used in these regions. And it was tried to prepare fresh TMR with corn stover for some local farms in a feed manufacturing company; during transportation or storage, the fresh TMR would ferment, and then fermented TMR were produced. It was indicated that the chop length of forage grass influenced the intake characteristics and digestibility as well as the fermentation quality of the silage and suggested that 50-mm-long Chinese wild rye was feasible in fresh TMR, and they considered the effects of forage size in fresh TMR on the characteristics of digestion in dairy cows (Bhandari et al. 2008).

FTMR is typically made using high-moisture food by-products, but aside from corn stover, few other food by-products are used in these regions (Cao et al. 2009). A feed manufacturing company attempted to prepare fresh TMR with corn stover for some local farms. However, during transportation or storage, the fresh TMR would ferment, resulting in FTMR. Studies have shown that the chop length of forage grasses affects both the intake characteristics and digestibility, as well as the fermentation quality of silage. It was suggested that using 50-mm-long Chinese wild rye in fresh TMR is feasible, and the impact of forage size on digestion characteristics in dairy cows was also considered (Liu et al. 2016).

The length of forage used in FTMR should consider both its digestive and fermentative characteristics (Lv et al. 2020). At present, it was clear that fermented TMR has a moderate amount of moisture and a higher lactic acid content (LAC) and produces high-quality TMR silage. FTMR products can be packaged as bagged FTMR and canned FTMR.

Table 14.2 displays the chemical composition of fresh TMR and FTMR, which were prepared using a compound feed, ground corn grain, and soybean meal, with varying chopping lengths of Chinese wild rye or corn stover. FTMR contained lower OM, NFC, and NDF, but higher EE contents compared with fresh TMR. Comparisons of the different types of roughage in TMR showed no significant differences in OM, CP, EE, NDF, ADF, and GE. Except for FTMR with 50 mm

Table 14.2 Chemical composition of fresh or fermented total mixed ration

	DM	OM	CP	EE	NFC	NDF	ADF	GE
	($g\ kg^{-1}$)	($g\ kg^{-1}$ DM)						($MJ\ kg^{-1}$ DM)
Contrast of fresh and fermented TMR								
Fresh TMR	916‡	935†	140	25‡	187†	582†	379	18.6
FTMR	424†	924‡	141	32†	176‡	555‡	366	18.9
Contrast of TMR with Chinese wild rye and chopped corn stover								
Chinese wild rye	678	927	144	31	145§	607	383	19.4
10 mm	668	941	144	28	208†	554	371	18.4
30 mm	668	930	140	28	199†‡	550	370	18.5
50 mm	666	919	133	29	175‡	561	366	18.7
SEM	2.89	3.68	4.21	1.36	3.12	8.73	3.77	0.24
Significance of main effects and interactions								
Fermentation (Fe)	<0.01	0.005	0.063	<0.01	0.042	<0.01	0.270	0.089
Forage (Fo)	0.270	0.104	0.126	0.073	0.035	0.567	0.534	0.843
Fe × Fo	0.554	0.651	0.813	0.499	0.044	0.477	0.311	0.216

†, ‡, §Means within volume with symbols significantly differ ($P < 0.05$), $n = 6$
ADF acid detergent fibre, *CP* crude protein, *DM* dry matter, *EE* ether extract, *GE* gross energy, *NDF* neutral detergent fibre, *NFC*, nonfibrous carbohydrate = OM − CP − EE − NDF, *OM* organic matter, *SEM* standard error of the mean, *TMR* total mixed ration

corn stover, all others with corn stover had higher NFC than TMR with Chinese wild rye. The interaction of fermentation and roughage influenced NFC.

In sheep feeding trials (Table 14.3), replacing Chinese wild rye with corn stover in FTMR influenced both fermentation and roughage, affecting the apparent digestibility of EE and digestible CP (DCP). Although CP digestibility was not influenced by fermentation, it was influenced by forage type. aNDFom digestibility was influenced by fermentation but was not influenced by roughage. The apparent digestibility of EE and aNDFom for FTMR was higher than that for fresh TMR. ADFom also tended to be higher in FTMR than in fresh TMR.

In cattle feeding trials (Table 14.4), animals fed HCT (high-level corn stover) in the addition of FTMR showed significantly higher apparent digestibility of CP, EE, NDF, and ADF compared to those fed LCT (low-level corn stover). In the control groups, the apparent digestibility of CP and EE in animals fed LCT was higher than those fed HCT. It was observed that probiotic supplementation had an effect on EE digestibility. In the control groups, animals fed HCT had lower EE digestibility than those fed LCT, whereas the opposite trend was observed in the addition groups.

Growth performance in finishing beef cattle fed TMR containing different levels of corn stover supplemented with or without probiotics is shown in Table 14.5. In the FTMR addition groups, animals fed HCT showed significantly higher FBW and ADG compared to those in the control groups. Animals in the addition groups exhibited higher DM intake, lower IBW, and a better feed-to-gain ratio compared to those in the control groups, regardless of the corn stover level. Overall, with the supplementation of corn stover and probiotics in the TMR, the economic benefit for finishing cattle increased by ¥201.03 compared to the control groups.

Table 14.3 Nutrient digestibility and nutrient content in fresh or fermented total mixed ration fed to sheep

	Apparent digestibility (%)						
	DM	OM	CP	EE	NDF	ADF	GE
Fresh TMR	71.9	73.9	66.1	72.4‡	73.2‡	69.5	73.7
Fermented TMR	75.7	78.2	67.1	83.3†	78.5†	75.7	74.8
Chinese wild rye	69.7	71.6	67.8†‡	64.0‡	74.8	67.9	71.3
10 mm	77.8	79.9	70.0†	89.6†	80.0	76.6	76.7
30 mm	74.5	76.9	65.9†‡	81.4†	74.8	73.6	74.3
50 mm	73.1	75.9	63.1‡	76.3†‡	73.8	72.3	74.7
SEM	3.48	3.37	4.54	5.61	3.56	4.13	3.46
Fermentation (F)	0.147	0.086	0.174	0.015	0.044	0.050	0.127
Roughage (R)	0.183	0.143	0.039	0.003	0.324	0.243	0.141
F × R	0.570	0.687	0.457	0.375	0.254	0.691	0.572

†, ‡Means within volume with symbols significantly differ ($P < 0.05$), n = 6
ADF acid detergent fibre, *CP* crude protein, *DM* dry matter, *EE* ether extract, *GE* gross energy, *NDF* neutral detergent fibre, *NFC* nonfibrous carbohydrate = OM − CP − EE − NDF, *OM* organic matter, *SEM* standard error of the mean; TMR, total mixed ration

Table 14.4 Apparent digestibility in finishing beef cattle fed TMR containing different levels of corn stover supplemented with or without probiotics

	DM	OM	CP	EE	NDF	ADF
	g/kg	g/kg of DM				
Control						
High	628.2	702.2	587.4^{Bb}	489.4^{Bb}	459.4^{b}	430.0^{b}
Low	628.4	746.2	666.5^{A}	636.9^{A}	447.8	399.0
Addition						
High	645.7	786.4	760.6^{a}	724.5^{a}	598.6^{a}	634.4^{a}
Low	631.8	776.8	687.0	671.8	492.1	466.4
SEM	8.01	25.79	37.73	37.71	52.91	56.09
P value						
A	0.2282	0.0770	0.0342	0.0094	0. 0264	0.0103
S	0.4184	0.4931	0.0122	0.0323	0.2901	0.2685
A × S	0.4051	0.2891	0.2474	0.0314	0.5804	1.0000

DM dry matter, *OM* organic matter, *CP* crude protein, *EE* ether extract, *NDF* neutral detergent fibre, *ADF* acid detergent fibre, *SEM* standard error of the means, *A* addition of probiotic, *S* level of corn stover
Different lowercase letters in the same column indicate that there is significant difference between different treatments under the same corn stover proportion ($P < 0.05$), marked with different capital letters in the same column indicating that there was significant difference between different corn stover proportions under the same treatment ($P < 0.05$)

In conclusion, FTMR utilizes fermentation technology to process many unconventional resources that cannot be directly fed to animals. The success and effectiveness of fermentation are influenced by various factors, requiring appropriate formulation and additives for different feed recipes to ensure high-quality

Table 14.5 Growth performance in finishing beef cattle fed TMR containing different levels of corn stover supplemented with or without probiotics

	DMI	ADG	F:G	Feed cost	Sales income	Economic benefit
	kg/d		kg/kg	¥/head		
Control						
High	8.03	1.24[b]	6.48	983.0	2885.3	1902.3
Low	8.23	1.37	6.00	1040.0	2970.7	1930.7
Experimental						
High	8.25	1.56[a]	5.29	1263.9	3367.2	2103.3
Low	8.36	1.48	5.65	1309.4	3202.7	1893.3
SEM	0.909	0.046	0.795	–	–	–
P value						
A	0.8529	0.0269	0.4901	–	–	–
S	0.8725	0.7719	0.7847	–	–	–
A × S	0.9563	0.3717	0.8490	–	–	–

IBW initial body weight, *FBW* final body weight, *DMI* dry matter intake, *ADG* average daily gain, *F:G* feed to gain ratio, *SEM* standard error of the means, *A* addition of probiotic, *S* level of corn stover

Different lowercase letters in the same column indicate that there is significant difference between different treatments under the same corn stover proportion ($P < 0.05$), marked with different capital letters in the same column indicating that there was significant difference between different corn stover proportions under the same treatment ($P < 0.05$)

fermentation. As a novel feed formulation technology, FTMR has promising development potential for small- to medium-sized farms or family-operated farms. It allows effective use of unconventional resources in feed production, conserves roughage resources, reduces feeding costs, improves ruminant production performance, and enhances livestock product quality. It also addresses the issues of poor stability and commercialization challenges associated with traditional TMR (Shin et al. 2021). However, FTMR still faces challenges in production and application, such as labour-intensive packaging, time-consuming unwrapping, and inconvenient distribution. Therefore, further improvements in fermentation technology and related equipment are essential to fully realize the production potential of FTMR.

References

Bhandari SK, Li S, Ominski KH, Wittenberg KM, Plaizier JC. Effects of the chop lengths of alfalfa silage and oat silage on feed intake, milk production, feeding behavior, and rumen fermentation of dairy cows. J Dairy Sci. 2008;91:1942–58.

Cao Y, Takahashi T, Horiguchi K. Effects of addition of food by-products on the fermentation quality of a total mixed ration with whole crop rice and its digestibility, preference, and rumen fermentation in sheep. Anim Feed Sci Technol. 2009;151:1–11.

Cao Y, Cai Y, Takahashi T, Yoshida N, Tohno M, Uegaki R, et al. Effect of lactic acid bacteria inoculant and beet pulp addition on fermentation characteristics and in vitro ruminal digestion of vegetable residue silage. J Dairy Sci. 2011;94(8):3902–12.

Cao Y, Zang Y, Jiang Z, Han Y, Hou J, Liu H, et al. Fermentation quality and nutritive value of fresh and fermented total mixed rations containing Chinese wild rye or corn Stover. Grassl Sci. 2016;62(4):213–23.

Helander C, Nørgaard P, Zaralis K, Martinsson K, Murphy M, Nadeau E, et al. Effects of maize crop maturity at harvest and dietary inclusion rate of maize silage on feed intake and performance in lambs fed high-concentrate diets. Livest Sci. 2015;178:52–60.

Liu Q, Li X, Desta S, Seare T, Zhang J, Shao T. Effects of *Lactobacillus plantarum* and fibrolytic enzyme on the fermentation quality and in vitro digestibility of total mixed rations silage including rape straw. J Integr Agric. 2016;15(9):2087–96.

Long Y, Huang W, Han Y, Yuan C, Wang S, Wang D, et al. Effects of mushroom Hypsizygus marmoreus residue fermentation total mixed rations on feeding and rumination behavior of growing male lambs in Guizhou black goats. Feed Ind. 2023;44(1):65–73.

Lv R, Zhang L, Wang C. Effects of fermented total mixed ratio with different adding ratios rice husks on growth performance of Hainan black goats. Chin J Anim Sci. 2020;56(5):117–21.

Mahmoud SA, Abdel-Hafez A, Zaki MM, Saleh EA. Factors affecting the microbial and chemical composition of silage. Zentralbl Bakteriol Naturwiss. 1979;134(1):34–9.

Nishino N, Wada H, Yoshida M, Shiota H. Microbial counts, fermentation products, and aerobic stability of whole crop corn and a total mixed ration ensiled with and without inoculation of *Lactobacillus casei* or *Lactobacillus buchneri*. J Dairy Sci. 2004;87(8):2563–70.

Qiu X, Xi K, Xie X, Tian Y, Wu X, Zeng J, et al. Fermentation quality and aerobic stability of the total mixed ration after replacing winged bean with straw. Food Ferment Ind. 2019;45(24):98–104.

Shin M, Ban O, Jung Y, Yang Y, Kim Y. Genomic characterization and probiotic potential of *Lactobacillus casei* IDCC 3451 isolated from infant faeces. Lett Appl Microbiol. 2021;72(5):578–88.

Yang M, Cai Y, He Z. Study on fermentation quality, in vitro rumen fermentation of total mixed diet supplementation of persimmon peel and its application in beef cattle. China Feed. 2021;20:103–6.

Part IV
Advances in Silage Science and Technology

Chapter 15
Progress in Microbial Research Methods for Silage

Zhumei Du

Abstract In recent years, the research methods of silage microorganisms have made remarkable progress, from traditional culture methods to comprehensive analysis based on molecular biology and multi-omics techniques. 16S rRNA high-throughput sequencing, metagenomics and transcriptomics revealed the diversity and functional characteristics of silage microbial communities, while metabolomics analysed the dynamic changes of microbial metabolites and their effects on fermentation quality. In addition, ecological network analysis provides a new perspective for the study of microbial interactions and population dynamics. The combination of these methods laid a solid foundation for revealing the silage fermentation mechanism and optimizing the silage process.

Keywords Microbial diversity · Microbiome · Molecular taxonomy · Silage fermentation · 16S rRNA sequencing

1 Traditional Microbial Culture and Classification Methods

Silage fermentation is a complex process through a variety of microbial metabolic action of mutual collaboration or mutual inhibition, in which lactic acid bacteria (LAB) play a key role in the fermentation process. Microorganisms, such as clostridia, yeasts, moulds and other microorganisms, can also affect the quality of silage fermentation with the change of fermentation conditions (Cai et al. 2020; Du et al. 2022, 2024). Traditional microbial culture methods, such as solid and liquid cultures, can isolate and identify some microbial species in silage by selective media and culture conditions. The plate counting method is a traditional method to determine the number of viable microorganisms in the sample by using agar medium, which is based on the number of microorganisms contained in the sample; carry out a series of gradient dilutions, so that the microorganisms are dispersed into

Z. Du (✉)
Yangzhou University, Yangzhou, China

Y. Cai, K. Ataku (eds.), *Cultural History and Modern Production Technology of Silage*, https://doi.org/10.1007/978-981-96-5787-2_15

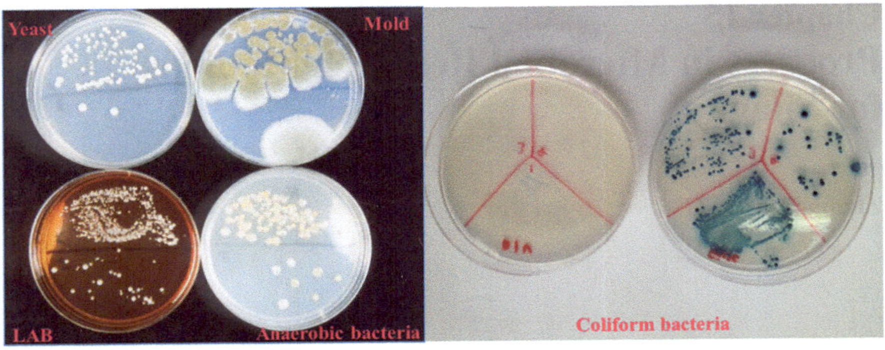

Fig. 15.1 Microbial colonies cultured in different media. (Photos by Cai)

individual cells; and then take a certain amount of diluted bacterial liquid inoculated onto the plate culture medium (Green and Goldman 2021). After incubation and under suitable temperature conditions, individual cells grow and multiply to form colonies. The number of microorganisms in the silage sample is calculated by observing the number of colonies formed. In the study of microorganisms related to silage fermentation (Du et al. 2024), the commonly used media for LAB, aerobic bacteria, coliform bacteria and yeasts (including moulds) are de Man, Rogosa and Sharpe agar (MRS, Difco Laboratories, Detroit, MI, USA), Nutrient agar (NA, Nissui-Seiyaku Co., Ltd., Tokyo, Japan), Blue Light agar (BLA, Nissui-Seiyaku) and Potato Dextrose agar (PDA, Nissui-Seiyaku).

Plate counting is a classic microbial enumeration method that is simple, inexpensive, and widely used (Cai 1999; Kozaki et al. 1992). This method is usually classified according to bacterial morphology and physiological and biochemical traits and plays an important role in the study of silage microbial ecology (Cai et al. 1998), population composition, quality management and feed safety. However, this method is limited in that it can only count the number of viable and culturable microorganisms. Currently, most microorganisms in nature are still unable to grow on the available media, and the plate count method does not accurately reflect the full picture of microbial ecology. Figure 15.1 shows colonies of lactic acid bacteria, aerobic bacteria, yeast and mould cultured by plate culture method using MRS, NA and PDA agar media.

2 The Application of Modern Molecular Biology Technology

Molecular taxonomy of microorganisms is a classification technique based on microbial genetic information and molecular signatures.

It identifies and classifies microorganisms by means of molecular biology, especially gene sequencing and analysis techniques. Though the traditional culture and classification method are difficult to comprehensively cover the microbial diversity in the silage environment, molecular taxonomy has become an indispensable and important tool in the current silage microbiology research due to its advantages of high throughput and high accuracy. The core of microbial molecular taxonomy lies in taxonomic identification, construction of evolutionary relationships and analysis of ecological functions through similarities and differences in gene sequences. These methods use signature genes (e.g. 16S rRNA, 18S rRNA, ITS region) as the basis of classification and are supplemented with computational biology analysis tools to construct evolutionary trees and classification systems of silage fermentation microorganisms.

2.1 *Molecular Taxonomy of Microorganisms*

16S rRNA Gene Sequence Analysis

16S rRNA gene sequence analysis is a commonly used method for molecular classification of bacteria (Ennahar et al. 2003). 16S rRNA gene is highly conserved in bacteria and archaea that contains several highly variable regions, which is suitable for classification and identification of bacteria and archaea (Cho et al. 2007). 16S rRNA gene sequence analysis can provide a basis for genus and even species identification (Kimura and Ohta 1972). The 16S rRNA gene sequencing is used to classify bacteria and archaea based on the homology and phylogenetic relationship of their 16S rRNA gene sequences (Pang et al. 2011). The analysis process involves extracting microbial DNA, amplifying specific regions of the 16S rRNA gene, sequencing and performing sequence comparisons. This is followed by annotation using databases (e.g. SILVA, Greengenes and RDP). The 16s RNA base sequence analysis is performed using the BLAST programme to search for 16S rDNA sequence similarities in the GenBank database, and then the sequence information is introduced into the CLUSTAL W software programme for assembly and comparison (Duan et al. 2008; Tan et al. 2010). By the current microbial classification standard, the 16S rDNA identity coefficient between two species is 97%, or less may be different species. Figure 15.2 shows the phylogenetic tree of LAB isolated from vegetable residue silage (Pholsen et al. 2016; Liu et al. 2014; Ni et al. 2015). Isolates NJ 414 and NJ 415 were the closest relatives to the type strain *Lactococcus fujiensis*, and the 16S rRNA sequences of all three were more than 99% similar to each other, so these two isolates were identified as *Lactococcus fujiensis* (Cai et al. 2011).

However, the 16S rDNA base sequences between some closely related strains were highly similar, and some strains could be as high as 99% or more, which could not be identified to the strain level based on the 16S rRNA gene base sequence analysis method, and therefore other microbial classification methods, such as DNA-DNA hybridization, were needed to continue further identification or classification (Ezaki et al. 1989). Figure 15.3 shows the analytical procedure for 16S rRNA sequencing and DNA-DNA hybridization. These methods are now widely used in the study of silage microbiology.

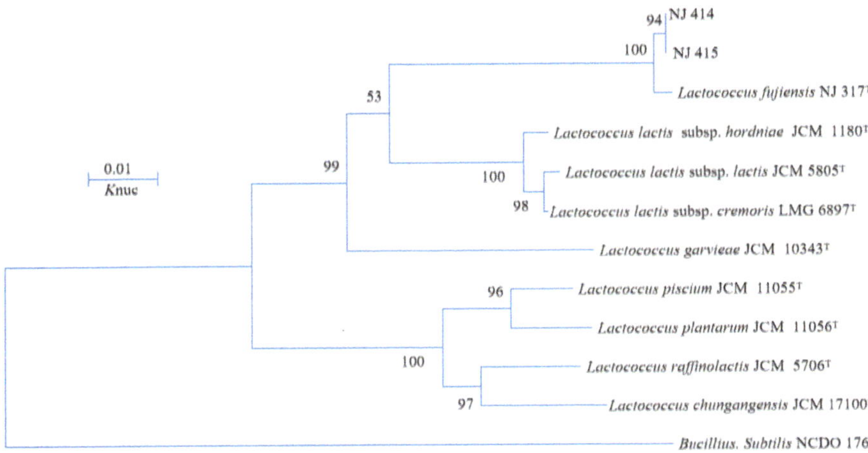

Fig. 15.2 Phylogenetic tree showing the relative positions of vegetable isolates NJ 414 and NJ 415 as inferred by the neighbour-joining method of complete 16S rDNA sequences. Bootstrap values for a total of 100 replicates are shown at the nodes of the tree. *B. subtilis* is used as an outgroup. The bar indicates 1% sequence divergence. Knuc, nucleotide substitution rates (isolate analysis data from Cai)

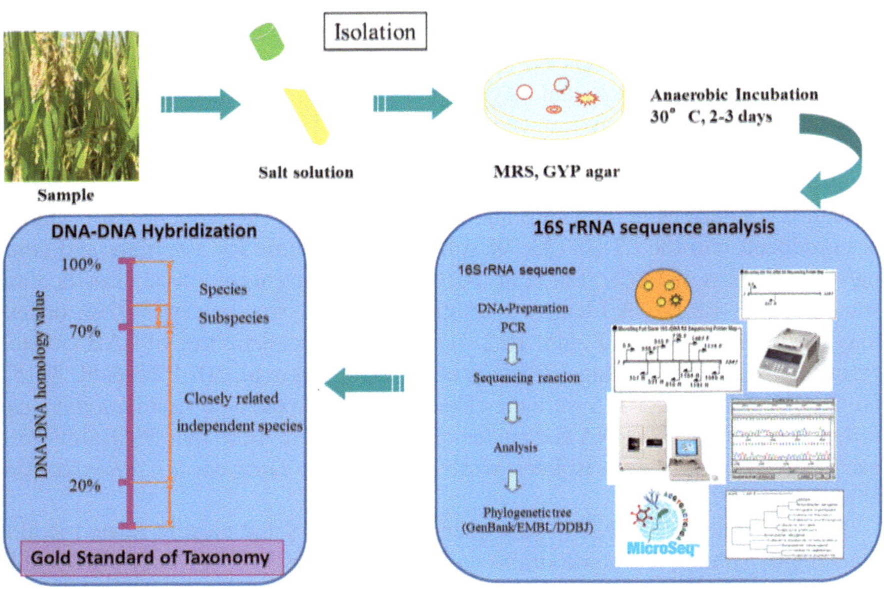

Fig. 15.3 16S rDNA sequence and DNA-DNA hybridization analysis

Sequence Analysis of 18S rRNA Gene
The 18S rRNA gene sequence is mainly used for the identification of eukaryotic microorganisms (e.g. fungi, protozoa) (Janson et al. 2000). Similar to the 16S rRNA gene sequence analysis method, the 18S rRNA gene also contains conserved and variant regions, which is suitable for the molecular classification of eukaryotic microorganisms (Al Sheikh et al. 2014). The analytical procedure is the same PCR amplification of the 18S rRNA gene and taxonomic annotation using databases (e.g. GenBank, SILVA). This method is suitable for diversity studies of eukaryotic microorganisms in silage environmental samples. However, this method is difficult for the identification of some low abundance species and may be interfered by other coexisting microorganisms.

Internal Transcribed Spacer (ITS) Region Sequence Analysis
The ITS region is located between two parts of the ribosomal genes of fungi, with a high degree of variability, and is the main marker gene for molecular classification of fungi. The ITS region is more conserved within the species and has a high degree of interspecies variability, which makes it suitable for fungal taxonomic identification. The process is to amplify the ITS region by PCR and annotate it with databases (e.g. UNITE, NCBI). It is characterized by high specificity and is particularly effective for species-level classification. However, this method suffers from PCR amplification bias and is not representative enough for some complex silage environmental samples.

Genome Analysis
Whole genome sequencing (WGS) provides richer information than single genes and allows high-resolution classification through the complete sequence of the genome. WGS not only identifies species but also provides information on gene function, metabolic pathways, etc. Its process includes sample DNA extraction, whole genome sequencing, assembly and annotation and finally classification and functional analysis. It is characterised by high resolution and can provide rich genetic information. However, this method has the disadvantages of high cost, complex data analysis and time-consuming processing and is less used in the field of silage science.

Macrogenome Sequencing
Macrogenome sequencing is suitable for the study of all microbial genomes in complex silage environments. It can directly obtain genomic information of different microorganisms in a community without isolating the cultured microorganisms. The process involves extracting all the DNA from an environmental sample, constructing a macrogenomic library, sequencing and bioinformatics analysis to obtain microbial taxonomic and functional information. It is characterised by its ability to describe the diversity and function of all microorganisms in the sample and is a powerful tool for classification of environmental microorganisms. However, this method has disadvantages such as the difficulty of genome splicing and the inability to accurately locate to species-level classification.

DNA-DNA Hybridization

DNA-DNA hybridization is a traditional molecular biology technique originally used to determine the relatedness of microbial species. DNA-DNA hybridization compares the genomes of two DNA molecules by measuring the degree of hybridization, i.e. the degree of pairing between the DNA strands of the two molecules under specific and similar conditions (Sang and Huang 2012). The process is that when the sequences of two DNA molecules are highly similar, they form a more stable double-stranded structure during hybridization and vice versa; hybridization is less efficient. The basic principle of this technique is based on the complementarity of DNA sequences, which allows the quantification of similarities between genomes from different species or strains. This method has become the ultimate standard for microbial identification and is also known in the industry as the gold standard for microbial strain identification. Going by the current Bergy classification criteria, DNA identity coefficients above 70% are considered to be the same species, and below 70% are considered to be different species (Fig. 15.3).

DNA-DNA hybridization was applied to the study of silage microorganisms in 1999, and the population structure and ecological mechanism of silage fermentation microorganisms were deeply explored, and a new species of lactic acid bacteria isolated from silage was published (Cai et al. 2012; Li et al. 2015; Tamaoka and Komagata 1984). Table 15.1 shows the results of DNA-DNA hybridization experiments of lactic acid bacteria isolated from vegetable residue silage. The G + C content of the isolates NJ414 and NJ415 was similar to that of the type strain *Lactococcus fujiensis* NJ 317, and the homology of DNA-DNA hybridization between the three

Table 15.1 DNA base composition and DNA relatedness of lactic acid bacteria isolated from vegetable residue silage

Strain	G + C content (mol %)	DNA-DNA reassociation (%) with NJ 317^T JCM 5805^T JCM 10343^T		
NJ 414	42.5	98.2	17.6	9.8
NJ 415	42.2	96.6	17.7	12.4
Lactococcus fujiensis NJ 317^T	42.1	100	15.4	15.6
Lactococcus lactis subsp. *lactis* JCM 5805^T	39.4	20.2	100	13.2
Lactococcus lactis subsp. *hordniae* JCM 1180^T	nd	11.7	85.6	10.4
Lactococcus lactis subsp. *cremoris* JCM 16167^T	nd	9.4	80.2	17.0
Lactococcus garvieae JCM 10343^T	44.3	12.3	8.4	100
Lactococcus piscium JCM 11055^T	38.5	8.6	7.8	18.6
Lactococcus plantarun JCM 11056^T	nd	10.3	10.3	8.6
Lactococcus raffinolactis JCM 5706^T	41.5	11.3	12.4	10.0

T type strain, *nd* no data, *JCM* Japan collection of microorganism (isolate analysis data from Cai)

of them was higher than 96.6% or more, and with the other type strains, it was lower than 17.6% below. Therefore, isolates NJ 414 and NJ 415 could be identified as the same strain as *Lactococcus fujiensis* (Cai et al. 2012).

3 Combination of Culture and Unculture Techniques

With the development of modern molecular techniques, the combination of culture and unculture methods has become an important trend in the study of silage microorganisms. This combination not only obtains data on the diversity of microorganisms but also provides pure culture of active strains for further functional studies.

3.1 Single-Cell Isolation and Culture

Single microbial cells from silage samples can be isolated and cultured by micromanipulation techniques or flow cytometric sorting, which is particularly suitable for isolation of low abundance or specialised microorganisms. Single cell isolation has helped to isolate some unique or undiscovered strains in silage research, providing a rich strain resource for silage microbiological research.

3.2 Enrichment Culture and High-Throughput Sequencing

Enrichment culture is a method for preferentially cultivation of certain types of microorganisms under specific environmental conditions, and then in combination with high-throughput sequencing technology, it can provide a deeper understanding of the community structure and function of the dominant microorganisms in the silage fermentation process (Liu et al. 2014; Du et al. 2021). This method overcomes the limitations of the traditional culture method and is able to obtain information on both cultured and uncultured microorganisms, resulting in a more comprehensive analysis of the silage microbial community. Figure 15.4 shows the schematic illustration of high-throughput sequencing and single-molecule real-time sequencing (Gao et al. 2022). In the study of silage microorganisms, high-throughput sequencing usually identifies to the genus level, whereas single-molecule real-time sequencing identifies to the species level, which better reflects the whole picture of microbial community structure in silage.

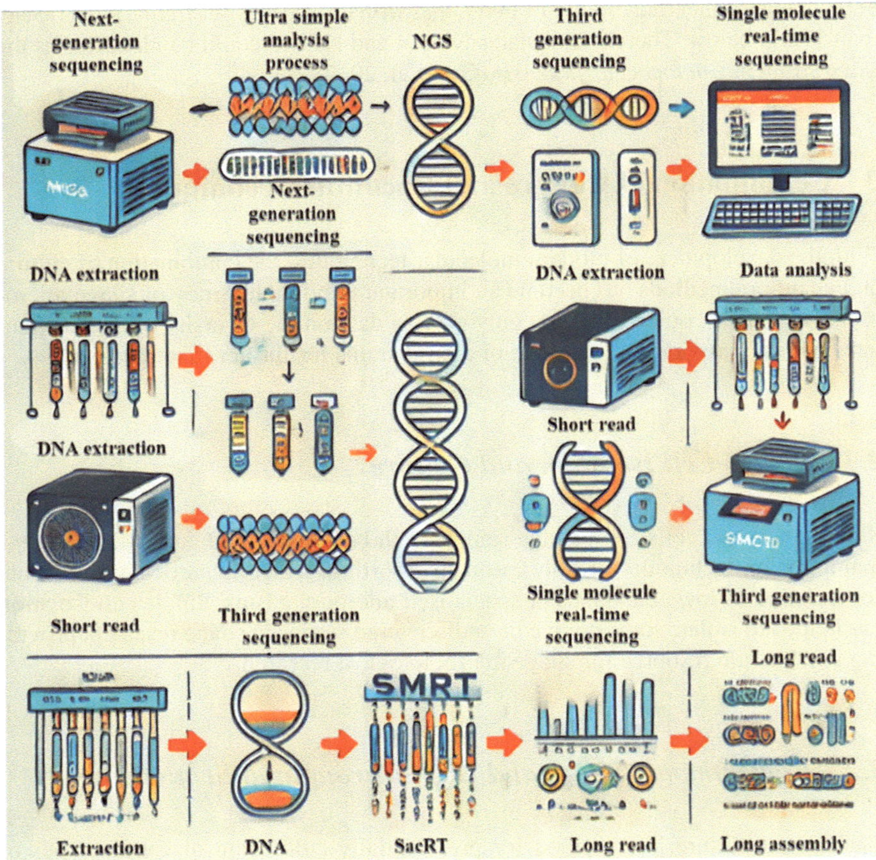

Fig. 15.4 High-throughput sequencing and single-molecule real-time sequencing created using AI mapping software

3.3 *Stable Isotope Probe*

Stable isotope probing is a technique that uses stable isotope labelled substrates to track the metabolic activity of microorganisms. By culturing microorganisms in silage samples and adding isotope-labelled substrates, it is possible to detect microorganisms active in specific metabolic pathways. Stable isotope probing technique helps researchers to identify metabolically active microbial populations during ensiling and reveals their specific functions in silage fermentation. Currently, this method of analysis is costly and has not been widely used in silage research.

4 Development of Data Analysis Techniques

With the popularity of unculture technologies such as microbiomics and metabolomics, advances in data analysis techniques play an important role in silage microbiology research. Complex microbial data in silage can be processed and interpreted using bioinformatics analysis methods.

Microbial community diversity analysis: Alpha diversity (species richness, evenness, etc.) and beta diversity (variation between samples) analysis are widely used in silage microbial community research. Through these indicators, the microbial diversity in silage can be quantitatively described, providing a reference for optimizing fermentation conditions.

Functional gene analysis: Functional gene analysis tools (e.g. KEGG, COG and EggNOG) based on macro-genomic data can help in identifying functional genes in microorganisms during silage making and infer microbial metabolic pathways. Functional gene analysis allows further optimization of the silage fermentation process, e.g. to enhance lactic acid fermentation or to inhibit the activity of spoilage microorganisms (Li et al. 2015).

Network analysis and machine learning: Co-occurrence network analysis can reveal the co-existence of microbial populations in the fermentation process and the association between microbes and fermentation products of silage. Machine learning methods (e.g. random forests, support vector machines) can be used in silage microbiology research to predict the function of key microbes and their populations as well as the optimization of the fermentation process.

5 Analysis Process and Data Interpretation of Microbial Molecular Taxonomy

5.1 Data Preprocessing

It includes removing low-quality sequences and filtering short sequences and false-positive results in PCR amplification products. Data preprocessing ensures the accuracy and representativeness of subsequent analyses.

5.2 Sequence Alignment and Annotation

The sequence data of the samples are aligned with known microbial genetic information using common databases (e.g. SILVA, Greengenes, NCBI GenBank) to identify specific strains. The annotation results after sequence comparison can help to construct the composition and taxonomic hierarchy of microbial communities, which can provide an important basis for the functional analysis of silage ecosystems.

5.3 Evolutionary Tree Construction

The microbial evolutionary tree was constructed by comparing similar sequences and using phylogenetic tree construction tools (e.g. MEGA, RAxML). Evolutionary trees can help to reveal the population structure and kinship of silage microorganisms, as well as speculating the evolutionary paths (Saitou and Nei 1987).

5.4 Community Diversity Analysis

By calculating alpha diversity (e.g. species richness, evenness) and beta diversity (e.g. inter-sample variation), the structural changes and diversity differences of silage microbial community can be analysed under different environmental or treatment conditions (Chow and Suttle 2015).

5.5 Interpretation and Application of Results

The ecological functions of microorganisms in silage environments are revealed by combining the analyses of community structure and functional gene annotation. The diversity of silage microorganisms and their potential functions can be interpreted by analysis data, which can provide a reference for the preparation of high-quality silage.

6 Research and Application of Silage Microorganisms

6.1 Optimization of Silage Fermentation Process

By the analysis of the microbial community structure under different fermentation conditions, the most favourable silage conditions for beneficial microorganisms, such as LAB, are determined. For example, by regulating the microbial population structure and metabolites, the acidity of silage can be moderately increased, inhibit unfavourable microorganisms, and increase the activity of LAB, thus prolonging the preservation time of silage (Badotti et al. 2012).

6.2 Screening for Advantageous Strains for Silage Fermentation

Selection of efficient lactobacilli strains in silage fermentation can increase the rate of acid production in fermentation, lower the pH and prevent the growth of harmful microorganisms (Ma et al. 2022). Using uncultivation techniques more acid-tolerant and adaptable strains can be identified, and these advantageous strains can be utilized in production to improve the fermentation quality of silage.

6.3 Monitoring Silage Safety

The use of high-throughput sequencing technologies allows the detection of potentially harmful microorganisms during ensiling and the inhibition of their growth by modulating fermentation conditions (Gihring et al. 2012). For example, by controlling anaerobic and acidic environments, mould growth is inhibited to ensure silage fermentation quality and feed safety.

Silage research has developed greatly along with advances in microbial culture and research methods. From the traditional selective culture methods to the application of modern high-throughput sequencing and macro-genomics, the continuous progress of these technologies has made the study of silage microorganisms more comprehensive and in-depth. In the future, with the application of single-cell sequencing, SIP technology and artificial intelligence in silage microbial research, the research in this field is expected to be further enhanced, providing more scientific data support for silage fermentation process optimization, strain screening and quality control.

References

Al Sheikh YA, Marie MAM, John J, Krishnappa LG, Dabwab KHM. Prevalence of 16S rRNA methylase genes among β-lactamase-producing Enterobacteriaceae clinical isolates in Saudi Arabia. Libyan J Med. 2014;9(1):24432.

Badotti F, Gomes FCO, Rosa CA, Hui Y. Handbook of plant-based fermented food and beverage technology. Boca Raton: CRC Press; 2012. p. 639–48.

Cai Y. Identification and characterization of *Enterococcus* species isolated from forage crops and their influence on silage fermentation. J Dairy Sci. 1999;82:2466–71.

Cai Y, Benno Y, Ogawa M, Ohmomo S, Kumai S, Nakase T. Influence of *Lactobacillus* spp. from an inoculant and of *Weissella* and *Leuconostoc* spp. from forage crops on silage fermentation. Appl Environ Microbiol. 1998;64:2982–7.

Cai Y, Yang J, Pang H, Kitahara M. *Lactococcus fujiensis* sp. nov., a lactic acid bacterium isolated from vegetable. Int J Syst Evol Microbiol. 2011;61(Pt 7):1590–4.

Cai Y, Pang H, Kitahara M, Ohkuma M. *Lactobacillus nasuensis* sp. nov., a lactic acid bacterium isolated from silage, and emended description of the genus *Lactobacillus*. Int J Syst Evol Microbiol. 2012;62(Pt_5):1140–4.

Cai Y, Du Z, Yamasaki S, Nguluve D, Tinga B, Oya T, et al. Community of natural lactic acid bacteria and silage fermentation of corn stover and sugarcane tops in Africa. Anim Biosci. 2020;33(8):1252.

Cho HS, Park SY, Ryu CM, Kim JF, Kim JG, Park SH. Interference of quorum sensing and virulence of the rice pathogen Burkholderia glumae by an engineered endophytic bacterium. FEMS Microbiol Ecol. 2007;60(1):14–23.

Chow CET, Suttle CA. Biogeography of viruses in the sea. Annu Rev Virol. 2015;2(1):41–66.

Du Z, Sun L, Lin Y, Yang F, Cai Y. The use of PacBio SMRT technology to explore the microbial network and fermentation characteristics of woody silage prepared with exogenous carbohydrate additives. J Appl Microbiol. 2021;131(5):2193–211.

Du Z, Lin Y, Sun L, Yang F, Cai Y. Microbial community structure, co-occurrence network and fermentation characteristics of woody plant silage. J Sci Food Agr. 2022;102(3):1193–204.

Du Z, Yamasaki S, Oya T, Nguluve D, Euridse D, Tinga B, et al. Microbial network and fermentation modulation of Napier grass and sugarcane top silage in southern Africa. Microbiol Spectr. 2024;12(1):e03032–23.

Duan Y, Tan Z, Wang Y, Li Z, Li Z, Qin G, et al. Identification and characterization of lactic acid bacteria isolated from Tibetan Qula cheese. J Gen Appl Microbiol. 2008;54(1):51–60.

Ennahar S, Cai Y, Fujita Y. Phylogenetic diversity of lactic acid bacteria associated with paddy rice silage as determined by 16S ribosomal DNA analysis. Appl Environ Microbiol. 2003;69:444–51.

Ezaki T, Hashimoto Y, Yabuuchi E. Fluorometric deoxyribonucleic acid-deoxyribonucleic acid hybridization in microdilution wells as an alternative to membrane filter hybridization in which radioisotopes are used to determine genetic relatedness among bacterial strains. Int J Syst Bacteriol. 1989;39:224–9.

Gao P, Wu C, Zhang J, Wang S, Huang Y, Dong Y, et al. Evaluation and optimization of microdrop digital PCR for detection of serotype A and B *Clostridium botulinum*. Front Microbiol. 2022;13:860992.

Gihring TM, Green SJ, Schadt CW. Massively parallel rRNA gene sequencing exacerbates the potential for biased community diversity comparisons due to variable library sizes. Environ Microbiol. 2012;14(2):285–90.

Green LH, Goldman E. Practical handbook of microbiology. Boca Raton: CRC Press; 2021.

Janson S, Gisselson LA, Salomon PS, Granéli E. Evidence for multiple species within the endoparasitic dinoflagellate *Amoebophrya ceratii* as based on 18S rRNA gene-sequence analysis. Parasitol Res. 2000;86:929–33.

Kimura M, Ohta T. On the stochastic model for estimation of mutation distance between homologous proteins. J Mol Evol. 1972;2:87–90.

Kozaki M, Uchimura T, Okada S. Experimental manual-for lactic acid bacteria, vol. 5. Tokyo: Asakurasyouten; 1992. p. 29–72.

Li D, Ni K, Pang H, Wang Y, Cai Y, Jin Q. Identification and antimicrobial activity detection of lactic acid bacteria isolated from corn stover silage. Anim Biosci. 2015;28(5):620.

Liu W, Pang H, Zhang H, Cai Y. Lactic acid bacteria: fundamentals and practice, vol. 10. Amsterdam: Springer; 2014. p. 978–94.

Ma H, Wang L, Yu H, Wang W, Wu G, Qin G, et al. Protease-producing lactic acid bacteria with antibacterial properties and their potential use in soybean meal fermentation. Chem Biol Technol Agric. 2022;9(1):40.

Ni K, Wang Y, Li D, Cai Y, Pang H. Characterization, identification and application of lactic acid bacteria isolated from forage paddy rice silage. PLoS One. 2015;10(3):e0121967.

Pang H, Qin G, Tan Z, Li Z, Wang Y, Cai Y. Natural populations of lactic acid bacteria associated with silage fermentation as determined by phenotype, 16S ribosomal RNA and recA gene analysis. Syst Appl Microbiol. 2011;34:235–41.

Pholsen S, Khota W, Pang H, Higgs D, Cai Y. Characterization and application of lactic acid bacteria for tropical silage preparation. Anim Sci J. 2016;87(10):1202–11.

Saitou N, Nei M. The neighbor-joining method: a new method for reconstructing phylogenetic trees. Mol Biol Evol. 1987;4:406–25.

Sang H, Huang JZ. A full scale approximation of covariance functions for large spatial data sets. J R Stat Soc B. 2012;74(1):111–32.

Tamaoka J, Komagata K. Determination of DNA base composition by reversed-phase high-performance liquid chromatography. FEMS Microbiol Lett. 1984;124:11–6.

Tan Z, Pang H, Duan Y, Qin G, Cai Y. 16S ribosomal DNA analysis and characterization of lactic acid bacteria associated with traditional Tibetan Qula cheese made from yak milk. Anim Sci J. 2010;81(6):706–13.

Chapter 16
Forage Rice Production

Hiroki Matsuyama and Jong Geun Kim

Abstract Whole crop rice silage (WCRS) is a type of cattle feed made by harvesting both rice grains and straws together and then ensiling them. In Japan and South Korea, significant efforts have been made in the production and utilization of WCRS to address the increasing crop rotation and idle rice fields, rising imported feed prices, and the need for safe livestock products. Since 2000, advancement in WCRS has included the development of specialized rice varieties, harvesting machinery, and silage feed additives, which have greatly improved its fermentation quality and nutritional value. As a result, the area planted with WCRS has rapidly expanded. This chapter focuses on the varieties of WCRS, the machinery used, and the additives specifically used in Japan and South Korea. It covers aspects of harvesting and preparation, feed characteristics, and the effects of WCRS on cattle feeding.

Keywords Cattle feeding · Feed characteristics · Harvester · Rice variety · Whole crop rice silage

1 Forage Rice Variety Breeding, Cultivation, Harvesting, Preparation, and Storage, Livestock Feeding Technology in Japan

Hiroki Matsuyama

Currently, efforts are underway in Japan to cultivate forage rice in fallow fields and abandoned paddy fields, harvest the hulled rice and stover together, ferment them into whole-crop rice silage (WCRS), and feed them to dairy and beef cattle. The

H. Matsuyama (✉)
Yamagata University, Tsuruoka, Japan
e-mail: matsuyama@tds1.tr.yamagata-u.ac.jp

J. G. Kim
Seoul National University, Pyeongchang, South Korea
e-mail: forage@snu.ac.kr

Y. Cai, K. Ataku (eds.), *Cultural History and Modern Production Technology of Silage*, https://doi.org/10.1007/978-981-96-5787-2_16

area planted with forage rice was less than 100 ha until 1999; however, by 2023, the most recent year for which data are available, this area exceeded 50,000 ha. The following can be attributed to this background.

1.1 Increase in Shifting Crop Fields and Idle Rice Paddies

As the Japanese diet has become more westernized, the annual per capita consumption of livestock products (milk and meat products) has increased from 58 to 137 kg over the past 40 years, whereas rice consumption has decreased from 112 to 61 kg. As a result, rice farmers have been forced to convert nearly one million hectare of paddy fields from food rice production to other crops.

1.1.1 Increasing Prices of Imported Feed

Over the past 40 years, the number of head per livestock farmer has increased from 3.4 to 61.5 for dairy cattle and from 1.3 to 32.2 for beef cattle. One major cause for this increase in the number of cattle is the existence of imported feed that is stably supplied at a low cost. However, the current sharp rise in global grain prices and the trend toward a shortage of tradable hay are putting pressure on livestock feeding. To stabilize management, the production of self-sufficient feed is essential.

1.1.2 Provision of Safe Livestock Products

Diseases suspected to be caused by imported feed, such as bovine spongiform encephalopathy (BSE) and foot-and-mouth disease, have highlighted the importance of self-feeding. Additionally, to provide consumers with safe livestock products and peace of mind, self-supplied feed with a clear production process is necessary.

Therefore, there are high expectations for forage rice that can utilize shifting crop fields and idle rice paddies to produce fodder. The production and utilization of WCRS had been considered since the 1970s, when the policy of reducing rice acreage for food crops was introduced, but did not take root immediately; since 2000, a national project has been developing a wide range of technologies for varieties, cultivation, harvesting, preparation, and feeding. In particular, the breeding of dedicated WCRS varieties, development of dedicated harvesters, and development of silage additives have substantially improved the fermentation quality and feed value of WCRS, leading to the rapid expansion of WCRS production and utilization. In addition, technical manual has been published in Japan to support the spread of WCRS production and utilization (Japan Grassland Agriculture and Forage Seed Association 2020). This chapter introduces the dedicated WCRS breeds, machinery, and additives involved in harvesting and preparation, forage characteristics, and feeding effects on cattle in Japan.

1.2 Rice Varieties for WCRS

WCRS rice varieties suitable for cultivation in various regions, from Hokkaido to Kyushu, have been bred and grown throughout Japan. These WCRS rice varieties have a higher total dry matter yield and total digestible nutrient (TDN) than food rice varieties and are less prone to collapse, even when grown with high fertilizer. Among these stover varieties are extremely short-eared stover varieties with an extremely low percentage of unhulled rice. Stem and leaf varieties are mostly late maturing and can be used from Kanto to western Japan, whereas conventional varieties are used from Hokkaido to Tohoku. Stem and leaf varieties often exceed 1.0 m in rod length but are highly resistant to downfall, which significantly reduces their quality as forage (Fig. 16.1). In addition, because low-cost production is required for feed, it is more important than food rice varieties to provide a wide range of disease and insect resistance to reduce the costs of disease and insect control.

Table 16.1 shows the name of the forage rice variety, rod length, yield, and TDN content. In terms of TDN yield, which is the most important factor for WCRS rice varieties, "Leafstar," "Tachiaoba," and "Tsukikatsuka" showed more than 1.17 tons/10a, all of which are stem-and-leaf type varieties with a low percentage of paddy and a large amount of stem and leaves. The extremely short panicle stem and leaf type varieties, "Tachiayaka," "Tachisuzuka," "Tsukisuzuka," and "Tsukikatsuka," have extremely short ears, resulting in extremely little paddy and high sugar content required for fermentation at harvest, making them highly suitable as rice varieties for WCRS. The leaf-stem type varieties all have a rod length of 1.0 m or more, and the grass length, including the tips of the stolons, is even longer; therefore, a harvester that can handle long rods may be necessary when harvesting by direct cutting. On the other hand, the conventional varieties with a high percentage of unhulled

Fig. 16.1 Conventional type variety "Kusahonami" and short panicle type variety "Tachisuzuka" (Photos by Matsuyama)

Table 16.1 Rod length, yield, and total digestible nutrients of forage rice varieties

			Yellow ripe stage	Full ripe stage			
Breed Name	Breeding area	Rod length (cm)	Total weight (t/10a)	Total weight (t/10a)	Brown rice weight (t/10a)	Estimated TDN content (%)	EstimatedTDN yield (t/10a)
Kitagenki	Hokkaido	76	1.17	1.91	0.85	59.4	0.70
Kitajyoubu	Hokkaido	77	1.53	2.11	0.76	59.0	0.91
Ushiyutaka	Aomori	96	1.42	2.28	0.84	49.4	0.68
Bekogonomi	Akita	79	1.17	1.55	0.69	62.1	0.73
Bekoaoba	Akita	70	1.37	1.77	0.73	61.9	0.85
Yumeaoba	Niigata	86	1.52	1.73	0.72	61.2	0.93
Tachihayate	Ibaraki	114	1.74	2.1	0.49	53.2	0.93
Hoshiaoba	Hiroshima	101	1.52	1.91	0.71	58.6	0.91
Momiroman	Ibaraki	89	1.80	2.12	0.82	61.0	1.10
Tachiayaka	Hiroshima	112	1.62	–	0.10	56.1	0.89
Kusahonami	Ibaraki	95	1.85	2.08	0.67	59.2	1.10
Leafstar	Ibaraki	109	1.92	2.14	0.42	61.0	1.17
Tachisuzuka	Hiroshima	121	1.87	–	0.23	52.7	0.99
Tsukisuzuka	Hiroshima	121	1.73	–	0.05	53.0	0.95
Mogumoguaoba	Fukuoka	104	1.92	2.16	0.75	57.3	1.10
Minamiyutaka	Miyazaki	101	1.29	1.47	0.31	–	–
Tachiaoba	Fukuoka	106	2.13	2.41	0.66	59.5	1.27
Tsukikotoka	Hiroshima	138	2.13	–	0.04	54.7	1.17

Technical manual for whole crop rice production and utilization, 7th edition (2020) partially modified
TDN total digestible nutrients

rice, such as Kitagenki, Bekogomomi, Bekoaoba, Yumeaoba, Momiroman, Hoshiaoba, Kusahonami, and Mogumoguaoba, have achieved high TDN yield in both unhulled rice and stem leaves.

The short panicle and high straw yield of varieties were clearly smaller in ear size than the conventional varieties and were specifically designed for WCRS (Matsushita et al. 2011, 2012; Nakagomi 2016). Paddies, which are less digestible in cattle, are significantly reduced, and the percentage of stover, which is more digestible, is increased. Sugar obtained through photosynthesis is translocated to the hull and stored as starch in brown rice; however, in the extremely short-earned stover-type variety with reduced hull, sugar accumulates in the stover, which is also called a high-sugar-content stover-type variety (Matsushita et al. 2012; Kusa et al. 2018). Such varieties not only improve digestibility for cows but also improve fermentation quality because of the abundance of sugar in the substrate during lactic acid fermentation (Kono et al. 2014; Kono 2016; Kusa et al. 2018) (Table 16.2).

Table 16.2 Fermentation quality of whole crop rice silage

		(% DM)			(% FM)		
Item	*n*	CP	TDN	pH	Lactic acid	Butyric acid	VBN/TN
Short panicle type	68	59.1 ± 3.6	6.3 ± 1.4	4.3 ± 0.5	0.94 ± 0.4	0.014 ± 0.04	4.01 ± 1.6
Conventional type	16	53.8 ± 6.2	7.3 ± 1.1	4.7 ± 0.3	0.53 ± 0.2	0.1 ± 0.1	5.21 ± 2.57

Technical manual for whole crop rice production and utilization, 7th edition (2020) partially modified

n sample number, *DM* dry matter, *FM* fresh matter, *CP* crude protein, *TDN* total digestible nutrient, *VBN* volatile basic nitrogen, *TN* total nitrogen

1.3 Harvest and Preparation of WCRS

It is difficult for forage rice plants to create anaerobic conditions because their stems have a rigid hollow structure that tends to increase the amount of residual oxygen (Cai et al. 2003). In addition, there were a few attached lactic acid bacteria, and many aerobic bacteria, molds, and yeasts were also attached. Conventional varieties contain fewer soluble sugars than corn and other forage crops, but a short panicle and high straw yield of varieties contain sufficient amounts. Similar to other forage crops, rice can achieve a good fermentation quality by following the basics of silage preparation (Cai et al. 2003; Kono et al. 2014). In addition, silage of consistently high quality can be prepared using additives according to the harvest time, moisture content, and other preparation conditions (Cai et al. 2003; Tohno et al. 2012).

The optimum time for harvesting conventional varieties is from the dough ripening stage to the yellow ripening stage (approximately 30 days after ear emergence). As the rate of rice husk discharge into the feces increases as growth advances (Shinde et al. 2008; Higuchi et al. 2009), it is important to combine varieties and crop types to avoid late harvesting and distribute the crop season. To prepare good quality WCRS, the moisture content of forage rice should be approximately 65%, and most forage rice that has reached the yellow maturity stage has a moisture content of 65% or less, allowing direct harvesting. However, if harvesting occurs at an extremely high moisture content, direct harvesting should be avoided, and moisture adjustment should be made through a pre-drying system. Ripening time was determined by the number of days after ear emergence and the ear condition. As the ripening of fodder rice is accelerated under high temperature and high light conditions and delayed under low temperature and low light conditions, weather conditions during the ripening period should be taken into consideration.

Unlike conventional varieties, the soluble non-nitrogen and sugar content in the stem and leaves and dry matter yield gradually increase after ear emergence, reaching their maximum values around 40 days after ear emergence, so the best time to harvest is 40 days after ear emergence (Nakagomi et al. 2022). Therefore, the best time for harvesting is 40 days after ear emergence or later. Even if early harvesting is unavoidable, the crop should be harvested 30 days after ear emergence. If the crop is harvested 30 days after ear emergence, the yield and fermentation quality will be reduced, and the advantage of using the extra short-earned leaf stalk-type variety will not be expected. According to the results of a survey of the short panicle and high straw yield of variety, the moisture content gradually decreased after the ear emergence date but remained approximately 65% from 40 to 70 days after ear emergence (Kobayashi et al. 2021). A long period of approximately 65%, which is the moisture range suitable for silage preparation, is effective in promoting good lactic acid fermentation. In contrast, until 20 days after ear emergence, the moisture content was above 75%, which was not suitable for direct harvesting. However, acid detergent fiber (ADF) does not increase easily after ear emergence, and digestibility does not decrease even if harvesting is delayed for very short-earning stover-type varieties (Kobayashi et al. 2021; Nakagomi et al. 2022). Therefore, the best time for harvesting is 40 to 70 days after ear emergence, when a particularly nutritious roughage (TDN content of approximately 60% in dry matter) is required, such as for lactating cows, and WCRS with TDN content of 50% to 55% in dry matter can be obtained at least until approximately 90 days after ear emergence (Kobayashi et al. 2021; Nakagomi et al. 2022). Therefore, it is advisable to avoid early harvesting of very short-earned stem leaf-type varieties and harvest them 40 days after ear emergence.

Harvesting and preparation of rice for WCRS can be broadly classified into a pre-drying system using machinery for pasture and a direct-cut system in which rice is harvested directly with a dedicated harvester or a general-purpose harvester (Takahashi et al. 2015; Urakawa 2017). In direct-cut systems, dedicated harvesters (combined harvesters) and general-purpose harvesters (roll bales) are used to produce roll bales in the field. In dry fields with large plots and good drainage conditions, efficient grass harvesting is possible using a machine, whereas in wet fields with poor drainage conditions, a harvester equipped with a crawler driving unit is required (Urakawa and Yoshimura 2003).

In a pre-drying system using grass machinery, forage rice is cut using a mower attached to a tractor, pre-dried in the sun, inverted by a tedder, collected by a rake, picked up by a roll baler, formed into roll bales, and released into the field. The rolled bales were sealed with a stretch film using a bale wrapper. The advantages of the pre-drying system are that livestock farmers' existing pasture machinery is used; thus, there is no need to install new machinery, sufficient soil fertility can be obtained during harvesting and preparation, and large plots can be operated with high efficiency. In addition, the pre-drying process enables moisture adjustment of WCRS and preparation of WCRS with low β-carotene content (Oya et al. 2005). However, disadvantages include quality deterioration owing to rainfall after harvesting and the possibility of soil contamination in wet fields.

The combine-type exclusive-purpose harvester (Urakawa and Yoshimura 2003) is based on a self-pulling combine and uses the harvesting section of the self-pulling combine as is; instead of a threshing section, it is equipped with a roll bale-forming chamber. The roll bale-forming chamber was equipped with an agitator in front of the roll bale-forming chamber, which cut the stalks and leaves with a disc cutter to homogenize the stalks and ears. Because the combined harvester uses rubber crawlers for the traveling section, it can work in wet paddy fields and is especially efficient even with rice plants that have grown over.

The general-purpose harvester (roll bale-type) (Shito et al. 2011; Tachibana et al. 2011) has a forage harvester mounted at the front of the machine and a roll baler for the shredded material at the rear. The current model has a rotary drum-type harvesting section of the forage harvester, which can be used for long-rod fodder rice, fodder corn, and other long and large crops. It can also harvest crops that have been sown in any row spacing. The cut grass is shredded in the shredding section and thrown into the storage hopper at the rear to form roll bales. The mowing section could be equipped with a device to add LAB. This machine can harvest long and large crops without changing the attachments, and the shredding process enables high-density packing. Similar to combine-type dedicated harvesters, rubber crawlers are used as traveling devices, making them easy to work in wet paddy fields.

Bale wrappers are machines that pick up roll bales released from harvesters in both pasture and direct-cut systems (Takahashi et al. 2015; Urakawa 2017). They seal roll bales in the field, seal them with wrapping films as they move, and unload them. There are two types of bale wrappers: crawler-type self-propelled bale wrappers (Urakawa et al. 2004) for wet fields and tractor-mounted bale wrappers for pasture grass, whereas the others have a vertical roll bale-placing device and a lift mechanism for truck loading. When forming a roll bale, it is necessary to increase the packing density of the roll bale to promote lactic acid fermentation and eliminate air from the rice. The target value of packing density should be 150 kg/m3 or more (Momose et al. 2005; Kawamoto et al. 2009). Before the operation, a preoperative inspection should be conducted to ensure that the belt of the harvester is not loose, and sufficient care should be taken to prevent the density from becoming too low. Sealing should be performed promptly after packing because aerobic microorganisms consume the monosaccharides contained in the rice material if they are exposed to air for a long period after packing. The more the rolls of the film, the better the long-term storage; however, for long-term storage beyond the summer following harvest, eight or more layers of film are desirable (Fig. 16.2). Rewrapping after transportation to a storage location is another effective method.

For production organizations that grow and harvest forage rice over a large area or contractors integrated with TMR centers that use WCRS as feedstock, silage preparation using large bunker silos may be advantageous because it reduces preparation costs compared with a roll bale system (Takahashi et al. 2015; Takahashi and Okada 2020). However, because rice straw are hollow and do not allow for adequate tread pressure, it is advisable to use a forage harvester capable of fine cutting and add a LAB (Tohno et al. 2012) when preparing silage in a bunker silo.

Under adverse conditions at the time of preparation, such as high moisture or low temperatures, or when there is a concern about deterioration after silo opening, the use of additives can promote fermentation and stabilize quality. When using additives, it is important to select the most appropriate additive based on the forage rice variety, harvest time, and quality problems anticipated at the time of feeding.

A variety of LAB were isolated from good-quality silage of each forage crop, and through experiments, a LAB strain, "*Lactobacillus plantarum* Chikuso-1" was

Fig. 16.2 Direct cutting system for forage rice using a combine harvester and a self-propelled bale wrapper (Photos by Matsuyama)

selected for its ability to improve the fermentation quality of WCRS (Cai et al. 2003). This LAB was freeze-dried and the additive "Chikuso-1" (Snow Brand Seedling Co. Ltd., Japan) was commercialized. Currently, "Chikusan-Lacto" (Snow Brand Seedling Co. Ltd., Japan) with three types of lactic acid bacteria "*Pediococcus inopinatus* IWT685," "*Lactococcus lactis* SBS-0011-S," and "*Lacticaseibacillus paracillus* SBS-0001" are being developed for use in conventional varieties of forage rice with low sugar content in the stems and leaves, in order to strongly promote lactic acid fermentation even from low sugar content. The actual method of adding *Lactobacillus lactis* involves dissolving it in water and spraying it on forage rice, while collecting it using an automatic adding device attached to a roll baler (Urakawa et al. 2003). "Chikusan-Lacto" suppresses butyric acid fermentation, which leads to a lower quality when the crop is harvested at higher moisture levels after rainfall or when early harvesting is required, such as during the milk-ripening period (Table 16.3).

"Chikusan-Aceto" (Snow Brand Seedling Co., Ltd., Japan) is an additive containing two types of LAB, "*Lentilactobacillus buchneri* IWT192" and "*Pediococcus inopinatus* IWT685", and when used in extremely short panicle (high sugar content) stover-type varieties, it promotes lactic acid fermentation while producing a moderate amount of acetic acid and other useful substances, and it can inhibit the growth of mold and yeast (Tohno et al. 2012; Tohno 2017; Fukuma et al. 2018). This inhibits aerobic deterioration and stabilizes the quality even after opening. Furthermore, "Chikusan-Aceto" contains lactic acid bacteria with excellent low-temperature growth potential, making it suitable not only for warm conditions but also for cold conditions such as 4 °C and also for harvesting and preparation operations for late and very late harvests where the morning and evening temperature drops overlap during the early stages of fermentation.

Table 16.3 Fermentation quality and microbiological analysis of silage ensiled for 60-days and 370-days

	Silage ensiled for 60 days				Silage ensiled for 370 days			
	Hamasari		Kusahonami		Hamasari		Kusahonami	
	Control	Chikuso-1	Control	Chikuso-1	Control	Chikuso-1	Control	Chikuso-1
Fermentation quality								
pH	5.04	3.98	5.18	4.15	4.65	3.75	4.87	3.95
DM(%)	32.37	29.42	32.11	33.54	31.27	30.63	33.33	30.25
Lactic acid (% FM)	0.24	1.20	0.14	0.93	0.65	1.68	nd	1.24
Acetic acid (% FM)	0.44	0.32	0.46	0.4	0.32	0.43	0.57	0.56
Butyric acid (% FM)	0.35	0.01	0.5	nd	0.28	nd	0.07	nd
Propionic acid (% FM)	0.04	nd	0.05	nd	nd	nd	0.07	nd
Ammonia N (g/kg FM)	0.79	0.24	0.91	0.32	0.21	0.17	0.65	0.24
Microorganism composition (log colony-forming units per gram of FM)								
Lactic acid bacteria	5.23	7.52	5.2	6.85	4.65	6.45	5.52	6.52
Aerobic bacteria	6.32	nd	5.5	nd	5.2	nd	4.5	3.21
Clostridia	4.3	nd	3.5	nd	4.3	nd	3.5	nd
Escherichia	nd	nd	3.2	3.5	nd	nd	3.2	nd
Bacilli	4.5	4.02	4.8	3.2	4.5	4.35	4.8	3.2
Yeast	3.8	4.2	4.2	4.8	3.8	4.2	4.2	3.5
Mould	nd	nd	4.6	nd	< 3 0	nd	7.5	nd

Cai et al. (2003)

DM dry matter, *nd* not detected, *FM* fresh matter

1.4 Feed Characteristics of WCRS

Rice WCRS vary widely in forage characteristics depending on the grass type of the cultivar grown. Edible varieties with high seed yields or forage varieties with both high seed and stover yields are described as "conventional varieties," while varieties with a significantly lower percentage of seed compared to conventional varieties are described as "extremely short-eared stover varieties."

From the ear emergence stage, non-fibrous carbohydrates (NFC) and soluble non-nitrogenous matter (NFE) in conventional varieties gradually increased as starch accumulated in the ear, whereas crude protein (CP), crude fiber, and neutral detergent fiber (NDFom) gradually decreased (Islam et al. 2004a; Hattori et al. 2005; Omura et al. 2007). Thus, WCRS prepared from conventional varieties at yellow maturity, the optimum harvest stage, has more NFC such as starch and, conversely, relatively less NDFom such as cellulose and hemicellulose than imported hay (Islam et al. 2004a; Hattori et al. 2005; Omura et al. 2007). CP is similar to or slightly less than imported grass hay but varies with the level of fertilizer applied during cultivation, and NDFom is higher in WCRS than in imported hay. WCRS is also characterized by a higher crude ash content than imported rice hay because of the high accumulation of silicic acid in the hulls and stover (Islam et al. 2004a; Hattori et al. 2005; Omura et al. 2007).

Compared to silages of grasses, such as Italian ryegrass, the short panicle and high straw yield of variety contains more soluble non-nitrogen and less crude fiber and protein, similar to conventional varieties (Kono, 2016; Kobayashi et al. 2021; Nakagomi et al. 2022). Despite the extremely small ears and extremely low starch-containing fruit, the composition of the short panicle and high straw yield of variety is similar to that of conventional varieties because large amounts of sugar and starch accumulate in the stover, which has a high NFC content (Kobayashi et al. 2021; Nakagomi et al. 2022).

In WCRS prepared from conventional varieties, NFE digestibility gradually increased as the ripening stage advanced and starch significantly accumulated in the ears (Islam et al. 2004b; National Agriculture and Food Research Organization 2009; Matsuyama 2009). In contrast, the digestibility of crude fiber gradually decreased owing to the accumulation of lignin in the stem and leaf parts, and the digestibility of CP tended to decrease as the ripening stage advanced. The TDN of the WCRS of conventional varieties increased as the ripening stage progressed, but remained the same or decreased slightly when reaching full ripeness compared to the yellow ripening stage. This was due to an increase in the amount of undigested grain and a decrease in the digestibility of the fiber components mentioned above (Shinde et al. 2008). In the WCRS of conventional varieties harvested at yellow maturity, the digestibility of CP, crude fat, and NFE was similar to or higher than that of imported grass hay; however, the digestibility of crude fiber was lower (National Agriculture and Food Research Organization 2009). The digestive tract passage rate of the WCRS tended to be slower (Yamamoto et al. 2008) (Table 16.4). Therefore, caution should be exercised when determining the amount of WCRS fed

Table 16.4 Chemical composition of whole crop rice silage

Item		Moisture (%)	Chemical composition (% DM)							
			Crude protein	Crude fat	NFE	NFC	Crude fiber	ADFom	NDFom	Crude ash
Whole crop rice silage (conventional variety)	Heading stage	75.6	7.2	2.9	41.8	9.2	31.8	39.7	64.4	16.3
	Milk ripe stage	71.8	7.3	2.6	46.4	13.9	27.1	35.3	59.6	16.6
	Dough ripe stage	63.6	6.4	2.4	48.1	18.7	26.6	33.6	55.9	16.6
	Yellow ripe stage	61.0	5.8	2.4	51.2	28.9	26.0	31.1	48.3	14.6
	Full ripe stage	59.0	5.4	2.6	53.2	–	24.5	–	–	14.3
Import hay	Sudangrass	10.3	8.6	1.7	47.9	12.6	32.6	38.8	67.9	9.2
	Oat	12.0	6.4	1.7	53.4	24.6	32.2	36.6	61.0	6.3
	Italian ryegrass	9.4	6.2	1.5	53.9	20.7	32.2	42.0	65.4	6.2
	Timothy	11.1	7.6	2.1	49.2	17.0	34.3	37.9	66.6	6.7
	Bermuda grass	9.1	8.9	1.6	56.9	12.7	24.7	31.6	68.9	7.9
	Tall fescue	12.1	6.7	1.2	47.8	14.6	37.6	40.5	70.8	6.7

Standard tables of feed composition in Japan (2009) partially modified. Sudangrass (30% < CF < 35%), Oat (30% < CF < 35%), Timothy (32% < CF < 37%)
DM dry matter, *NFC* non fiber carbohydrates, *NFE* nnitrogen free extracts, *ADFom* acid detergent fiber, *NDFom* neutral detergent fiber

as a simple replacement for grass hay, as it may suppress dry matter intake (DMI), especially in lactating cows.

Studies conducted by Japanese research institutes have shown that the TDN of WCRS (mostly conventional varieties) during the yellow ripening stage ranges from 40% to 60% (Matsuyama et al. 2005; Matsuyama et al. 2006; Higuchi et al. 2009; Higuchi et al. 2015). In particular, low TDN can be attributed to preharvest downfall, late cutting, pest damage, weed contamination, and extremely poor fermentation quality (Matsuyama et al. 2013) (Table 16.5). The best time to harvest conventional varieties is from the standpoint of nutritional value, from the "glue ripening" stage to the "yellow ripening" stage, when the ears are full.

Conventional varieties have high acid detergent fiber (ADF) and silicate contents, which are indigestible fiber components, and their fiber digestibility is low quality compared to that of other pastures. On the other hand, the short panicle and high straw yield of variety has low ADF content and a crude fiber digestibility of approximately 60%, which is significantly higher than the 50% of conventional varieties. This results in a higher TDN content. And the TDN content of the extra-short stover variety in dry matter at yellow maturity is 58%–61%, about 5 percentage points higher than that of the conventional variety at the same yellow maturity

Table 16.5 Digestibility and nutritive value of whole crop rice silage

		Digestibility				Nutritive values	
		(% DM)				(% DM)	(MJ/kg DM)
Item		Crude protein	Crude fat	NFE	Crude fiber	TDN	ME
Whole crop rice silage (conventional variety)	Heading stage	55	63	53	67	51.5	7.72
	Milk ripe stage	69	60	55	60	50.3	7.51
	Dough ripe stage	56	64	68	51	53.2	8.03
	Yellow ripe stage	54	60	66	53	54.0	8.16
	Full ripe stage	50	60	66	50	53.6	8.10
Import hay	Sudangrass	51	38	56	60	52.2	7.85
	Oat	38	40	58	66	56.2	8.55
	Italian ryegrass	60	53	68	69	64.4	10.00
	Timothy	51	50	58	57	54.4	8.24
	Bermuda grass	75	51	52	64	54.0	8.16

Standard tables of feed composition in Japan (2009) partially modified
Sudangrass (30% < CF < 35%), Oat (30% < CF < 35%), Timothy (32% < CF < 37%)
DM dry matter, *NFE* nitrogen free extracts, *TDN* total digestible nutrients, *ME* metabolizable energy

(Kawamoto et al. 2010; Kono et al. 2014; Kono 2016). The nutritional value of conventional varieties declines as the growth stage progresses because the content of ADF, a non-digestible fiber, increases linearly from 20 days after ear emergence. However, in the case of the short panicle and high straw yield of variety, ADF remained low level after 20 days post-ear emergence, resulting in a very slow decline in dry matter digestibility after ear emergence (Kawamoto et al. 2010; Kono et al. 2014; Kono 2016).

On voluntary intake of WCRS, many studies have reported that the preference for WCRS in cattle is similar to or better than that for grass silage and hay (Ishida 2000; Islam et al. 2004a, b, c; Hosoda et al. 2005; Yamamoto et al. 2005; Takahashi et al. 2007; Matsuyama 2009; Shinde 2010). There are also reports that preference varies among WCRS depending on variety, harvest time (ripening period), harvest method, preparation method, etc., but the results are not uniform. In general, it can be said that, as with silages of other forage crops, if the WCRS has good fermentation quality, it is certainly well-liked by cattle. Furthermore, silages with a good fermentation quality showed less nutrient loss before preparation. On the other hand, WCRS with poor fermentation quality are not only less palatable but can also be detrimental to livestock health, for example, due to the accumulation of mold toxins from mold outbreaks. Therefore, it is important to follow the basics of silage preparation and use additives, such as lactic acid bacteria, according to harvest conditions to achieve good fermentation quality, not only for palatability but also to increase livestock productivity.

It has been reported that the voluntary intake of WCRS prepared with conventional breeds is affected by the time of harvest (ripening period). In a study conducted on beef cattle (Japanese Black and Japanese Shorthorn), the voluntary intake of WCRS increased from the flowering period as the cattle matured (Hashimoto and Fukagawa 2005; Nakanishi 2006; Shinoda et al. 2007). However, after the dough ripening period, the feeding amount almost reached an equilibrium state. At that time, the voluntary intake of WCRS (60% moisture content) was 20–25 kg of raw material equivalent (8–10 kg of dry matter equivalent) per day, which was comparable to silages of Italian ryegrass and sorghum (Hashimoto and Fukagawa 2005; Nakanishi 2006; Shinoda et al. 2007). The voluntary intake of WCRS in lactating cows is affected by milk yield, the percentage of concentrate in the feed ration, and the number of days since calving (lactation stage). In a study that measured the voluntary intake of WCRS at different lactation stages by varying the proportion of concentrate in the diet, cows consumed 12.5 to 30.5 kg (5.0 to 12.2 kg dry matter equivalent) of WCRS (60% moisture content) per day and were capable of eating about 20 kg (8 kg dry matter equivalent) of concentrate per day under all conditions (Ishida 2000; Matsuyama 2009; Sinde et al. 2008, 2012a, b).

The roughage value index (RVI) is a measure of the physical characteristics of a feed. This index was calculated by measuring the chewing time of cows fed a ratio of interest (the sum of feeding and recursion times). RVI is closely related to homeostasis of the first stomach environment. RVI is closely related to homeostasis of the first stomach environment because the amount of saliva secreted, which has a buffering capacity against the organic acids produced in the first stomach, is affected by chewing time. The RVI of the WCRS of conventional varieties was 82 min/kg, which is similar to that of timothy hay and Sudan grass hay and higher than that of alfalfa hay and corn silage (Matsuyama 2009). Thus, the physical properties of WCRS prepared with conventional varieties were high among the roughages. In lactating cows, chewing time, as indicated by RVI, is closely related to milk fat percentage, and to maintain a milk fat percentage of 3.5%, the RVI of the fed ration must be at least 30.5 min/kg. Three types of TMR contain 30% each of WCRS, corn silage, and Italian ryegrass silage. In a study in which lactating cows were fed three TMRs containing 30% each of WCRS, corn silage, and Italian ryegrass silage, the RVI of each was 34.3 min/kg, 27.3 min/kg, and 36.9 min/kg, respectively (Yamamoto et al. 2008). These results indicated that the RVI of WCRS was higher than that of corn silage and similar to that of Italian ryegrass silage. The RVI was also affected by the shape of the feed. In the WCRS, the longer the cutting length, the higher the RVI, and the chewing effect of the chewed feed decreased the anthesis of the undigested offspring (Shinde and Iwashimizu 2008). However, if the cutting length is too long, the time required for digestion suppresses the dry matter intake, resulting in a lower milk yield (Shinde et al. 2008). According to previous studies, the optimum cutting length is approximately 3.0 cm to ensure adequate RVI and to consider DMI and the amounts of undigested cotyledons excreted (Table 16.6).

When cattle are fed WCRS prepared from conventional varieties with a high percentage of grain, a noticeable amount of undigested grain is expelled in the feces. This phenomenon occurs when the whole-crop feeding of forage crops with offspring, such as corn and wheat, is used, including rice. Generally, as grains ripen, the outer skin of the grain and the grain themselves harden, making digestion difficult for cattle (Shinoda et al. 2002). When fodder rice seedlings are directly immersed in the first stomach of cattle (without being chewed by cattle), they are almost undigested, and even if chemical treatments such as alkali and enzymes are applied, the passage rate in the first stomach after 48 hours is only about 10% (Shinoda et al. 2002). However, physical treatments, such as crushing and grinding, result in approximately 60%, and brown rice with the husks removed results in more than 80% (Shinoda et al. 2002). This indicates that indigestible hulls prevented the digestion of WCRS grains. The percentage of undigested grain in the WCRS prepared with conventional varieties depended on dry matter intake and the percentage of WCRS in the feed ration (Shinde et al. 2005). It is reported to average about 10%

Table 16.6 Roughage value index of whole crop rice silage

Item	Roughage value index (min./kg DM)
Whole crop rice silage	82
Imported timothy hay	79
Imported sudanese grass hay	77
Imported alfalfa Hay	47
Corn silage	66

Standard tables of feed composition in Japan (2009)
DM dry matter

in beef breeding heifers, fattening cows, and growing cows and can reach 50% in lactating cows. The high percentage of undigested grain, especially in lactating cows, is due to the high intake of dry matter, which accelerates the rate at which feed passes through the digestive tract and prevents digestion of the grain. A remedy for this is to reduce unnecessary excretion of undigested grain by adjusting the amount of roughage containing the WCRS and the cutting length to promote digestion of the grain by ensuring adequate chewing time.

The potassium content of WCRS prepared with conventional varieties was low compared to that of other diets and is considered to be a suitable feed for dairy cows during the dry period to prevent hypocalcemia and milk fever during calving (Table 16.7). However, it has become clear that the potassium content is higher in forage rice grown in fields where potassium has accumulated, such as through the continuous use of cattle manure, and that there are strains and varieties of rice with a high potassium absorption capacity. Because there is a concern that the potassium content may be higher in WCRS prepared from these sources, it is recommended that feed analysis should be conducted in advance to confirm the potassium content.

Nitrate-nitrogen levels in the WCRS prepared with conventional varieties were lower than those in Sudan grass hay and other silages (Table 16.8). When feed containing high levels of nitrate is ingested, nitrate is reduced to nitrite in the first stomach, which reacts with hemoglobin to form methemoglobin after being transferred into the bloodstream. This results in nitrate poisoning, which reduces the ability to transport and exchange oxygen and carbon dioxide, and causes oxygen deprivation. Because forage rice is grown in an environment where it is difficult to accumulate nitrate, and forage rice itself has characteristics that make it difficult to accumulate nitrate, the nitrate-nitrogen concentration in the WCRS is low.

WCRS contains more β-carotene and vitamin E (α-tocopherol) than imported hay, particularly direct-cut WCRS (Matsuyama 2009). β-carotene and α-tocopherol are abundant in the leaf part and gradually decrease with maturity in forage rice as a whole, especially in conventional varieties with a high ear proportion (Table 16.8).

Table 16.7 Mineral and nitrate nitrogen of whole crop rice silage

		Mineral (% DM)				(ppm)
Item		K (Potassium)	Ca (Calcium)	Mg (Magnesium)	P (Phosphorus)	HNO_3-Nitrogen)
Whole crop rice silage (conventional variety, yellow ripe stage)		1.20	0.23	0.11	0.19	80
Import hay	Timothy	1.63	0.31	0.12	0.18	157
	Sudanese grass	2.42	0.43	0.32	0.22	1109
	Oat	1.52	0.22	0.14	0.17	185
Silage	Corn	1.65	0.24	0.13	0.25	369
	Sorghum	2.13	0.33	0.17	0.26	1343

Standard tables of feed composition in Japan (2009)
DM dry matter

Table 16.8 β-carotene and α-tocopherol of whole crop rice silage

Item		(mg/kg DM)	
		β-carotene	α-tocopherol
Whole crop rice silage (conventional variety)	Milk ripe stage	36 (7.3–92)	145 (20–385)
	Dough ripe stage	51 (2.1–100)	–
	Yellow ripe stage	32 (9.6–144)	154 (17–618)
	Full ripe stage	–	50 (6–154)
Hay	Timothy	9 (2.6–20)	14
	Orchard grass	13 (5.3–20)	40
	Bermuda grass	15 (0.8–36)	31
	Rice straw	4 (0.5–9.3)	21
Silage	Corn	42	39
	Sorghum	46.0	–

Standard tables of feed composition in Japan (2009)
Whole crop rice silage is prepared without pre-drying
DM dry matter

This is especially true for conventional varieties with high ear content. In addition, both β-carotene and α-tocopherol are decomposed or oxidized by ultraviolet light, heat, and oxygen, so they are significantly reduced by pre-drying treatment, and the β-carotene content of WCRS pre-dried for more than 2 days becomes as low as that of dried rice straw (Hiraoka et al. 2004; Oya et al. 2005; Knaya 2007). A feeding method that promotes fat accumulation in the skeletal muscle tissue by controlling vitamin A in the blood is commonly practiced in Black Japanese fattening cattle. Therefore, when feeding rice WCRS to beef cattle for vitamin A-controlled fattening, there have been cases where efforts have been made to produce rice WCRS with reduced β-carotene content by methods such as pre-drying and delaying the harvest time (Knaya 2007).

1.5 Effects of Feeding WCRS in Beef and Dairy Cattle

It has been reported that α-tocopherol has a physiological effect of inhibiting the synthesis of lipid peroxide in the body (antioxidant capacity) and that feeding WCRS to fattening cows increases α-tocopherol accumulation in the meat and maintains good meat color, while feeding WCRS to lactating cows increases α-tocopherol concentration in the milk and inhibits milk fat in lactating cows and feeding WCRS increases α-tocopherol concentration in milk and inhibits oxidation of milk fat.

α-Tocopherol has physiological effects that inhibit the synthesis of lipid peroxides in the body (antioxidant capacity) and is also known as a functional substance involved in fertility. Therefore, α-tocopherol is used as a feed additive in dairy cattle feed management. However, the potency (biological activity) of natural α-tocopherol in WCRS is higher than that of synthetic α-tocopherol in feed additives, so WCRS is advantageous over feed additives when increasing α-tocopherol concentration in blood and raw milk. When lactating cows fed imported oat hay were started on WCRS, α-tocopherol concentration in both blood and milk increased, indicating that α-tocopherol contained in WCRS was transferred into the body of lactating cows. In order to increase the α-tocopherol concentration in raw milk, it is necessary to continue feeding WCRS for a certain period of time (Matsuyama 2009).

The low TBARS levels in the blood and milk of dairy cows fed with WCRS or imported Timothy hay and incubated with an oxidation-accelerating reagent for a certain period indicated that oxidation resistance was enhanced (Matsuyama 2009). The effects of α-tocopherol in dairy cows have been investigated using feed additives, and it has been reported that α-tocopherol inhibits oxidation of milk fat and oxidative odor in raw milk and improves immune function, disease prevention, and fertility in sound livestock. It is possible that α-tocopherol in WCRS can reproduce the effects observed with feed additives, although it depends on future research.

α-Tocopherol in feed is absorbed from the digestive tract and then accumulates in beef. The results of previous studies on α-tocopherol accumulation and oxidation inhibition in beef have shown that it is necessary to accumulate about 3 mg of α-tocopherol per kg of beef in order to demonstrate antioxidant capacity (Yamada 2009, 2016; Nakanishi et al. 2018). When consumers purchase beef, meat color is a major consideration. During refrigeration, meat gradually turns brown, which is caused by the oxidation of myoglobin (the red pigment in beef) to a brown pigment called metmyoglobin. In addition, beef preferred by consumers is characterized by its good flavor when eaten. However, when beef is oxidized, an oxidized odor is produced, and the flavor deteriorates. The results of a study in which WCRS was fed to Japanese black cattle for the entire period of fattening showed that α-tocopherol content in the beef increased and discoloration of the beef during refrigeration was suppressed compared to the control cattle that were conventionally fattened with dried rice straw (Yamada 2009, 2016; Nakanishi et al. 2018). Furthermore, the increase in thiobarbituric acid reactive substances (TBARS), an indicator of beef oxidation, was lower in beef with increased α-tocopherol content fed WCRS than in beef in the control condition during cold storage.

1.6 Conclusion

In the future, it is expected that technology related to WCRS will be disseminated, and livestock products produced by feeding WCRS will be branded. Currently, Japanese agriculture is required to improve food and feed self-sufficiency and shift to environmentally friendly agriculture. The remaining technical issues should be studied and promoted in cooperation with other Asian countries, as feed production issues are unique to temperate monsoon regions.

2 Forage Rice Production and Cattle Feeding in South Korea

Jong Geun Kim

Rice is one of the important grains worldwide and is used as a food crop in many countries. It is particularly widely consumed in most Asian countries. Approximately 505.27 million tons of rice are produced globally, with about 90% of this production taking place in Asia (Shahbandeh 2008).

Statistical data related to rice production in South Korea have been compiled since 1965, and Fig. 16.3 shows the rice cultivation area and production levels over the years. In 1965, the rice cultivation area was 1.2 million hectares, reaching its peak at 1.3 million hectares in 1987. However, this area has steadily decreased, reaching 727 thousand hectares by 2022, reflecting a decrease of 42.3%. The development of new rice varieties through rice breeding programs, initiated in the mid-1960s, led to the widespread cultivation of an improved variety called "Tongil" in the early 1970s. This variety was recognized as a groundbreaking solution to South Korea's food shortage problem. Since 1976, rice production has consistently exceeded five million tons, with a record high of 6.053 million tons achieved in 1988. Rice production nearly doubled over 33 years, from 2.959 million tons in 1955 to 6.053 million tons in 1988 (Kosis 2023).

In South Korea, the per capita rice consumption was 128 kg in 1985; however, this figure has seen a steady decline over the past three decades, reaching 57.7 kg per capita by 2020, a reduction of approximately 55%. Although per capita rice consumption in South Korea is still relatively high compared to other East Asian countries where rice is a staple food, there is a forecast that this decreasing trend will continue (Fig. 16.4).

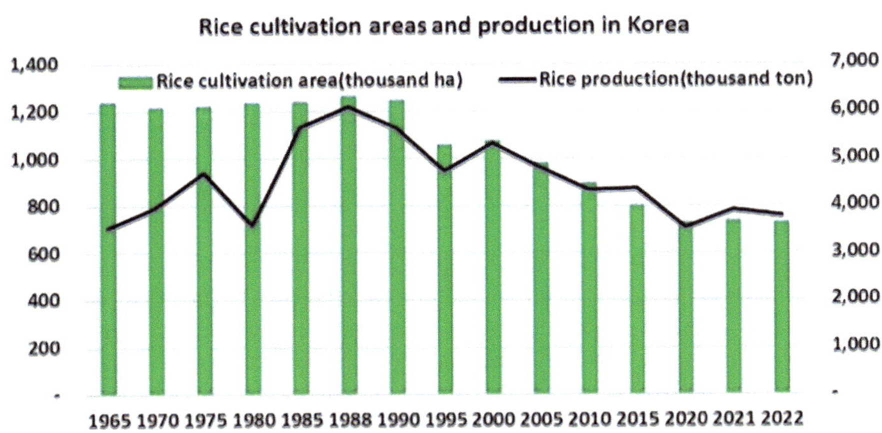

Fig. 16.3 Rice cultivation areas and production in Korea

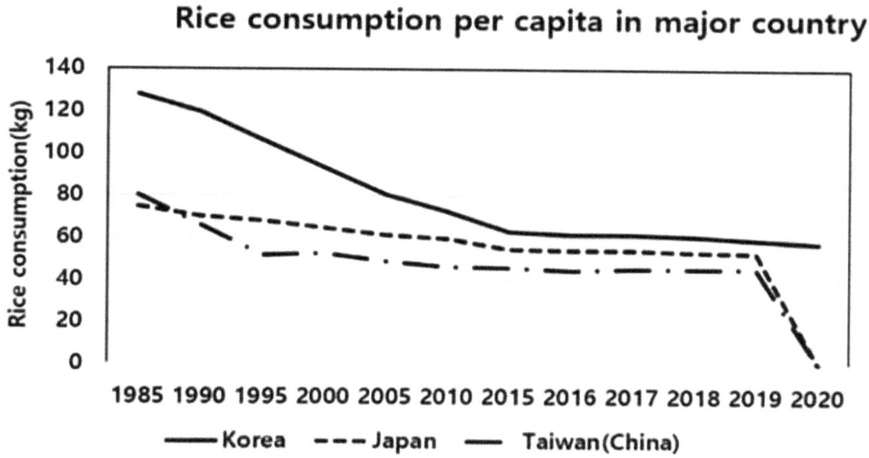

Fig. 16.4 Rice consumption per capita in East Asia countries

Alternative approaches are being suggested to repurpose paddy fields when there is excess rice production. However, in South Korea, paddy fields have various public functions, and simply eliminating them cannot be the only solution. First, during rice cultivation, paddy fields help retain water, which aids in conserving water and regulating floods during heavy rainfall. Additionally, cultivating rice in paddy fields helps prevent soil erosion and generates oxygen. Moreover, there is a recent trend to utilize paddy fields for amenity purposes, which provide emotional benefits.

In South Korea, the total number of cattle, including Hanwoo (Korean native cattle) and dairy cows, is approximately four million. These livestock require a total of 5.22 million tons of feed annually. Domestically produced feed accounts for about 82.7% of the total demand (approximately 4.32 million tons), with the remaining one million tons met through imported hay. The government regulates the import quantity through a quota system with reduced tariffs (MAFRA 2022). However, farmers have increasingly advocated for an increase in the amount of imported hay, which is not an easy task. Furthermore, it is expected that the hay market will see an increase in imports when markets in countries like Australia, Canada, and the United States open up from 2024 onward (Jeong 2021) (Fig. 16.5).

Therefore, to address the shortage of forage resources, regulate rice production, and maintain the diverse public functions of paddy fields, it is desirable to utilize paddy fields for the production of whole crop rice. In this context, various research findings on whole crop rice silage, as reported in recent studies, have been outlined.

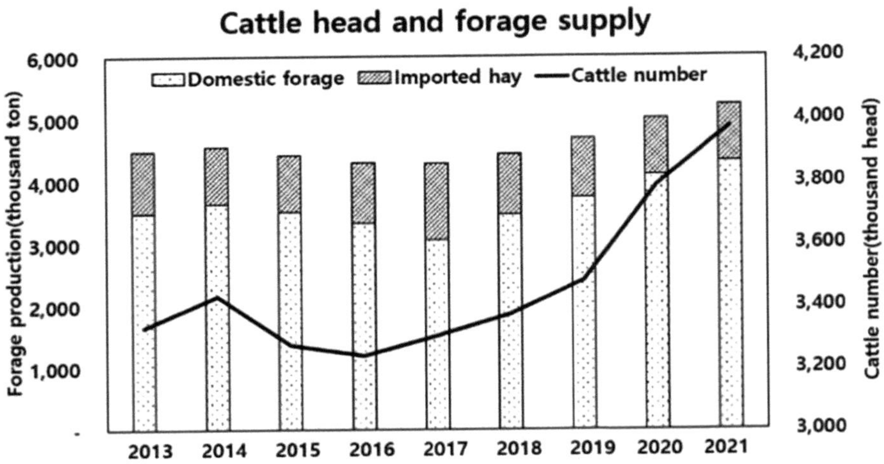

Fig. 16.5 Number of cattle and forage supply in Korea

2.1 Productivity and Forage Quality of Whole Crop Rice

2.1.1 Productivity

The utilization of rice for feed can be pursued in various ways. While using dedicated rice varieties for cultivation is one approach, it is also important to explore the option of utilizing rice intended for human consumption as feed, especially when there is a concern about overproduction based on production forecasts. In the absence of dedicated rice varieties initially, evaluations were conducted using rice intended for human consumption, and the average productivity was observed to be 10 tons/ha in terms of dry matter (DM) yield. When referring to the harvest period (dough stage) for rice used as feed, the DM content was found to be 32.8%, and the fresh yield exceeded 37 tons/ha, while the DM yield exceeded 12 tons/ha (Table 16.9).

2.1.2 Forage Quality

The nutritional value of rice used for feed, particularly in terms of CP, was found to be 8.4% at dough stage. The ADF, NDF, and TDN recorded were 35.3%, 63.3%, and 60.9%, respectively (Table 16.10). Although the feed value of rice intended for human consumption showed lower figures compared to newly developed varieties, there was no significant difference between the general sample and the feed crop. Therefore, in cases where rice production exceeds annual consumption (over 800,000 to 1,000,000 tons), utilizing it as feed based on production forecasts is considered a desirable policy for rice production adjustment.

Table 16.9 Dry matter and forage productivity of food rice "Chucheong" (NIAS 2006)

			Yield (kg/ha)		
Harvest time	Harvest date	DM (%)	FM	DM	TDN
Heading	18 Aug.	22.5	26,723	6017	3520
Flowering	29 Aug.	26.5	29,789	7865	4670
Milk	15 Sep.	30.8	35,004	10,756	6489
Dough	18 Sep.	32.8	37,427	12,256	7473
Yellow ripen	5 Oct.	38.2	33,958	12,979	8078
Fully ripen	16 Oct.	43.8	27,700	12,178	7705
Mean	–	32.4	31,767	10,342	6323

[a] *FM* fresh matter, *DM* dry matter, *TDN* total digestible nutrient

Table 16.10 Forage quality of edible rice "Chucheong" (NIAS 2006)

Harvest time	CP (%)	ADF (%)	NDF (%)	TDN (%)
Heading	9.5	38.5	68.4	58.5
Flowering	9.2	37.3	66.7	59.4
Milk	8.7	36.1	64.0	60.3
Dough	8.4	35.3	63.3	60.9
Yellow ripen	7.8	33.7	61.0	62.2
Fully ripen	7.3	32.4	59.6	63.2
Mean	8.5	35.5	63.8	60.8

2.1.3 Improvement of Silage Quality

To ensure the proper storage of large-scale feed production, it is important to consider effective preservation methods, with hay and silage being the most commonly used. In Korea, silage is predominantly used due to its resistance to adverse weather conditions as frequent rainfall during the harvest season poses challenges for hay production.

2.1.4 Silage Quality of Edible Rice Variety

The silage quality of edible rice varieties is presented in Table 16.11. The pH level, which is an indicator of silage quality, was slightly higher at 4.61 during the heading stage. However, the lactic acid content was higher at 2.84%, indicating improved fermentation. Additionally, lower acetic acid and butyric acid were also recorded, which is indicative of a superior silage quality. In contrast, the silage quality during the flowering and ripening stages was lower, suggesting that the optimal harvest stage of feed rice for silage is the milk to early dough stages.

Table 16.11 Silage quality of edible rice variety "Chucheong" (NIAS 2006)

Harvest time	pH	Organic acid (% DM) Acetic acid	Butyric acid	Lactic acid	Flieg's score	Quality grade
Heading	4.50	0.77	0.25	2.32	48	3
Flowering	4.67	0.82	0.16	2.29	50	3
Milk	4.72	0.63	0.11	2.66	64	2
Dough	4.61	0.47	0.09	2.84	74	2
Yellow ripen	4.92	0.54	0.14	2.52	64	2
Fully ripen	5.15	0.34	0.24	2.19	57	3

Table 16.12 Fermentation quality of whole crop rice silage (SNU 16 ~ 18)

Varieties	pH	Organic acid (% DM) Acetic acid	Butyric acid	Lactic acid
Suweon 605	4.82	0.74	0.21	2.21
Yeongwoo	4.37	0.32	0.04	2.84
Mogwoo	4.63	0.56	0.19	2.62
Average LSD (0.05)	4.61 0.25	0.54 0.18	0.15 0.07	2.56 0.37

2.1.5 Silage Quality of Whole Crop Rice Variety

In the evaluation test of new whole crop rice varieties for silage production, the pH level was significantly lower in the Yeongwoo variety, while Suweon 605 exhibited a relatively higher pH level. In terms of organic acids in silage, the Yeongwoo variety had lower acetic acid and butyric acid coupled with the highest lactic acid content. Based on these findings, the silage quality of the Yeongwoo variety is superior (Table 16.12).

2.2 Livestock Utilization of Whole Crop Rice Silage

2.2.1 Hanwoo (Korean Native Cattle)

To meet the nutritional requirements of Hanwoo cattle using whole crop rice, the trend of using total mixed ration (TMR) is becoming more common. However, many Hanwoo farms still rely on concentrates and straw for feeding. In the future, the use of TMR is expected to increase. The utilization of whole crop rice in TMR feed is particularly desirable, and two studies evaluating the effects of feeding TMR containing whole crop rice silage are discussed below.

2.2.2 TMR Using Whole Crop Rice Silage in the Central Region (Pyeongteak)

The composition ratio of the TMR was determined based on the nutritional requirements at each growth stage. Compared to the control group, the whole crop rice TMR (WCR-TMR) included approximately 45% whole crop rice silage during the growing stage and was gradually reduced to 35%, 15%, and 9% during the fattening-preparation stage (Kim et al. 2021).

Table 16.13 shows the growth characteristics of Hanwoo cattle. The initial weight was adjusted to an average of 284 kg, and there was no significant difference in weight up to the pre-fattening stage. However, during the fattening and finishing stages, the weight of the WCR-TMR feeding group was significantly higher than that of the control group. At 29 months of age, the weight was 631 kg/head in the control group, whereas it was 647 kg/head in the WCR-TMR feeding group, which was significantly higher ($P < 0.05$).

In terms of average daily gain (ADG), the WCR-TMR feeding group showed a significant increase during the growing and pre-fattening stages, but there was no difference during the fattening and finishing stages (Table 16.14). Over the entire period, the control group had an average daily gain of 0.71 kg/day/head, while the WCR-TMR feeding group achieved an average daily gain of 0.75 kg/day/head, which was significantly higher ($P < 0.05$). In a study by Kim et al. (2019) on Hanwoo cattle fed TMR primarily consisting of whole crop rice (WCR 13.00–15.04% inclusion), the average daily gain was reported to be 0.78 kg/day throughout the

Table 16.13 Effect of WCR-based TMR on growth performance of Korean native cattle (NIAS 2006)

Item	Control	WCR-TMR	SE
Body weight (kg)			
Initial (age of 9 Mon.)	284^{a}	284^{a}	6.19
12 mon.	364^{a}	375^{a}	7.78
18 mon.	453^{b}	473^{a}	8.32
22 mon.	581^{b}	603^{a}	9.96
29 mon.	631^{a}	647^{a}	10.41
Initial~12 mon. 13 ~ 18 mon.			
19 ~ 22 mon. 23 ~ 29 mon.			
Overall period			
Average daily gain (kg)	0.85^{b}	0.98^{a}	0.041
	0.72^{b}	0.79^{a}	0.012
	0.70^{a}	0.71^{a}	0.004
	0.59^{a}	0.52^{a}	0.027
	0.71^{b}	0.75^{a}	0.034

SE standard error

^{a,b}Values with different superscripts in the same row differ at $P < 0.05$

Table 16.14 Effect of WCR-TMR feeding on carcass characteristics of Korean native cattle (NIAS 2006)

Item	Control	WCR-TMR	SE
Quantity grade			
Back fat thickness (mm)	11.7[a]	9.3[b]	1.2
Rib eye area (cm^2)	85[a]	84[a]	0.5
Carcass weight (kg)	398[a]	397[a]	0.2
Meat yield index	65.55	67.12	0.78
Grade (A:B:C)	3:3:1	5:2:0	–

SE standard error
[a,b]Values with different superscripts in the same row differ at $P < 0.05$

feeding period. This was higher than the control group's 0.66 kg/day/head. Consistent with the present study, no significant difference in average daily gain was observed during the finishing stage.

Table 16.2.14 also presents the carcass and meat quality characteristics of the WCR-TMR feeding group. Backfat thickness was significantly greater in the control group (11.7 mm), compared to the WCR-TMR (9.3 mm). However, there were no significant differences in rib eye area, carcass weight, or meat yield index. Regarding meat grading, the WCR-TMR feeding had five heads graded as A and two heads graded as B, while the control group had three heads graded as A, three heads graded as B, and one head graded as C, indicating a higher occurrence of A-grade carcasses in the WCR-TMR group. However, in a study by Jin et al. (2012) on TMR with barley silage as the main ingredient, the control group (rice straw-based) showed a higher occurrence of A-grade carcasses, differing from the results of the present study.

In terms of meat quality, the marbling score was significantly higher in the WCR-TMR group (7.1), while there were no significant differences in meat color, fat color, texture, or maturity between the two groups. Similarly, Kim et al. (2019) observed no significant differences in meat quality parameters in their study on whole crop rice silage feeding (Table 16.15). However, in terms of meat quality grading, the WCR-TMR group showed a higher occurrence of grade 1^{++}, with 100% of the samples classified as grade 1 or higher, compared to 85.7% in the control group. Despite these findings, it is difficult to draw definite conclusions about the feed type on the proportion of meat graded as 1 or higher, based on these results.

2.2.3 TMR Using Whole Crop Rice Silage in Southern Region (Sancheong)

The research results on TMR feed, primarily using whole crop rice silage, conducted in the southern region of South Korea are summarized in Table 16.16. At 8 months of age, the control group had an initial weight of 292 kg, while the WCR-TMR group had an initial weight of 251 kg. Although the control group exhibited significantly higher weight up to the growing stage, there was no significant

Table 16.15 Effect of WCR-based TMR feeding on meat quality of Korean native cattle (SNU 16–18)

Item	Control	WCR-TMR	SE[f]
Quality grade			
Marbling score (1–9)	5.6[b]	7.1[a]	0.75
Meat color (1–7)	4.7[a]	4.7[a]	0.02
Fat color (1–7)	2.9[a]	3.2[a]	0.15
Texture (1 = soft, 9 = firm)	1.1[a]	1.0[a]	0.04
Maturity (1 = young, 9 = old)	2.1[a]	2.1[a]	0.01
Grade (1^{++}:1^{+}:1:2:3)	1:3:2:0:1	3:3:1:0:0	-

[a,b]Values with different superscripts in the same row differ at $P < 0.05$

Table 16.16 Effect of WCR feeding on growth performance of Hanwoo steers (SNU 16–18)

Item	Control	WCR-TMR	SEM	$Pr > F$
Body weight (kg)				
– Initial (8 months of age)	292 ± 12.01[a]	251 ± 8.63[b]	8.837	0.016
– 14 months of age	392 ± 10.67[a]	374 ± 4.89[b]	6.120	0.047
– 21 months of age	585 ± 15.00	604 ± 11.15	9.356	0.326
– Final (30 months of age)	726 ± 17.77[b]	765 ± 14.42[a]	12.135	0.012
– Total gain	434 ± 14.95[b]	514 ± 18.09[a]	15.252	0.004
Average daily gain (kg)				
– Growing period	0.67 ± 0.04[b]	0.82 ± 0.06[a]	0.041	0.045
– Early fattening period	0.76 ± 0.03[b]	0.90 ± 0.03[a]	0.030	0.011
– Late fattening period	0.57 ± 0.02	0.64 ± 0.04	0.026	0.131
– Whole period	0.66 ± 0.02[b]	0.78 ± 0.02[a]	0.023	0.005

WCR whole crop rice silage, *SEM* standard error of means
[a,b]Values with different superscripts in the same row differ at $P < 0.05$

difference in weight between the two groups at the end of the pre-fattening stage ($P > 0.05$). However, at 30 months of age at the time of slaughter, the WCR-TMR group had a significantly higher weight. The total weight gain from the start of the experiment to slaughter was 514 kg in the WCR-TMR group, significantly higher than the 434 kg observed in the control group ($P < 0.05$). Kim et al. (2010) also reported a slightly higher final weight for the WCR-TMR group compared to the control group, although the difference was not statistically significant.

Regarding average daily gain (ADG), the WCR-TMR group showed significantly higher value from the growing stage until slaughter, but there was no significant difference in ADG during the finishing stage ($P > 0.05$). Notably, the average daily gain during the growing and pre-fattening stages was significantly higher in the WCR-TMR group, consistent with the findings of Kim et al. (2010). Over the entire period, the average daily gain was 0.78 kg/head/day in the WCR-TMR group, 18% higher than the control group's 0.66 kg/head/day. Particularly during the growing stage, the WCR-TMR feeding group showed a 22% improvement in average daily gain.

Table 16.17 Effect of WCR-TMR feeding on carcass characteristics of Hanwoo steers (SNU 16–18)

Item	Control	WCR-TMR	SEM	*Pr* > F
Marketing weight (kg)	726 ± 17.77^b	765 ± 14.42^a	12.135	0.012
Cold carcass weight (kg)	417.8 ± 12.98^b	450.4 ± 10.38^a	9.068	0.046
Back fat thickness (mm)	11.75 ± 1.84	13.00 ± 1.26	1.094	0.586
Longissimus dorsi area (cm^2)	88.00 ± 3.04	89.88 ± 4.64	2.694	0.740
Yield index	65.87 ± 1.21	64.30 ± 0.86	0.748	0.311
Yield grade (A:B:C)	2:4:2	2:4:2	–	–

**WCR* whole crop rice silage, *SEM* standard error of means
^{a,b}Values with different superscripts in the same row differ at $P < 0.05$

Table 16.18 Effect of WCR-based TMR feeding on meat quality of Hanwoo steers (SNU 16–18)

Item	Control	WCR-TMR	SEM	*Pr* > F
Marbling score (1–9)	4.00 ± 0.53	4.13 ± 0.58	0.381	0.876
Meat color (1–7)	4.75 ± 0.16	4.75 ± 0.25	0.144	1.000
Fat color (1–7)	3.13 ± 0.22	2.88 ± 0.29	0.182	0.512
Texture (1 = soft, 3 = firm)	1.25 ± 0.16	1.50 ± 0.18	0.125	0.334
Maturity (1 = young, 9 = old)	2.00 ± 0.00	2.00 ± 0.18	0.091	1.000
Quality grade(1^{++}:1^{+}:1:2:3)	0:2:4:2:0	0:2:4:2:0	–	–

WCR whole crop rice silage, *SEM* standard error of means
^{a,b}Values with different superscripts in the same row differ at $P < 0.05$

Table 16.17 illustrates the changes in carcass characteristics associated with the WCR-TMR feeding. The carcass weight was significantly higher in the WCR-TMR group, with a slaughter weight of 450.4 kg compared to 417.8 kg in the control group. Despite this difference, there was no significant variation in backfat thickness, rib eye area, or meat yield index between the two groups ($P > 0.05$).

Both the control and WCR-TMR feeding groups achieved the same meat grading results, with two heads graded as A, four heads graded as B, and two heads graded as C. However, Kim et al. (2010) reported different grading outcomes in their study: the control group had a grading distribution of A:B:C = 3:3:1, while the WCR-TMR group had a distribution of 5:2:0, indicating an improvement in meat grading for the WCR-TMR group.

In terms of intramuscular fat, levels were slightly higher in the WCR-TMR group, but this difference was not statistically significant. Additionally, there were no significant differences in meat color, maturity, fat color, or between the two groups. The meat quality grade also showed no significant difference between the two groups, with two heads graded as 1+, four heads as 1, and two heads as 2 for both treatments (Table 16.18). Notably, no meat was graded as 1++ in this experiment. In a similar study, Hwangbo et al. (2008) observed the positive effects from feeding barley silage, including an increased occurrence of grades higher than 1+. Further findings indicated that feeding whole crop barley silage influenced rumen development during the growing stage, increased rib eye area during the pre-fattening stage, and enhanced intramuscular fat deposition during the post-fattening period due to adequate energy supply (Kim and Seo 2006; Mowat and Slumskie 1971).

2.3 Dairy Cattle (Holstein)

The specification test for using rice as a feed for dairy cattle was conducted as a substitution test for corn silage and rye silage (Tables 16.19 and 16.20).

2.4 Substitution Test for Corn Silage

In the substitution test for corn silage, the rice silage was compared with corn silage. The rice silage feeding group produced an average of 25.6 kg of milk per head per day (98% of the corn silage group's production), slightly lower than the 26.1 kg per head per day observed in the corn silage feeding group. There was no significant

Table 16.19 Milk yield and composition in Holsteins fed either corn silage or whole crop rice silage in TMR

Item	Control (corn silage)	Whole crop rice silage
Milk yield (kg/d)	26.1	25.6
Milk composition		
– Fat, %	3.47	3.42
– Protein, %	3.25	2.95
– Lactose, %	4.83	4.80
– Total solids, %	12.68	12.30
– MUN, mg/dl	14.46	16.68
SCC, thousand/ml	214.07	180.04

SCC somatic cell count

Table 16.20 Milk yield and composition in Holsteins fed either rye silage or whole crop rice silage in TMR

Item	Rye-TMR	WCR-TMR
Milk yield (kg/d)	22.0	25.7
Milk composition		
– Fat, %	2.96	2.99
– Protein, %	3.15	3.27
– Lactose, %	3.56	3.73
– Total solids, %	8.55	8.75
SCC, thousand/ml	263	211

SCC somatic cell count

difference in the milk components such as milk fat, milk protein, lactose, and total solids. These results suggest that the nutritional value of rice silage is comparable to that of corn silage, with potentially higher feed efficiency.

2.5 Substitution Test for Rye Silage

In Table 16.20, rice silage-based TMR was compared with rye silage-based TMR. The rice silage-based TMR feeding group exhibited higher milk production compared to the rye silage-based TMR feeding group, although no significant differences were noted in the milk components. Notably, a decrease in somatic cell count, which is an indicator of udder health, was observed in rice silage-based TMR group. This decrease was also seen in the comparison with corn silage.

2.6 Conclusion

In several East Asian countries, including Korea, Japan, and Taiwan-China, rice consumption has been declining due to shift in dietary habits. This trend has led to increased rice stocks, creating challenges for the government. Rather than reducing rice production by eliminating paddy fields, an alternative solution is to repurpose these fields for cultivating rice intended for animal feed. Feed rice, known for its high nutritional value, can serve as livestock feed and help regulate overall rice production.

Despite the potential benefits of using feed rice to manage rice production, the Korean government has been cautious in actively pursuing this approach. In contrast, Japan has long implemented policies to support the cultivation of feed rice. As external pressures intensify, it is hoped that Korea will develop and expand effective policies to enhance the cultivation and utilization of feed rice.

References

Cai Y, Fujita Y, Murai M, Ogawa M, Yoshida N, Kitamura A, et al. Application of lactic acid bacteria (*lactobacillus plantarum* Chikuso-1) for silage preparation of forage paddy rice. Jpn J Grassl Sci. 2003;49:477–85.

Fukuma T, Kono Y, Kanda N, Shinde S. Effect of lactic acid bacteria addition on fermentation quality and aerobic degradation of high sugar content forage rice 'Tachisuzuka' whole crop silage. Livestock Technol Res Cent. 2018;17:1–6.

Hattori I, Sato K, Kobayashi R. Chemical composition of major forage rice varieties grown in the Kyushu region. Kyushu J Grassl Sci. 2005;35:26–31.

Higuchi K, Takusari N, Nonaka I, Tajima K, Yabumoto Y, Tomohisa Tomaru T, et al. Effects of grain filling, variety and cutting length of whole crop rice silage on nutrient utilization, ruminal fermentation and chewing time in dry cows. Bull Natl Inst Livest Grassl Sci. 2009;9:1–14.

Higuchi K, Takusari N, Nonaka I, Tjima K, Tomaru T, Ohtani F, et al. Estimation of metabolizable energy requirement for maintenance by both energy balance trials and integration of energy costs for activity in dry cows fed whole crop rice silages. Nihon Chikusan Gakkaiho. 2015;86:169–77.

Hiraoka H, Yamamoto Y, Urakawa S, Yamada H, Kand Y, Karita S, et al. Change of the β-carotene content of feed rice plant (*Oryza sativa* L.) with advancing growth stages and storage period of whole crop silage. Jpn J Grassl Sci. 2004;50:71–3.

Hosoda K, Nishida T, Ishida M, Matsuyama H, Yoshida N. Intake, digestibility and milk production by Holstein cows in early lactation fed 'Hoshiaoba' developed as a new breed of rice for whole crop silage. Jpn J Grassl Sci. 2005;51:48–54.

Hwangbo S, Choi SH, Kim SW, Kim WH, Son DS, Jo IH. Effects of dietary concentrate levels based on whole-crop barley silage on growth and meat quality in growing Korean black goats. J Anim Sci Tech. 2008;50:527–34.

Ishida M. Preliminary observations on milk production and feed utilization of dairy cows fed forage rice 'Kantoshi-206' roll bale silage. Kanto J Anim Sci. 2000;50:14–21.

Islam MR, Ishida M, Ando S, Nishida T, Yoshida N, Arakawa M. Effect of variety and stage of maturity on nutritive value of whole crop rice, yield, botanical fractions, silage fermentability and chemical composition. Anim Biosci. 2004a;17:183–92.

Islam MR, Ishida M, Ando S, Nishida T, Yoshida N, Arakawa M. Effect of variety and stage of maturity on nutritive value of whole crop rice silage for ruminants: in situ dry matter and nitrogen degradability and estimation of metabolizable energy and metabolizable protein. Anim Biosci. 2004b;17:1541–52.

Islam MR, Ishida M, Ando S, Nishida T, Yoshida N. Estimation of nutritive value of whole crop rice silage and its effect on milk production performance by dairy cows. Anim Biosci. 2004c;17:1383–9.

Japan Grassland Agriculture, Forage Seed Association. Technical manual for whole crop rice production and utilization. 7th ed; 2020. p. 1–214.

Jeong JS. The current status and problems of domestic forage production and distribution. In: Proceedings of 2021 symposium and conference of Korean Society of Grassland and Forage Science. Pyeongchang; 2021. p. 23–42.

Jin GL, Kim JK, Qin WZ, Jeong J, Jang SS, Sohn YS, et al. Effect of feeding whole crop barley silage- or whole crop rye silage based-TMR and duration of TMR feeding on growth, feed cost and meat characteristics of Hanwoo steers. J Anim Sci Tech. 2012;54(2):111–24.

Kawamoto H, Yamaguchi H, Komatsu T, Tanaka O, Oshibe A. Effect of chopping and high-density ensiling on the silage fermentation of forage paddy rice (Oryza sativa L.). Jpn J Grassl Sci. 2009;54:323–7.

Kawamoto H, Yamaguchi H, Sekiya H, Oshibe A, Nakagomi K. Chemical composition and total digestible nutrients content of spikelet decreased strain for forage paddy rice silage. Jpn J Grassl Sci. 2010;56:144–8.

Kim WH, Seo S. Cultivation and utilization of barley as the main winter crop in paddy field. Proceedings of Korean Society of Grassland and Forage Science Conference. 2006:37–57.

Kim JG, Park HS, Seo S, Yoon SH, Kim MJ, Lim YC. Effect of TMR using whole crop rice silage on carcass quantity and quality of Korean native cattle. In: Proceedings of the 56th International Congress of Meat Science and Technology. Jeju, Korea; 2010. p. 168.

Kim JG, Zhao GQ, Liu C, Wei SN, Kim HJ, Kim KH, et al. Feeding effects of whole crop rice based TMR on growth performance and carcass characteristics of Hanwoo steers. J Korean Soc Grassl Forage Sci. 2019;39(2):97–104.

Kim JG, Cheong EC, Li YF, Kim HJ, Farhad A, Kim MJ. Effect of TMR feed mixed with whole crop rice on growth performance and meat quality of Hanwoo steers. J Korean Soc Grassl Forage Sci. 2021;41(4):267–72.

Knaya C. β-carotene, a-tocopherol contents of forage rice plant (*Oryza sativa* L.) and silage making for fattening beef cattle. Jpn J Grassl Sci. 2007;53:167–71.

Kobayashi H, Nakagomi K, Senda M. Comparison of dry matter yield and nutritive components in two cultivars of forage rice with different heading dates. Jpn J Crop Sci. 2021;90:83–91.

Kono Y. Short panicle paddy rice cultivar (specialized varieties of high digestibility and high sugar content for whole-crop silage use) of feed characteristics and feeding to cattle. Livest Tech. 2016;7:21–6.

Kono S, Shinde S, Kanda N, Shirota K, Fukuma T, Tsukazaki Y. Nutrient value and ruminal degradability of whole crop silage prepared from Tachisuzuka as short panicle paddy rice cultivar. Jpn J Grassl Sci. 2014;60:91–6.

Kosis. Rice production in Korea. Korean Statistical Information Service; 2023.

Kusa K, Uegaki R, Kimura T. Effects of nitrogen fertilizer application and harvesting stage on the chemical composition and fermentation quality of the rice cultivar Tachisuzuka for whole-crop silage. Jpn J Grassl Sci. 2018;64:7–17.

MAFRA. The current situation of forage increase production and supplementation policy. Ministry of Agriculture, Food and Rural Affairs; 2022.

Matsushita K, Iida S, Ideta O, Sunohara Y, Maeda H, Tamura Y, et al. 'Tachisuzuka', a new rice cultivar with high straw yield and high sugar content for whole-crop silage use. Breeding Sci. 2011;61:86–92.

Matsushita K, Iida S, Ideta O, Sunohara Y, Maeda H, Tamura Y, et al. Tachisuzuka, a new rice cultivar with high straw yield, high digestibility and high sugar content for whole-crop silage use. Bull NARO Western Reg Agr Res Cent. 2012;11:1–13.

Matsuyama H. Feeding of whole crop rice silage to dairy cows and its effects. Livest Tech. 2009;10:9–14.

Matsuyama H, Shioya S, Ishida M, Nishida T, Hosoda K, Eruden B, et al. Feed characteristics of rice plants (*Oryza sativa* L. cv. Hamasari, Yumetoiro, Hokuriku-184) as whole crop silage. Jpn J Grassl Sci. 2005;51:289–95.

Matsuyama H, Shioya S, Nishida T, Hosoda K, Eruden B, Yoshida N, et al. Nutritive value of rice plants (*Oryza sativa* L. cv. Kusayutaka, Hamasari, Kusahonami) as whole crop silage. Jpn J Grassl Sci. 2006;51:385–9.

Matsuyama H, Shioya S, Hosoda K, Eruden B, Ishikawa T, Ishida M. Feed characteristic of rice plant (*Oryza sativa*) as whole crop silage with insect damage straight swifts (*Parnara guttata*). Jpn J Grassl Sci. 2013;59:33–7.

Momose Y, Hara T, Tsuchiya M, Nakazawa N. Effects of chopping length on the loading density and fermentation quality of whole crop rice silage. Jpn J Grassl Sci. 2005;51:190–4.

Mowat DM, Slumskie RA. Barley silage, ground whole plant barley and corn silage for finishing beef cattle. Can J Animal Sci. 1971;51:201–7.

Nakagomi K. Breeding trends in specialized varieties of high digestibility and high sugar content for whole-crop silage use. Livest Techn. 2016;7:9–12.

Nakagomi K, Fujimoto H, Ishioka G, Sasahara H, Shigemune A, Ideta O. Changes in the feed composition content of short-panicle rice varieties after heading in the western region of Japan. Jpn J Crop Sci. 2022;91:59–66.

Nakanishi Y. Rational feeding method of whole crop rice silage to beef breeding cattle. Kyushu J Grassl Sci. 2006;36:35–9.

Nakanishi N, Ide T, Ishizaki S, Yoshida N, Yamada T, Ishida M. Effects of feeding whole crop rice silage on meat quality of crossbred (Japanese black × Holstein) and Japanese brown cattle. Bull Natl Inst Livest Grassl Sci. 2018;18:17–25.

National Agriculture and Food Research Organization. Standard tables of feed composition. 2009. (in Japanese).

Omura H, Ishihara K, Machida Y, Taniguchi S, Matsuda M, Kuroe H. Variation over time in the composition of forage components and β-carotene content of forage rice hay by forage rice cultivar. Kyushu J Grassl Sci. 2007;37:37–42.

Oya Y, Yamashita D, Yamasaki K, Sato K. Kinetics of β-carotene in forage rice hay utilization system for feeding fattening cattle. Kyushu J Grassl Sci. 2005;35:7–15.

Shahbandeh M. Total global rice consumption. 2008;9:2022/23.

Shinde S. Intake, feeding of rice whole crop silage for lactating dairy cow. Jpn J Grassl Sci. 2010;55:365–72.

Shinde S, Iwashimizu T. The influence of the cutoff lengths of rice whole crop silage for the paddy excretive amount on lactating cow. Bull Hiroshima Prefectural Technol Res Inst Livestock Technol Res Cent. 2008;15:9–13.
Shinde S, Shirota K, Nagao K. Effects of rice whole crop silage proportion in total mixed ration on milk production performance and chewing activity of lactating cow. Kansai J Animal Sci. 2005;156:7–14.
Shinde S, Shirota K, Nagao K. The influence of the harvesting time of rice whole crop silage for the paddy excretive amount on lactating cow. Bull Hiroshima Prefectural Technol Res Inst Livestock Technol Res Cent. 2008;15:1–7.
Shinde S, Kand N, Ito K. Feeding trial on farm of fermented total mixed ration composed of rice whole crop silage. Bull Hiroshima Prefectural Technol Res Inst Livestock Technol Res Cent. 2012a;16:6–14.
Shinde S, Yoshimura T, Ito K. Effect of cutting length for milk production of rice whole crop silage composed total mixed ration. Bull Hiroshima Prefectural Technol Res Inst Livestock Technol Res Cent. 2012b;16:29–36.
Shinoda M, Kushibiki S, Shinguu H, Ueda Y, Murai M, Takita T. Effect of cultivars and treatments of rice on digestion in rumen. Tohoku Agric Res. 2002;55:125–6.
Shinoda M, Kushibiki S, Shinguu H, Ueda Y, Thono E, Murai M. Effects of feeding whole crop rice silage during the later fattening stage on growth, blood profiles and carcass traits of Japanese black steers. Nihon Chikusan Gakkaiho. 2007;78:201–8.
Shito H, Tachibana Y, Kawaide T, Okajima H, Kitanaka T, Syouda M, et al. Improvement of suitability for farm work and calculation of minimum harvesting area for introducing a self-propelled harvesting roll baler. Jpn J Grassl Sci. 2011;57:13–20.
Tachibana Y, Shito H, Kawaide T, Takahashi K, Okajima H, Kitanaka T, et al. Development of forage rice plant harvesting functions for self-propelled harvesting roll baler. Jpn J Grassl Sci. 2011;57:21–6.
Takahashi K, Okada S. Fine cutting harvesting and preparation technology of extremely short ear rice for WCS. Jpn J Grassl Sci. 2020;66:46–9.
Takahashi T, Maehara M, Zhang Y, Motobayashi T, Ishii Y, Kanda S, et al. Effect of feeding of whole crop rice silage on milk production, rumen fermentation, blood metabolites and feeding behavior of lactating cows. Nihon Chikusan Gakkaiho. 2007;78:45–55.
Takahashi K, Kubota J, Teramoto I, Takahashi H, Horie T, Tsunekawa I, et al. Study of low-cost harvesting and silage making system for whole-crop silage rice. Jpn Soc Agric Mach Food Eng. 2015;77:105–12.
Tohno M. Development of lactobacillus addition 'Chikuso-2' for high quality rice WCS preparation and fresh TMR preparation. Livest Tech. 2017;6:10–5.
Tohno M, Kobayashi H, Tajima K, Uegaki R. Strain-dependent effects of inoculation of *lactobacillus plantarum* subsp. *plantarum* on fermentation quality of paddy rice (*Oryza sativa* L. subsp. *japonica*) silage. FEMS Microbiol Lett. 2012;337:112–9.
Urakawa S. Producing for rice whole crop silage and rice grain. Jpn J Grassl Sci. 2017;63:38–42.
Urakawa S, Yoshimura Y. Development of the cutting roll baler for the rice whole crop silage. Jpn J Grassl Sci. 2003;49:43–8.
Urakawa S, Yoshimura Y, Hiraoka K, Yamamoto Y. Development of the addition device attached in the cutting roll baler for the rice whole crop silage. Jpn J Grassl Sci. 2003;49:254–7.
Urakawa S, Yoshimura Y, Hiraoka K, Okumura M. Improvement of a self-propelled bale wrapper for high-efficiency operation. Jpn J Grassl Sci. 2004;50:304–9.
Yamada T. Feeding and meat quality evaluation of forage rice to beef cattle. Livest Tech. 2009;10:15–9.
Yamada T. Feeding technology of whole crop rice silage for beef cattle. Livest Tech. 2016;3:11–3.
Yamamoto Y, Mizutani M, Inui K, Urakawa S, Hiraoka H, Goto M. Feed intake and lactation of dairy cow fed on total mixed ration of whole rice crop silage. Jpn J Grassl Sci. 2005;51:40–7.
Yamamoto Y, Mizutani M, Inui K, Urakawa S, Hiraoka H, Goto M. Effects of different types of roughage on digestion of whole crop rice silage in total mixed ration. Jpn J Grassl Sci. 2008;54:12–8.

Chapter 17
Regional Feed Resources and Sustainable Livestock Production

Ping Li, Jinsong Yang, Chuncheng Xu, Yanhong Yan, and Toshiyuki Ohtani

Abstract This chapter summarizes the use of regional feed resources in silage production and animal husbandry. In the context of silage production, it is essential to carefully select appropriate corn varieties for plateau forage silage. This selection must be accompanied by precise regulation of temperature, humidity, and fermentation conditions. The silage process for fruit residues, tea grounds, and bamboo can utilize either natural lactic acid bacteria or additional bacterial agents. Local agricultural by-products can be combined with other materials for silage production. The "maize-soybean strip intercropping" system yields silage maize biomass comparable to monoculture, while also providing additional soybean yield. High-quality silage significantly improves feed quality, optimizes nutritional composition, enhances nutrient intake of animals, and effectively promotes animal growth and production performance. This has important implications for the sustainable development of the livestock industry, as it increases feed conversion efficiency, reduces breeding costs, and enhances both the economic and ecological benefits of the livestock sector.

Keywords Livestock production · Regional feed resource · Silage Fermentation · Sustainable development

P. Li (✉)
Guizhou University, Guiyang, China
e-mail: lip@gzu.edu.cn

J. Yang
Hainan University, Haikou, China

C. Xu
China Agricultural University, Beijing, China
e-mail: xucc@cau.edu.cn

Y. Yan
Sichuan Agricultural University, Chengdu, China

T. Ohtani
Shizuoka Prefectural Research Institute of Animal Industry, Kikugawa, Japan
e-mail: fujikinka@sp-exp.pref.shizuoka.jp

Y. Cai, K. Ataku (eds.), *Cultural History and Modern Production Technology of Silage*, https://doi.org/10.1007/978-981-96-5787-2_17

1 Silage Preparation of Plateau Grassland Resources and Sustainable Livestock Production

Ping Li

In recent years, with the continuous growth in people's economic statues, people's demand for high-quality livestock products such as meat, eggs, and milk has gradually increased, and animal production has been vigorously developed. The increment of animal production is inevitably accompanied by an increase in the demand for forage, and high-quality forage resources have become an important factor to ensure the growth of animal production in China. Grazing lands are the main source of feed for animals on the plateau. The productivity of livestock raised in plateau areas is reduced due to problems such as insufficient and low-quality feed in winter and early spring. This calls for the development of new sources of forage and feed to meet the needs of plateau areas.

The preparation of silage is an acidic fermentation process, dominated by lactic acid bacteria (LAB) under anaerobic conditions using forage with appropriate moisture content. LAB converts water-soluble carbohydrates (WSC) into organic acids (mainly lactic acid) to reduce the pH of silage, thereby inhibiting microbial activity and nutrient losses and thus forming silage. Silage has gradually become the main feed source for livestock in plateau areas. There are many raw materials for making silage. Currently, the most widely used raw material for silage is corn. Other raw materials include tall grasses such as oats and pennisetum, common forages such as alfalfa and ryegrass, and some unconventional silage raw materials such as *Broussonetia papyrifera* (Fig. 17.1). Silage is one of the most effective and lowest-cost methods for long-term preservation of nutrients in green forage. Silage can also improve the quality of structural carbohydrates in plants and degrade some cellulose and hemicellulose, thereby improving feed digestibility.

1.1 Evaluation and Influencing Factors of Silage

The quality of silage depends on the raw materials used, silage storage condition, and method used, and the nutritional value of silage is also closely related to the above mentioned factors. As the key to ensuring a stable feed supply for livestock throughout the year, the improvement of silage quality has many important meanings: At the economic level, high-quality silage can reduce feed waste and lower breeding costs. High-quality silage can enhance the digestion and nutrient absorption of feed, reduce the amount of feed fed, and at the same time reduce the treatment cost of livestock diseases caused by feed quality problems; in terms of environmental protection, high-quality silage helps reduce greenhouse gas emissions. Good silage production and preservation processes can reduce the production of greenhouse gases such as methane. From the perspective of breeding benefits,

Fig. 17.1 Common silage raw material crops

high-quality silage can improve the production performance of livestock, such as improving weight gain of beef and mutton cattle, increasing the milk yield and milk fat of dairy cows, thereby increasing income of farmers. In addition, LAB in silage also have probiotic effects on the intestinal health of ruminant, can improve the intestinal microecology of animals, produce various antibacterial substances such as small molecule antimicrobial peptides and bacteriocins, and inhibit harmful pathogenic bacteria in silage and animal intestines; It can also synthesize a variety of nutrients such as folic acid, vitamin B1, and vitamin B2, reduce the morbidity of livestock, and promote animal growth and improve production performance.

1.2 High-Quality Silage Is Usually Identified by the Following Factors

1.2.1 Color

The color of high-quality silage should be close to that of the raw materials used. For example, silage corn should be yellow-green or green. If the silage color turns black or brown, it usually indicates poor silage quality.

1.2.2 Smell

It has a sour and fragrant smell, similar to the smell of pickles, which is a normal smell produced by LAB fermentation. There should be no pungent ammonia smell, rotten smell, or musty smell. If there is a strong ammonia smell, it may mean excessive protein degradation; a rotten smell may indicate that the silage has severely deteriorated.

1.2.3 Texture

The structure is intact, the stems and leaves remain clearly visible, and the texture is soft and moist. There should be no sticky or dry feeling. When fetched with hand, there should not be much water remaining on the hand, but there should be a slight damp feeling.

1.2.4 pH Value

Generally speaking, the pH value of high-quality silage should be between 3.8 and 4.2. The lower the pH value, the more sufficient the silage fermentation, and the activity of harmful microorganisms is inhibited.

1.2.5 Dry Matter Content

The appropriate dry matter (DM) content is usually between 30% and 35%. High DM content may lead to poor silage fermentation, and too low content is prone to produce exudate, resulting in the loss of nutrients.

1.2.6 Nutritional Component

It should contain higher nutritional components such as crude protein and energy, and the loss of nutritional components is small compared with silage raw materials. For example, the crude protein content in silage corn should be maintained at a certain level to meet the nutritional needs of livestock.

1.2.7 Aerobic Stability

High-quality silage should be able to remain stable for a long time after being exposed to air and is not prone to heating and deterioration. This means that the beneficial microorganisms in the silage can inhibit the growth of harmful microorganisms and reduce the secondary fermentation of silage. These indicators are comprehensively evaluated to determine whether the quality of silage is high quality, thereby providing a good feed source for livestock and ensuring the healthy growth and high productivity from livestock.

1.3 Sustainable Production of Plateau Animal Husbandry

In recent years, some grasslands on the Qinghai-Tibet Plateau have shown varying degrees of degradation due to overgrazing, rodent damage, natural disasters, and other reasons, which is not conducive to the sustainable production of plateau animals. The cold season in the Qinghai-Tibet Plateau region lasts for a long time with low temperatures. The yield and nutritional components of natural grassland forage in the withered grass period are significantly lower than those in the warm season. Under the purely natural grazing mode, the demand for forage resources in the grassland animal husbandry during the cold season increases significantly. In the research on the sustainable development strategy of agriculture in Tibet, it is pointed out that the prominent problems in the development of animal husbandry in Tibet are overgrazing of pastures in plateau areas and insufficient forage resources. The development and utilization of high-quality and high-yield forage for silage in high-altitude pastoral areas can improve the current situation of insufficient forage resources and unbalanced seasonal supply on the Qinghai-Tibet Plateau, shorten the slaughtering cycle of plateau herbivorous livestock, improve breeding benefits, reduce the pressure on grassland livestock carrying capacity, and enhance the social, economic, and ecological benefits in the Qinghai-Tibet Plateau region.

Currently, the selection, planting, and processing of silage corn varieties in high-altitude areas have become an important way to solve the problem of forage resources in plateau areas, which can effectively guarantee the supply of forage for herbivorous livestock on the plateau, improve the utilization efficiency of grassland and the income of farmers and herdsmen in planting, and alleviate the pressure of livestock carrying capacity on pastures and the contradiction between grass and livestock. Some researchers have planted and assessed silage corn varieties that adapt to the alpine and cold climate of the plateau. Yao et al. (2022) assessed high-quality silage corn varieties suitable for planting in the plateau area in the Ganzi Tibetan Autonomous Prefecture with an average altitude of 3300 meters. The results showed that among the varieties, Zhongyu Grass No. 3 had a higher yield, adapted to the alpine and cold climate conditions, and recorded higher plant height, leaf length, and stem diameter. Yang et al. (2017) carried out a demonstration and promotion experiment of silage corn at an altitude of over 3200 meters in the plateau dam area of Diqing Plateau. The results showed that among the experimental varieties, Zhongyu 335 had the best total yield and individual plant yield and had good adaptability in the high-altitude environment. In addition, some researchers used spring corn and summer corn in the Tibet region as materials to produce silage and determine the fermentation quality and microbial diversity of the silage. The results showed that the quantity of lactic acid and WSC in the summer silage corn group was higher than those in the spring corn group, while the dry matter content and pH value were lower. Therefore, the quality of summer corn silage fermentation was better.

The demand for forages for livestock and poultry production is relatively large. Long-term uncontrolled harvesting of fresh forage can easily lead to damage to the natural environment and ecology. The development of forage silage technology effectively ensures the supply of high-quality forage in livestock and poultry breeding, and its application in forage is conducive to retaining the nutrients in the raw materials to the greatest extent; prolonging the preservation time of forage, improving the palatability of forage and increasing the digestion, absorption rate, and utilization rate of feed by livestock and poultry; and improving the economic and social benefits of forage. Therefore, silage is the choice of most farmers. Silage for feeding livestock and poultry reduces the uncontrolled, excessive, and destructive harvesting of fresh forage by farmers, effectively protects the ecological environment, promotes the harmonious development of forage protection and clearance breeding, and also lays a good development foundation for the industrialization development of forage.

2 Fruit Residue Silage and Fermentation Quality

Jinsong Yang

Banana, pineapple, and papaya are popular tropical fruits in the world. According to FAO subregional production statistics, their production per year worldwide was 119.83 million tons, 28.15 million tons, and 1.40 million tons in 2020, respectively. Commercial processing of tropical fruits produces large quantities of residues every year in many countries, including China, but some residues are dumped into landfills, whereas others are used as compost, which leads to wasted resources and possible environmental problems because of unsuitable disposal. Demand for efficient use of food by-products has increased due to economic and environmental concerns. Defective bananas, banana by-products, pineapple peels, and papaya peels are high in nutrients such as vitamins, minerals, and fiber and can be potential feed resources for livestock. However, these fruit residues are highly perishable because of their high moisture content. LAB is naturally present on the surface of some fruits. They are responsible for silage fermentation and also influence fermentation quality (Yang et al. 2016, 2022).

In this study, LAB from fruit residues were screened, isolated, and identified, focusing on strains that are most likely to play an important role in the fermentation process. The identification of the isolated strains was performed using 16S rDNA sequence analysis and at the molecular level. In order to effectively utilize fruit residues as animal feed, their chemical composition and silage fermentation characteristics were also studied (Yang et al. 2016, 2022).

The fruit residues of banana, pineapple, and papaya were collected from a local commercial food factory and a plantation base. Natural populations of LAB and silage fermentation of fruit residue were studied. Eighty-two strains of LAB were isolated from fruit residues such as banana leaf and stem, pineapple peel, and papaya peel (Yang et al. 2016). Eleven strains of LAB were isolated from the fruits and flowers of defective bananas. These LAB were identified and characterized. Pineapple and papaya peels contained higher crude protein (11.5–13.8% of dry matter) and water-soluble carbohydrate (WSC) (16.8–22.4%), but lower acid detergent fiber contents (21.2–26.4%) than banana stems and leaves (8.2% crude protein, 42.8% acid detergent fiber, and 5.1% WSC). Defective bananas contain 74.8–76.3% moisture, 7.2–8.2% crude protein, 5.9–6.5% ether extract, and 25.3–27.8% neutral detergent fiber on a dry matter (DM) basis. Strains isolated from fruit residue were identified as *Lactiplantibacillus plantarum* (54.9% of the total isolates), *Lactobacillus paraplantarum* (3.6%), *Lactobacillus nagelii* (8.5%), *Lactobacillus perolens* (4.9%), *Lactobacillus cas*ei (11%), *Lactobacillus fermentum* (9.8%), and *Enterococcus gallinarum* (7.3%), respectively (Yang et al. 2016). Strains isolated from the banana fruits and flowers were identified as *Lactiplantibacillus plantarum.*

Lactiplantibacillus plantarum was the dominate members of the LAB population in those fruit residues. These silages prepared by using a small-scale fermentation system were well preserved, with a low pH and relatively high content of lactic acid. The study suggests that the fruit residues contain abundant LAB species and nutrients and that they could be well preserved by making silage, which is a potential good fruit protein source for livestock diets.

As shown in Tables 17.1 and 17.2, these fruit residues have high number of epiphytic *lactobacilli* (10^5 cfu/g of FM). During the silage fermentation, the *lactobacilli* could produce sufficient lactic acid to reduce pH and inhibit the growth of harmful bacteria, and the resulting silage was of good quality.

The results confirm that fruit residue and defective bananas are nutritious, can be used to prepare high-quality silage, and have good potential as a feed source for ruminants. This is of great significance for effectively utilizing biological resources, reducing feed costs, alleviating feed shortages, and promoting the sustainable development of animal husbandry.

Table 17.1 Microbiological analysis of fruit residues

	Counts of viable microorganisms (cfu g^{-1} of FM)					
Sample	LAB	Bacilli	Coliform bacteria	Aerobic bacteria	Mold	Yeast
Banana stem and leaves	4.5 ± 0.8^b	4.6 ± 0.7^a	4.4 ± 1.2^a	6.7 ± 2.2^a	<2.0	3.4 ± 0.4^c
Defective banana	4.2 ± 1.8^b	2.5 ± 0.6^c	3.4 ± 1.5^b	6.4 ± 2.0^a	<2.0	3.5 ± 1.2^c
Banana flower	3.3 ± 1.4^c	<2.0	3.0 ± 1.2^c	3.5 ± 1.5^c	<2.0	<2.0
Pineapple peel	3.5 ± 0.5^c	2.0 ± 0.4^d	3.7 ± 1.0^b	6.6 ± 1.6^a	<2.0	5.2 ± 0.6^a
Papaya peel	5.8 ± 1.3^a	3.2 ± 0.5^b	3.6 ± 0.8^b	5.4 ± 1.1^b	<2.0	3.9 ± 0.5^b

cfu colony-forming unit, *FM* fresh matter, *LAB* lactic acid bacteria, *ND* not detected
$^{a, b, c, d}$Means within columns with different superscripts differ ($P < 0.05$)

Table 17.2 Chemical composition and silage quality of fruit residues

Item	Banana stem and leaves	Defective banana	Pineapple peel	Papaya peel
Chemical composition				
Moisture (%)	85.5 ± 1.1[b]	75.4 ± 0.8[c]	94.3 ± 2.2[a]	93.0 ± 1.6[a]
Organic matter (% DM)	85.8 ± 1.7[c]	93.8 ± 1.2[a]	87.3 ± 0.7[b]	88.2 ± 1.5[b]
Crude protein (% DM)	8.2 ± 0.5[c]	8.1 ± 0.8[c]	13.8 ± 0.9[a]	11.5 ± 0.7[b]
Ether extracts (% DM)	2.3 ± 0.1[b]	6.4 ± 0.6[a]	2.1 ± 0.3[b]	1.5 ± 0.2[c]
NDF (% DM)	44.3 ± 1.1[a]	26.9 ± 1.2[c]	28.5 ± 0.6[b]	22.2 ± 0.5[d]
ADF (% DM)	42.8 ± 1.1[a]	17.8 ± 0.8[d]	26.4 ± 0.7[b]	21.2 ± 0.6[c]
WSC (% DM)	5.1 ± 0.2[d]	18.9 ± 0.8[b]	22.4 ± 0.8[a]	16.8 ± 0.7[c]
Silage quality				
Moisture content (%)	86.7 ± 0.8[c]	75.1 ± 1.2[d]	94.5 ± 1.1[a]	93.3 ± 1.6[b]
pH	4.3 ± 0.1[a]	3.7 ± 0.6[b]	3.4 ± 0.3[b]	3.4 ± 0.2[b]
Lactic acid (% DM)	2.7 ± 0.2[d]	11.7 ± 0.7[c]	18.9 ± 1.2[b]	21.6 ± 0.5[a]
Acetic acid (% DM)	6.9 ± 0.2[b]	6.2 ± 0.2[c]	7.2 ± 0.5[a]	6.8 ± 0.3[b]
Butyric acid (% DM)	ND	ND	ND	ND
Propionic acid (% DM)	ND	ND	ND	ND
Ammonia-N (% DM)	4.1 ± 0.4[a]	3.6 ± 0.5[b]	3.3 ± 0.2[b]	3.2 ± 0.2[b]

Silage was stored for 60 days; data is the average of three fruit residue samples
DM dry matter, *NDF* neutral detergent fiber, *ADF* acid detergent fiber, *WSC* water-soluble carbohydrate, *ND* not detected
[a]Means within rows with different superscripts differ ($P < 0.05$)

3 Tea Ground Silage and Nutrient

Chuncheng Xu

Tea, made from *Camellia sinensis* leaves, is one of the most consumed non-alcoholic beverages in the world. With the prevalence of tea beverages, the demand for tea production has increased. As the country of origin of tea, China has the largest tea planting area and highest tea production. In 2022, its tea plantation area reached 3.39 million hectares, and tea production reached 3.34 million tons (Lei et al. 2024). With the increase in demand for tea products and tea drinks, large amounts of tea grounds are produced. As the major tea-producing and tea-consuming country, China has a lot of tea grounds and is exploring how to effectively utilize these tea grounds (Guo et al. 2021, 2023). Although a small amount of tea grounds are converted into raw compost material, most are generally buried. There is increasing demand for efficient use of food by-products due to economic and environmental concerns (Xu et al. 2007).

Table 17.3 Chemical composition of different tea grounds

Item	Green tea ground	Barley tea ground
DM (%)	25.0	29.2
Organic matter (% DM)	97.0	96.9
Crude protein (% DM)	31.1	12.6
Neutral detergent fiber (% DM)	41.0	34.8
Acid detergent fiber (% DM)	26.1	23.8
Ether extract (% DM)	5.7	3.1
Water-soluble carbohydrate (% DM)	0.3	0.1

DM dry matter

3.1 Feed Characteristics of Tea Grounds

Tea grounds refer to the solid organic waste generated from the production and processing of tea and tea beverages. Tea leaf has a useful amount of amino acids, proteins, vitamins, tannins, and polyphenols (Cai et al. 2001; Xu et al. 2007). Most of these nutrients remain in the tea grounds after extraction to make tea drinks. Studies have shown that tea grounds contain a variety of valuable nutrients (Wang et al. 2011; Xu et al. 2008b, 2008c). In general, tea grounds contain 1–2% of tea polyphenols, 0.1–0.2% of caffeine, 30% of crude protein (CP), and 16–18% of crude fiber on dry matter (DM) basis after extraction (Xu et al. 2003, 2007, 2008a). In addition, the composition of lysine and methionine in tea grounds is 1.5–2% and 0.5–0.7%, respectively, indicating high potential utilization value. All these suggest that tea grounds have a potential as a feed resource (Table 17.3).

Because of high-moisture content, tea grounds readily deteriorate after being released as a by-product from tea drink companies. Therefore, it must be used as soon as possible if fed directly, whereas direct feeding exhibits poor palatability due to the tannin components existing in tea grounds. Furthermore, drying process will consume a lot of energy if used as dry feed (Wang and Xu 2013). Therefore, ensiling is one of the suitable ways to preserve high-moisture tea grounds.

3.2 Fermentation Quality and Nutrition Value of Tea Ground Silage

Preservation by ensiling is highly dependent on lactic acid fermentation. Due to the low water-soluble carbohydrates (WSC) and low lactic acid bacteria (LAB), tea grounds are very difficult to ensiling alone (Cai et al. 2001; Xu et al. 2008a). Therefore, tea grounds are usually ensiled with additives or in mixed silage with other feed materials. The addition of *Lactiplantibacillus plantarum* FG1 and commercial acremonium cellulase in tea grounds greatly improved the fermentation quality and retained the nutrient components (Table 17.4). Barley tea grounds treated with mixtures of *Lactiplantibacillus plantarum* Chikuso-1 and acremonium

Table 17.4 Fermentation quality, chemical composition, apparent digestibility, and nutrient content of green tea ground silage

	FG1 + AUS	Control
Fermentation profile		
pH	3.49[c]	4.98[a]
Lactic acid (%FM)	1.55 [c]	0.23[a]
Acetic acid (%FM)	0.52	0.50
Propionic acid (%FM)	Nd	Nd
Butyric acid (%FM)	Nd	Nd
Ammonia N (%FM)	0.03	0.05
Chemical composition		
Dry matter (%)	25.2	25.5
Organic matter (%DM)	96.7	96.9
Crude protein (%DM)	31.3	31.0
Ether extract (%DM)	6.1	6.0
Acid detergent fiber (%DM)	21.5[b]	23.2[a]
Apparent digestibility		
Dry matter (%)	70.6 ± 2.8	–
Crude protein (%)	74.6 ± 4.4	–
Ether extract (%)	50.7 ± 2.9	–
Acid detergent fiber (%)	50.2 ± 2.2	–
Energy (%)	67.4 ± 1.9	–
Nutrient content		
Total digestible nutrients (%DM)	71.1 ± 0.3	–
Digestible energy (MJ/kg DM)	13.4 ± 0.4	–

FG1 Lactiplantibacillus plantarum, *AUS acremonium* cellulase, *FM* fresh matter, *DM* dry matter
[a, b, c] values with different superscript letters differ ($P < 0.01$)

cellulase, formic acid, and sodium hydroxide (NaOH) were also well preserved as indicated by enhanced lactic acid fermentation or restricted fermentation of almost all microorganisms (Xu et al. 2008a). In addition, it has been found that the preparation of mixed silage is effective for green tea grounds to improve fermentation quality. Green tea grounds could be ensiled successfully without bacterial inoculants when mixed with materials containing sufficient sugars (Nishino et al. 2007). Besides forage silage, the addition of green tea grounds could also enhance the lactic acid fermentation of byproducts-mixed silage when there are insufficient materials for lactic acid production (Kondo et al. 2004, 2006). Researches indicated that neither green tea grounds-associated LAB nor tea polyphenols account for the enhancement of lactic acid fermentation. Nasehi et al. (2018) conducted the experiment on the effect of different levels of green tea grounds and found that the lambs' average daily gain and the feed conversion ratio have been improved by the addition of 20 g/kg green tea grounds in the diet. The CP digestibility and N retention were significantly higher than other treatments. Moreover, methane emission has been reduced, which may be attributed to the tea saponins and phenolic compounds in the green tea grounds. Tea ground silage might also be a protein supplement as it contains 22–35% of crude protein.

Many researchers have studied the nutritive value of tea ground and tea ground silage so as to fully assess the feeding effect of tea ground silage treated with different additives or ensiled as mixed silage. The nutritive value of green tea grounds is thought to be equivalent to that of brewers' grains. Digestibility of CP, ether extract (EE), and acid detergent fiber (ADF) for green tea ground silage are 74.6%, 50.7%, and 50.2%, and the estimated TDN, digestible CP (DCP), and digestible energy (DE) are 71.1%, 23.9%, and 13.4 MJ/kg on DM basis, respectively (Xu et al. 2003). However, digestibility of CP, organic cell wall, and ADF for BTG are 32.5%, 28.8%, and 21.2%, and the estimated TDN, DCP, DE, and metabolizable energy are 71.0%, 2.9%, 12.9 MJ/kg, and 10.7 MJ/kg on DM basis, respectively. It is suggested that the nutritive value of barley tea ground silage is about 80% of that of barley and they can be used as a potential feed ingredient (Xu et al. 2003). Xu et al. (2008c) suggested that barley tea ground silage treated with NaOH had the highest ruminal degradability of DM and CP, followed by LAB and acremonium cellulase treatment. However, formic acid treatment had no effect on the DM and CP disappearance of silage. Wang et al. (2011) evaluated the fermentation characteristics and protein degradation dynamics of wet green tea ground silage treated with different additives and also found that formic acid treatment had higher NH_3-N and free-amino nitrogen (FAA-N) content than the control ($P < 0.05$). Previous study indicated that ensiling may degrade part of the components of tea catechins, while not affecting total phenols, antioxidative activity, and digestibility of green tea grounds (Nishino et al. 2007). It is inconsistent with conclusions that tannins could prevent protein degradation during forage ensiling and studies that the addition of green tea grounds could increase ruminal gas production and decrease the DM loss (Kondo et al. 2006). Nishida et al. (2006) suggested that feeding diets containing 20% of green tea ground silage on dietary DM had no negative impact on ruminal fermentation, but increased the plasma antioxidative activity and vitamin E concentration.

3.3 *Fermentation Quality and Nutritional Value of Tea Ground TMR Silage*

Ensiling is suitable for preserving high-moisture tea grounds. However, due to the high content of tea polyphenols, the palatability of tea ground silage is usually poor. If ensiled with dry feeds as a total mixed ration (TMR), the risk of effluent production would be minimized, and the time for mixing prior to feeding could be omitted. In addition, unpalatable by-products could be incorporated into TMR if their odors and flavors could be altered by silage fermentation. According to Xu et al. (2007, 2008a), the use of green tea grounds to produce TMR silage can effectively preserve the nutritional components of tea grounds and solve the problem of juice discharge in high moisture silage. Meanwhile, green tea beverage grounds can be used as a substitute for some grain concentrate feed, which will not affect the palatability, dry matter intake (DMI), and digestibility of nutrients. It has been shown that the substitution of green tea grounds for wet brewers' grains at different ratios of TMR could be well preserved with high lactic acid content and low pH and NH3-N content (Xu et al. 2007). Besides, TMR silages prepared with barley tea grounds at ratios of 10%, 20%, and 30% DM were also well preserved (Xu et al. 2004). Green tea grounds contain a lot of protein and functional components, such as tannin, caffeine, beta-carotene, and vitamin E (Cai et al. 2001). Therefore, it can be used as an additive to improve low-quality silage instead of the current expensive protein supplements. Previous study has indicated that condensed tannin (CT) at levels of 25–40 g/kg DM can protect amino acids to increase the absorption in the small intestine of ruminants, while deteriorating intake and digestibility at >55 g/kg DM (Min et al. 2003). All these suggest that many possible results will occur on the utilization of tea grounds, and the proper addition of tea grounds needs further study. Many researches have investigated the feeding value of different application rates of tea grounds. Xu et al. (2004) found that with the increasing proportions of barley tea grounds in TMR silage, digestibility of DM, CP, and TDN in TMR silages containing 10% and 20% barley tea grounds were significantly higher than that of 30% green tea grounds. Therefore, the ideal mixing proportion of the barley tea grounds for TMR silage is 10% to 20% on DM basis. Progressive increases of green tea grounds (0%, 5%, 10%, and 15% on DM basis) substituted for wet brewers' grains had no effect on the voluntary feed intake of TMR silage, while digestibility of DM, organic matter, CP, and energy were slightly lower than the control, and the N intake and fecal N increased. No differences among treatments were observed in urinary N, retention N, pH level, and total VFA concentration. However, the rumen NH_3-N contents tend to decrease with the increasing of green tea grounds. Therefore, the high green tea ground level of 15% of diet DM can be recommended for silage-based TMR (Xu et al. 2007).

3.4 Conclusion

As mentioned above, alternatives or additives for animal feed are necessary to relieve the feed resource pressure. Currently, ensiling of forages has become a promising method for preservation of feedstuffs and promoting livestock production. Tea grounds are abundant and available all year round. However, because of the high tannin contents, the palatability of tea grounds is relatively poor. Besides, the high moisture content of tea grounds limits its direct use as feedstuff. Therefore, ensiling could be one of the suitable ways to preserve high-moisture tea grounds. In view of the nutritive properties of tea grounds, there is a widespread interest in its use as a nutrient supplement and an effective animal feed source. Overall, tea grounds as silage and TMR silage have high nutrient values and exhibit positive advantages on feed intake, digestibility, gas production, and rumen fermentation for ruminants.

4 Mixed Sowing of Feed Corn and Soybeans and Their Silage Preparation

Yanhong Yan

Maize (*Zea mays*) and soybean (*Glycine max*) are China's main feed crops. To achieve self-sufficiency in maize and soybeans through monoculture, nearly 100 million hectares (ha) of land would be required in China. China annually imports nearly 100 million tons of soybeans, accounting for 75% of the total grain imports, with approximately 80% of these imports used for animal feed. The intercropping of maize and soybean is one of the most common cultivation patterns that not only efficiently utilizes resources but also enhances land productivity (Yang et al. 2024). The "maize-soybean strip intercropping" leverages the biological characteristics and border effects of both crops and increases light energy utilization effectively. This cropping system not only achieves silage maize biomass yields comparable to monoculture but also gets an additional soybean yield (Yan et al. 2015). Whole plant maize-soybean silage combines high-energy and high-protein feed, rich in starch and nitrogenous substances primarily in the form of protein. This system supports the transition from grain to forage production and promotes the sustainable and high-quality development of integrated crop-livestock farming systems.

Maize-soybean intercropping, through the scientific implementation of "selecting the compact varieties," "increased light exposure through widening space between maize and soybean rows," and "reducing plant spacing to ensure plant density," offers significant agricultural advantages and economic value. (1) In the maize-soybean strip intercropping systems, the use of compact or semi-compact maize varieties (e.g., Zhongyu No. 3) and shade-tolerant soybean varieties (e.g., Qihuang No. 34) is more conducive to sustaining high yield of both crops (Yang et al. 2017). (2) Maize-soybean intercropping systems can exploit the complementary spatial and temporal ecological niches of both crops, narrowing the row spacing of maize and soybean compared to monoculture, respectively, while expanding the inter-row spacing between maize and soybean rows. Compared to the monoculture planting of soybean and maize, the light resource utilization efficiency increases by 1.50 times and 1.16 times, respectively (Liu et al. 2018). In this arrangement, the maize strip consists of two rows planted in a wide-narrow row pattern, where the narrow row spacing within the maize strip is 0.4 m, and the wide row spacing is between 1.9 m and 2.5 m. Within the wide rows, soybeans are planted in strips of two to four rows, with an inter-row spacing of 0.3 m. The spacing between the maize strips and the soybean strips is maintained at 0.6–0.7 m (Yang et al. 2014) (Fig. 17.2) (3). The plant spacing within the maize strip is significantly reduced to achieve the planting density of pure maize, while the plant spacing within the soybean strip is moderately reduced to achieve over 70% of the pure planting density (Chen et al. 2023). Different intercropping patterns and appropriate seeding densities significantly affect yield and nutritional value. Studies have shown that

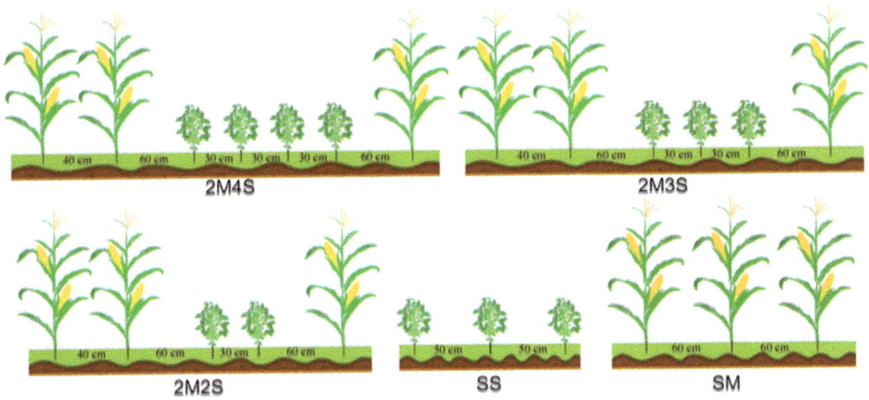

Fig. 17.2 Different strip intercropping patterns of maize-soybean. SS, soybean; SM, maize; 2M2S, 2 maize rows plus 2 soybean rows strip intercropping; 2M3S, 2 maize rows plus 3 soybean rows strip intercropping; 2M4S, 2 maize rows plus 4 soybean rows strip intercropping. Same as the below

intercropping maize and soybean in row ratios of 2:2, 2:3, and 2:4 resulted in land equivalent ratios exceeding 1.7, with the highest at 1.8 for the 2:3 and 2:4 row ratios (Zeng et al. 2022) (Table 17.5 and Table 17.6). Additionally, research has indicated that planting with a 2:3 row ratio, soybean and maize densities of 150,000 plants ha-1 and 75,000 plants ha-1, respectively, with soybean sown 9 days later than maize optimizes both silage quality and biomass yield (Li et al. 2022).

Maize, widely utilized as silage material, boasts several advantages such as tall stalks, high nutritional value, and excellent palatability. It is rich in water-soluble carbohydrates (WSC), making it highly suitable for ensiling (Han et al. 2018). However, maize silage possesses lower protein content, varying from 5.5% to 7.0% of dry matter (DM) (China standard NY/T 342004), and is unable to meet the protein demand of livestock (Qu et al. 2013). Soybeans are rich in protein and vitamins, whereas the high buffering capacity often leads to the degradation of protein nitrogen (PN) into non-protein nitrogen (NPN), resulting in nutrient loss and high levels of butyric acid (Batista et al. 2019). Ensiling is the primary processing method for the feed utilization of maize-soybean strip intercropping system. Whole-plant maize-soybean mixed silage effectively combines the advantages of individual whole plant of maize and soybean silages, improving the quality of the silage feed. This approach allows for simultaneous harvesting and ensiling, saving time and labor, maximizing whole-plant utilization, increasing forage supply rates, and reducing environmental pollution from improper straw use. Harvesting soybeans at the pod-filling to early maturity stages (R6-R7) and maize at the 1/2 to 2/3 milk line stages ensures optimal moisture content and digestible nutrients, thereby increasing the success rate of mixed silage. The stubble height of both crops should be maintained above 20 cm during harvesting to obtain high-quality roughage and to avoid introducing butyric acid bacteria from the soil, which can produce butyric acid

Table 17.5 Land equivalent ratios and biomass of maize and soybean as affected by different planting patterns in 2018

Treatment	Maize fresh weight	Soybean fresh weight	Total fresh weight	Maize dry weight	Soybean dry weight	Total dry weight	Land equivalent ratio
	t ha-1						
2M2S	65.26	10.34	75.60	23.36	3.40	26.76	1.74
2M3S	63.05	11.84	74.90	23.23	3.98	27.20	1.80
2M4S	59.39	12.58	71.98	21.58	4.17	25.75	1.80
SM	62.49	–	62.49	21.49	–	21.49	–
SS	–	14.94	14.94	–	4.73	4.73	–

Data from Zeng et al. (2022)

Table 17.6 Land equivalent ratios and biomass of maize and soybean as affected by different planting patterns in 2019

Treatment	Maize fresh weight	Soybean fresh weight	Total fresh weight	Maize dry weight	Soybean dry weight	Total dry weight	Land equivalent ratio
	t ha-1						
2M2S	63.40	11.14	74.54	20.75	3.42	24.17	1.76
2M3S	60.84	12.44	73.29	21.77	3.90	25.68	1.80
2M4S	57.32	13.38	70.70	19.66	4.06	23.72	1.80
SM	60.10	–	60.10	20.88	–	20.88	–
SS	–	15.83	15.83	–	3.73	3.73	–

Data from Zeng et al. (2022)

during the ensiling fermentation process, leading to significantly reductions in nitrate nitrogen content. The cutting length of the crops also affects the degree of compaction, which is crucial for ensuring normal fermentation in the rumen. The recommended cutting length for maize and soybeans is 1.0 to 1.9 cm. And it is also crucial to use the crushing equipment for corn and soybean grains during silage processing. For high-quality maize and soybean mixed silage, LAB inoculants should be added. The effectiveness of LAB inoculants is specific and does not guarantee favorable fermentation results in all environments. Therefore, selecting homotypic and heterotypic LAB mixed inoculants that are well-suited to local conditions is essential. The filling density should be greater than 750 kg.m^{-3}, and the weight of the compaction equipment should be over 15 tons. Faster filling and compaction are preferable. Ensiling can be done using silos or plastic bags. Regularly inspecting the plastic bags and cover films for any damage and promptly repairing any leaks is crucial.

Mixed ensiling of maize and soybeans can significantly improve the fermentation quality of silage, resulting in high-quality silage feed. Compared to monoculture, maize-soybean strip intercropping promotes the growth of epiphytic LAB attached to the raw materials (LAB $>10^5$ cfu/g fresh weight in intercropped materials) (Zeng et al. 2020). In maize-soybean mixed silage, *Weissella* and *Lactobacillus*

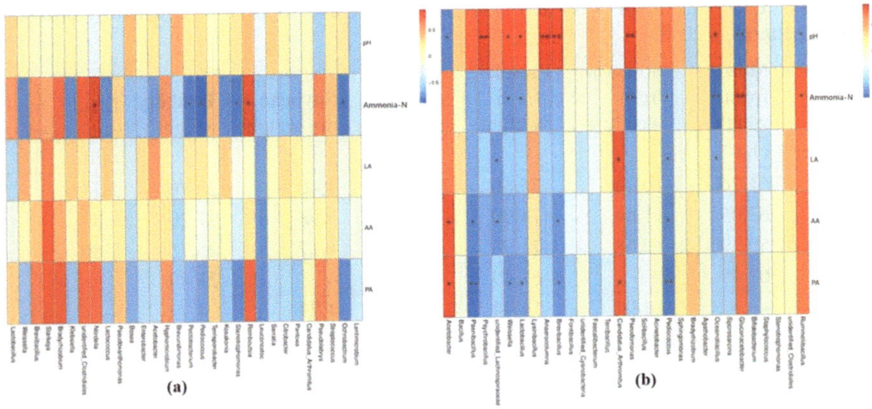

Fig. 17.3 Correlation analysis between bacterial community and organic acid, pH, and ammonia-N in maize-soybean mixed silage: (**a**) 60 days of ensiling; (**b**) 7 days of aerobic exposure. *LA* lactic acid content, *AA* acetic acid content, *PA* propionic acid content (Data and Fig. From Yan)

predominated; the relative abundance of *Lactobacillus* was higher than monoculture silages. Besides, low relative abundance bacteria such as *Pseudomonas, Gluconacetobacter*, and *Oceanobacillus* were associated with an increase in LA, WSC, and crude protein (CP) content and a decrease in ammonia nitrogen (NH_3-N/TN) content, which was able to improve the overall silage quality (Zeng et al. 2020) (Fig. 17.3). The presence of yeast and fungi during the ensiling process poses risks to the palatability, nutritional value, and aerobic stability of silage, leading to potential spoilage and toxin accumulation. Therefore, selecting high-quality LAB with antifungal activity and the ability to degrade mycotoxins is crucial for improving silage quality. *Lactobacillus rhamnosus* LG608 (CGMCC No.1.60089) reportedly can improve aerobic stability and efficiently degrade aflatoxin, which can significantly improve silage quality (Li 2024). This additive enhanced silage quality and aerobic stability, reduced dry matter loss (DML) during aerobic exposure of maize and soybean mixed silage, and reduced *mycotoxin* accumulation (e.g., OTA, DON, AFB1, FB, ZEN) by maintaining a higher relative abundance of *Lactobacillus* at the initial stage of aerobic exposure (Li 2024) (Fig. 17.4). Besides, maize-soybean mixed silage also promotes in vitro rumen fermentation in dairy cows, which enhances dry matter degradation rates and improves production performance (Kang et al. 2017).

In conclusion, the technology of maize-soybean mixed silage not only enhances the nutritional value of the feed but also significantly improves livestock production performance. Through scientific planting and management practices, along with the application of LAB additives, the ensiling process can be further optimized, contributing to the sustainable and high-quality development of modern herbivorous livestock farming.

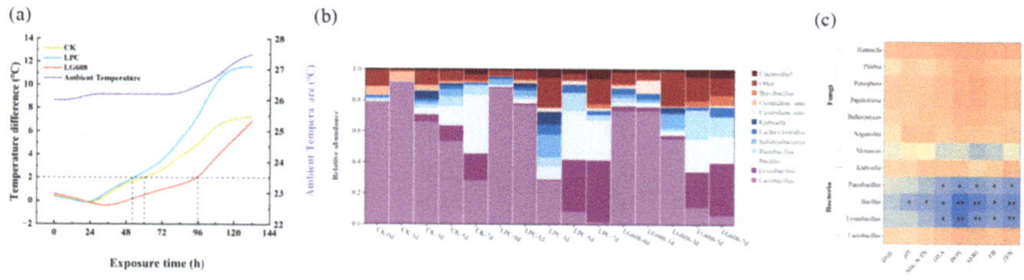

Fig. 17.4 Temperature changes (**a**), relative abundance of bacteria at the genus level (**b**), and correlation analysis of microorganism with DML, fermentation quality, and mycotoxins (**c**) in corn and soybean mixed silage prepared with lactic acid bacteria inoculants during aerobic exposure. *CK* maize and soybean ensiled without additives, *LPC* maize and soybean ensiled with commercial *Lactiplantibacillus plantarum, LPC* LG608, maize and soybean ensiled with *Lactobacillus rhamnosus* LG608, *DML* dry matter loss, NH_3*-N/TN* ammonia nitrogen/total nitrogen, *OTA* ochratoxin A, *DON* deoxynivalenol, *AFB1* aflatoxin B1, *FB* fumaricin, *ZEN* zearalenone (data and fig. From Yan)

5 Application of Bamboo Silage in Animal Production

Toshiyuki Ohtani

In many parts of Japan, neglected bamboo forests dominated by moso bamboo (*Phyllostachys heterocycla f. pubescent)*, a perennial evergreen plant belonging to the subfamily Madake of the *Poaceae* family, are expanding. Although bamboo is rich in dietary fiber and contains many functional ingredients such as vitamin E, it is rarely used in the livestock industry because it becomes stronger and more difficult to process after the period when it is called bamboo shoots.

Bamboo is said to vary in moisture content, nutrients, and functional components depending on the location and growing season. Several model areas in Shizuoka Prefecture showed slight variation in these components with respect to location and growing season (Table 17.7). Table 17.8 shows the chemical composition of bamboo harvested after 1 year. The crude protein was higher (11.3%) in the upper part of the bamboo (leaves) than the middle part (middle part of the stem) and the lower part (rod), recording 1.8 and 2.7%, respectively. This indicates that the bamboo plant's leafy parts can be used as feed for cattle and other herbivores.

On the other hand, the middle part of bamboo stems cannot be used as livestock feed in its original form because of its hard fibers. In addition, if the bamboo is not sufficiently ground, wisps are produced, which damage the oral cavity and stomach of livestock, and are also not palatable. Therefore, it is desirable that the hardy parts of the bamboo are processed into powder for use as animal feed. The finer the particle size, the easier it is for livestock to eat and transport, but the higher the production cost will be.

We have determined that a particle size of about 500 microns is optimal for the powdered bamboo, based on production efficiency, production cost, and livestock preference (Fig. 17.5).

Table 17.7 Chemical composition of bamboo in different regions

Area	DM (%)	Crude protein (% DM)	Crude fat (% DM)	Crude ash (% DM)	ADF (% DM)	NDF (% DM)
East	47.8	1.7	0.3	2.0	66.5	92.0
Center	38.0	3.6	0.3	2.8	65.6	91.2
West	42.9	2.9	0.8	3.2	65.8	91.4

Samples were collected in June in Shizuoka, Japan
DM dry matter, *ADF* acidic detergent fiber, *NDF* neutral detergent fiber

Table 17.8 Chemical composition of various parts of bamboo

Sample	DM (%)	Crude protein (% DM)	Crude fat (% DM)	Crude ash (% DM)	ADF (% DM)	NDF (% DM)
Leaf	63.7	11.3	2.3	4.3	47.6	77.4
Middle stem	38.0	1.8	0.6	1.6	68.1	89.6
Bottom stem	42.9	2.7	0.5	2.7	67.1	88.1

DM dry matter, *ADF* acidic detergent fiber, *NDF* neutral detergent fiber

Livestock feed utilization always requires a required quantity (quantitative) and stable composition (qualitative) regardless of the season (timed). This means that even if a large amount of material is available in the spring if it is not available in the summer or winter, it cannot be used for livestock feed. In addition, considering the efficiency of on-site bamboo harvesting and transportation, it is better to convert bamboo into powder in the spring than in the summer.

Bamboo powder (raw powder) at this time of the year (spring) can be used directly as livestock feed with rich nutrients.

However, as shown in Table 17.8, bamboo powder produced from the middle part of bamboo stems is not suitable for long-term storage because of its high moisture (about 40%) and easy deterioration (Ohtani and Iwasawa 2007). Lactic acid fermentation was attempted using Chikuso-1 (*Lactobacillus plantarum*), an inoculant used for conditioning grass silage, and other isolated strains. Finally, strain RO50 (*Lactococcus lactis* subsp. *lactis*) that produced two bacteriocins was successfully selected to inhibit harmful microorganisms and improve the fermentation quality of bamboo powder silage.

The LAC fermentation test of bamboo powder (bamboo silage) was started with a small-scale experimental system. One-year-old bamboo was cut down to stubble height of 1 m from the ground in early May (spring season) and immediately processed into powder (500 microns). Small portions (500 g) of this bamboo powder were fermented with LAB Chikuso-1 and RO50 after which they were degassed and sealed in aluminum packs and fermented for 60 days (Fig. 17.6).

It was found that the addition of Inoculant Chikuso-1 and RO50 separately lowered the pH more quickly than the control where no LAB was added (Iwasawa et al. 2007). Silage produced from lactic acid treated grass produced good quality silage in a short time and also stored for a long time.

Fig. 17.5 Moso bamboo forest (left), "Panda crusher" (center), and powder (right) 500 microns (Photos by Cai)

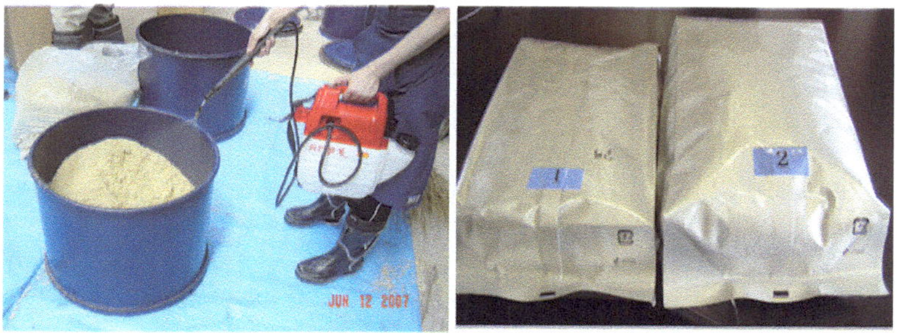

Fig. 17.6 Preparation of LAB-inoculated bamboo silage using a small-scale fermentation system (Photos by Cai and Ohtani)

We investigated the characteristics of the components of the bamboo-derived bioactive material (hereafter referred to as "material") using the lactic acid fermentation technique with the addition of RO50.

Bamboo contains 44.7 (IU/Kg) of functional components such as α-tocopherol (vitamin E), but the materials may lose these functional components due to lactic acid fermentation. Therefore, we analyzed the antioxidant capacity of materials stored at room temperature for about 60 days after silage preparation. The results showed that the antioxidant capacity of the raw bamboo powder was 4764 (luminescence count) and that of the material was 3318 (luminescence count), indicating a slight loss of components, but the antioxidant capacity of the material was still strong even after lactic acid fermentation (Table 17.9). In addition, a study of the free amino acid content of the material showed that it produced the highest amount of γ-aminobutyric acid (GABA), which has anti-stress effects such as sedative effects and lowering blood pressure, at approximately 600 mg/100 g, more than other components such as glycine (Gly) and alanine (Ala).

Therefore, it was considered that the material could be preserved for a long period of time by lactic acid fermentation, as well as maintain its nutritional and functional components and produce new components such as GABA.

Next, in the livestock feeding test, the effect of lactic acid fermentation and the possibility that components such as α-tocopherol could be transferred to products such as eggs and produce new eggs and chicken meat with functional properties

Table 17.9 Comparison of antioxidant capacity of bamboo and other raw materials

Material	Antioxidant power
Raw bamboo powder	4764
Raw material	3318
Antioxidant powdered tea	9147
Tea extract	1762
Lemon juice	916

Unit, luminescence count

Fig. 17.7 Chicken feed experiment of bamboo silage (Photo by Ohtani)

were investigated using chickens (egg-laying chickens and meat chickens) when the material was used in their feed.

As mentioned above, materials are rich in nutrients such as crude protein, but normal chicken feed (concentrate feed) is mainly composed of corn, soybean meal, and wheat, and compared to these feeds, it is not possible to replace all chicken feed with materials because of their low energy (TDN). Therefore, in order to examine the amount of materials added that would not reduce productivity, such as chicken preference, egg production, and weight loss, we investigated the amount of materials added to chicken feed at 2.5%, 5%, and 10% (Fig. 17.7) (Matsui et al. 2009). As a result, it was found that 2.5% of the chicken feed was appropriate and that adding more than this resulted in a decrease in the number of eggs laid by egg-laying hens and a decrease in the body weight of meat hens at the time of shipment and a decrease in productivity compared to the control (no bamboo added) (Table 17.10).

On the other hand, since the material was found to contain α-tocopherol, which is fat-soluble, the egg-laying hens (Rhode Island Red breed) were used to investigate the migration of the material in chicken eggs (egg yolk). During the study period, the hens were fed ad libitum, and free water was offered. The results showed that α-tocopherol in egg yolk was about 3.2 mg/100 g in the material-added area, which was significantly higher than that in the control area, about 2.9 mg/100 g

Table 17.10 Supplemental amount of bamboo silage and production performance of broilers

Bamboo silage addition level (%)	Shipping weight(g)	Feed intake (g/day/bird)	Feed requirement	Training rate (%)
0	2,681[a]	104.5[a]	1.87	100
2.5	2,670[a]	104.6[ab]	1.91	99
5	2,573[b]	105.1[bc]	19.6	100
10	2,438[c]	102.5[c]	20.1	100

[a, b, c]Significant difference between different signs ($P < 0.05$)

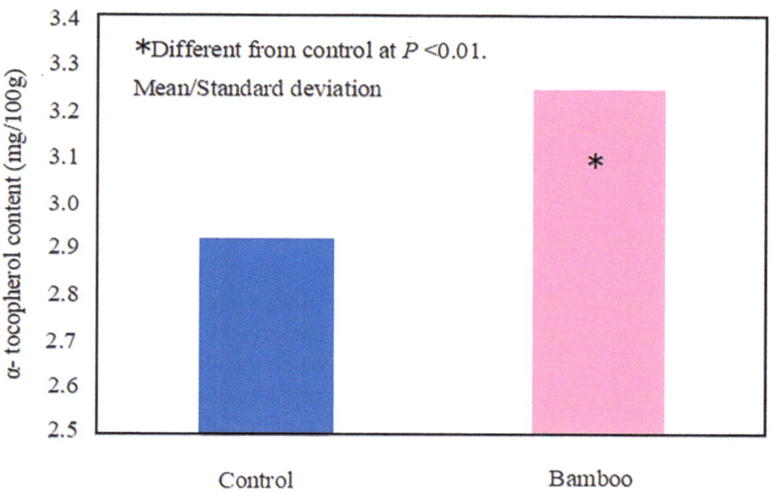

Fig. 17.8 Alpha-tocopherol content in egg yolk

(Fig. 17.8). Chickens did not show any decrease in laying performance or other effects on productivity during the study period (Matsui et al. 2010).

Next, a feeding test was conducted to determine the effect on meat quality of chickens (Chunky breed: 240 females). After 21 days of age, the chickens were fed a meat chicken finishing diet (CP 18.0%, ME 3200 kcal/Kg) with the appropriate amount of material added (2.5%) until 50 days of age. Chickens were fed ad libitum and free watering. After the test, the chicken meat was immediately vacuumed and refrigerated at −20 °C. At a later date, the meat was thawed at 4 °C and stored in a refrigerator. Later, after thawing at 4 °C, the chicken breasts were boiled for a certain period of time, and the results were evaluated by a questionnaire. The results showed that the chicken meat in the material-added group had a higher overall evaluation of tenderness, juiciness, flavor, and aroma than the chicken meat in the non-added group and was evaluated as "delicious chicken meat" with a good flavor (Fig. 17.9). The meat color was measured using a colorimeter on thigh meat (semitendinosus muscle) and shallow breast muscle (meat of the breast), but was not affected by the addition of materials (Matsui et al. 2010).

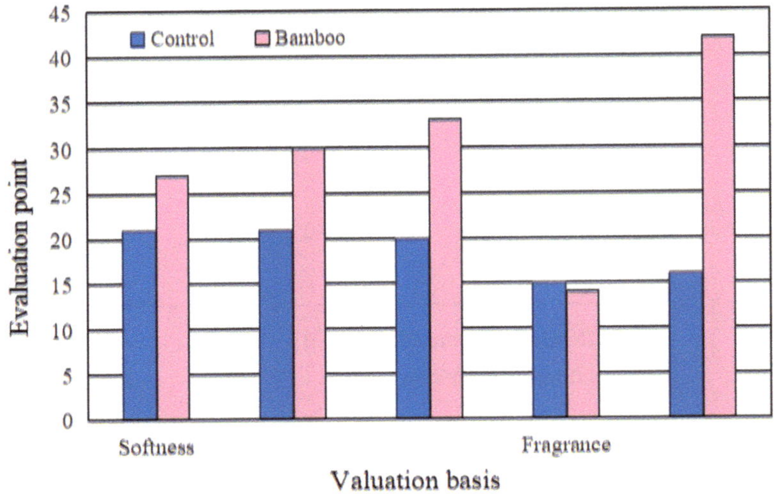

Fig. 17.9 Sensory test of chicken

In addition, changes in the intestinal microflora were investigated using meat chickens to determine the number of major bacteria in the intestines under the appropriate amount of added material (2.5%). When the chicken was disassembled, the contents of the cecum were collected anaerobically and immediately applied to the culture media for each bacterium, and the number of colonies was measured. *Lactobacillus* was investigated as a change in useful bacteria. On the other hand, the number of *Enterobacteriaceae*, *Staphylococcus*, and *Streptococcus* was investigated as changes in harmful bacteria. The results showed that there was no change in the number of bad bacteria, *Enterobacteriaceae* and *Streptococcus*, but a significant decrease in the number of *Staphylococcus*, indicating that feeding the materials improved the intestinal tract (Fig. 17.10). This is thought to be due to the non-digestible component of the material, dietary fiber, and the probiotic effect of the LAB strain RO50, which improved the intestinal flora and regulated the intestinal tract (Matsui et al. 2010).

Next, the physiologically active effect of lactic acid fermentation was investigated for 35 days using broiler chicken to determine the enhancement of liquid immune response under the appropriate amount of material (2.5%). The enhancement of liquid immune response was measured by measuring the agglutination titer against sheep erythrocytes and *Brucella melitensis* (BM). The results showed that there was no significant difference in sheep erythrocyte antibody titer, an indicator of T cell-dependent immunity, throughout the study period, but BM antibody titer, an indicator of T cell-independent liquid immunity, remained significantly higher from 14 days after inoculation to the end of the study (35 days) in the material-added group, confirming the enhancement effect of liquid immune response (Fig. 17.11). This indicates that the probiotic effect of the RO50 material is expected to enhance immunity and reduce the incidence of various diseases in chickens (Matsui et al. 2010).

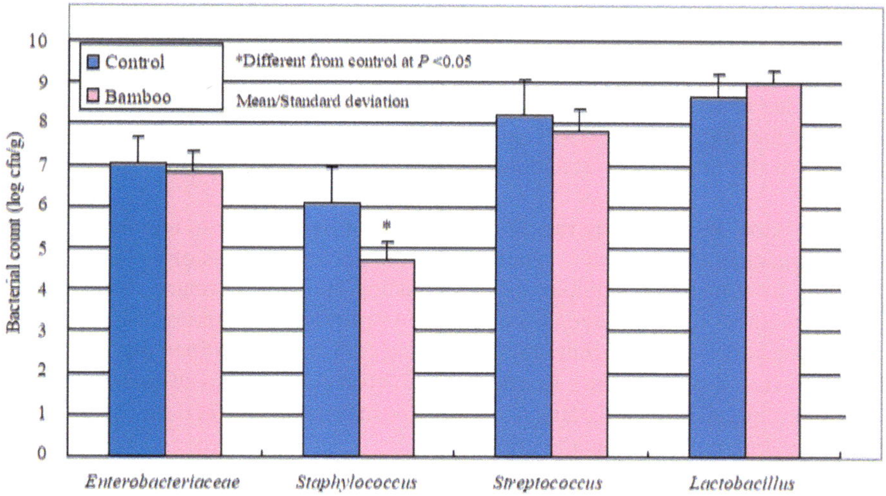

Fig. 17.10 Intestinal flora

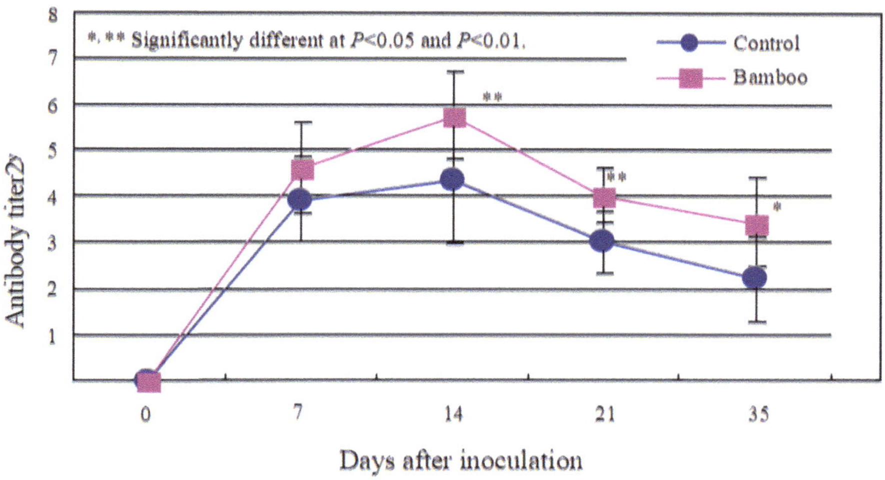

Fig. 17.11 Enhancing effect on liquid immunity

The results indicate that incorporating bamboo powder silage into chicken feed at a 2.5% inclusion level does not reduce egg production or lead to weight loss in chickens. Additionally, it may enhance the production of eggs with various beneficial properties for egg-laying hens, as well as improve the flavor of meat from broiler chickens. This could position bamboo as a key ingredient in creating a regional "brand animal product" for livestock feed. This could lead to the development of regional "branded livestock products" using bamboo as livestock feed. The use of bamboo as livestock feed will lead to the creation of a new bamboo industry and a drastic solution to the problem of neglected bamboo forests.

Currently, TMR is a mixture of several materials such as roughage (dried or grass silage), various concentrated feeds (grains and meal), and by-products such as food residues and is widely used for cattle feed. Cattle have a good taste for the "sourness" of lactic acid fermentation, and if our material introduced here could be used as one of the ingredients in TMR, more bamboo would be consumed, which would help solve the problem of abandoned bamboo groves and prevent the rising cost of feed.

In addition, as previously mentioned, the use of livestock feed requires both the correct quantity and consistent quality of ingredients, regardless of the season. It is essential to implement technologies for adjusting and storing materials on a large scale, such as roll-bale silage. We also need to provide the appropriate amount of feed to the cattle. In our recent chicken feeding trial, we determined this amount to be 2.5%. It will be necessary to establish the correct quantities of materials to be added through additional feeding trials and to examine their effects on productivity, including milk composition and milk yield.

On the other hand, bamboo is inexhaustible in Japan. Reducing the labor and transportation costs associated with harvesting bamboo will be important to promote its use as livestock feed.

In some areas of Kyushu, where the livestock industry is thriving, the problem of neglected bamboo forests is serious, as is the soaring cost of livestock feed. The upper part (leaves) of the harvested bamboo is used by private TMR centers as one of the materials for cattle feed because of its high nutritional value, and the middle part (middle part of the stem) and lower part (rod) are mainly composted and used as fertilizer and soil improvement in agriculture. As a result, the growth and taste of vegetables are said to be very good.

We hope that such joint efforts by the government and private sectors will spread throughout Japan to solve the problems of rising livestock feed prices and neglected bamboo forests.

Lastly, we have introduced the effectiveness of the use of lactic acid fermented bamboo silage (bamboo silage), made by reducing the granularity of bamboo to about 500 microns, in chicken feed. In particular, there is concern about outbreaks of serious diseases such as chicken influenza in chickens and foot-and-mouth disease in cattle. For this reason, we hope that the opinions of veterinarians and other experts on diseases will be taken into account when using bamboo as feed for livestock, including chickens, in the future.

References

Batista VV, Adami PF, Moraes KF, Oligini CL, Giacomel L. Row arrangements of maize and soybean intercrop on silage quality and grain yield. J Agr Sci. 2019;11(2):286.

Cai Y, Masuda N, Fujita Y, Kawamoto H, Ando S. Development of a new method for preparation and conservation of tea grounds silage. Anim Sci J. 2001;72(10):536–41.

Chen X, Zhao Y, Zhang H. Analysis of soybean and maize strip compound planting technology. J Chin Agric Mechanization. 2023;44:48–25.

Guo S, Awasthi M, Wang Y, Xu P. Current understanding in conversion and application of tea waste biomass: a review. Bioresour Technol. 2021;338:125530.
Guo X, Xu D, Li F, Bai J, Su R. Current approaches on the roles of lactic acid bacteria in crop silage. Microbial Biotechnol. 2023;16:67–87.
Han H, Li C, Yu Y, Xu Z, Li Q, Tang G, et al. Identification of lactic acid bacteria in the feces of dairy cows fed whole crop maize silage to assess the survival of silage bacteria in the gut. Anim Sci J. 2018;89(1):97–104.
Iwasawa T, Matsui S, Yokogoshi H, Cai Y. Seeking of biologically active substances in bamboo silage and their application to feeding broiler chicks. Bull Shizuoka Prefectural Res Inst Anim Ind Swine Poultry Res Cent. 2007;1:37–43.
Kang J, Song J, Marbun T, Kwon CH, Kim EJ. Effect of intercropped maize and soybean silage on nutritive values, in vitro ruminal fermentation, and milk production of Holstein dairy cows. J Korean Soc Grassl Forage Sci. 2017;37(3):216–22.
Kondo M, Kita K, Yokota HO. Evaluation of fermentation characteristics and nutritive value of green tea waste ensiled with byproducts mixture for ruminants. Anim Biosci. 2006;19(4):533–40.
Kondo M, Naoki N, Kazumi K, Yokota H. Enhanced lactic acid fermentation of silage by the addition of green tea waste. J Sci Food Agr. 2004;84(7):728–34.
Lei X, Na B, Zhou T, Qian Y, Xie Y, Zheng Y, et al. Effects of dried tea residues of different processing techniques on the nutritional parameters, fermentation quality, and bacterial structure of silaged alfalfa. Microorganisms. 2024;12:889.
Li J. Effect of lactic acid bacteria LG608 on aerobic stability and mycotoxins of corn and soybean mixed silage. Chengdu: Sichuan Agricultural University; 2024.
Li Q, Zeng T, Hu Y, Du Z, Liu Y, Jin M, et al. Effects of soybean density and sowing time on the yield and quality of mixed silage in maize-soybean strip intercropping system. Fermentation. 2022;8(4):140.
Liu X, Rahman T, Song C, Su B, Yang F, Yang W. Relationships among light distribution, radiation use efficiency, and land equivalent ratio in maize-soybean strip intercropping. Field Crop Res. 2018;224:91–101.
Matsui S, Cai Y, Oishi S, Yokogoshi H. Application of bamboo silage to poultry feeding. Bull Shizuoka Prefectural Res Inst Anim Ind Swine Poultry Res Cent. 2009:27–42.
Matsui S, Cai Y, Oishi S, Yokogoshi H. Application of bamboo silage to poultry feeding (II). 2010;3:27–33.
Min BR, Barry TN, Attwood GT, McNabb WC. The effect of condensed tannins on the nutrition and health of ruminants fed fresh temperate forages: a review. Anim Feed Sci Technol. 2003;106(1):3–19.
Nasehi M, Torbatinejad NM, Rezaie M, Ghoorchi T. Effects of partial substitution of alfalfa hay with green tea waste on growth performance and in vitro methane emission of fat-tailed lambs. Small Rumin Res. 2018;168:52–9.
Nishida T, Eruden B, Hosoda K, Matsuyama H, Nakagawa K, Miyazawa T, et al. Effects of green tea (Camellia sinensis) waste silage and polyethylene glycol on ruminal fermentation and blood components in cattle. Anim Biosci. 2006;19(12):1728–36.
Nishino N, Kawai T, Kondo M. Changes during ensilage in fermentation products, tea catechins, antioxidative activity, and in vitro gas production of green tea waste stored with or without dried beet pulp. J Sci Food Agr. 2007;87(9):1639–44.
Ohtani T, Iwasawa T. Preparation and conservation of bamboo powder silage and its fermentation quality. Bull Shizuoka Swine Poultry Exp Stn. 2007;17:53–7.
Qu Y, Jiang W, Yin G, Wei C, Bao J. Effects of feeding maize-lablab bean mixture silages on nutrient apparent digestibility and performance of dairy cows. Anim Biosci. 2013;26(4):509–16.
Wang HL, Xu CC. Utilization of tea grounds as feedstuff for ruminants. J Anim Sci Biotechnol. 2013;4:54.
Wang RR, Wang HL, Liu X, Xu CC. Effects of different additives on fermentation characteristics and protein degradation of green tea grounds silage. Anim Biosci. 2011;24(5):616–22.

Xu C, Cai Y, Fujita Y, Kawamoto H, Sato T, Masuda N. Chemical composition and nutritive value of tea grounds silage treated with lactic acid bacteria and Acremonium cellulose. Nihon Chikusan Gakkaiho. 2003;74:355–61.
Xu C, Cai Y, Fukasawa M, Matsuyama H, Moriya N. The effect of replacing brewers' grains with barley tea grounds in total mixed ration silage on feed intake, digestibility, and ruminal fermentation in wethers. Anim Sci J. 2008b;79:575–81.
Xu C, Cai Y, Moliya M, Eruden B, Hosoda K, Matsuyama H. Influence of replacing brewers' grains with green tea grounds on feed intake, digestibility, and ruminal fermentation characteristics of wethers. Anim Sci J. 2008a;79:226–33.
Xu C, Cai Y, Moriya N, Ogawa M. Nutritive value for ruminants of green tea grounds as a replacement of brewers' grains in totally mixed ration silage. Anim Feed Sci Technol. 2007;138:228–38.
Xu C, Cai Y, Murai M. Fermentation quality and nutritive value of total mixed ration silage with barley tea grounds. Nihon Chikusan Gakkaiho. 2004;75(2):185–91.
Xu C, Cai Y, Zhang J, Fukasawa M, Moriya N. Ensiling and subsequent ruminal degradation characteristics of barley tea grounds treated with contrasting additives. Anim Feed Sci Tech. 2008c;141:368–74.
Yan Y, Wan Y, Liu W, Wang X, Yong T, Yang W, et al. Influence of seed treatment with Uniconazole powder on soybean growth, photosynthesis, dry matter accumulation after flowering and yield in relay strip intercropping system. Plant Prod Sci. 2015;18:295–301.
Yang F, Huang S, Gao R, Liu W, Yong T, Wang X, et al. Growth of soybean seedlings in relay strip intercropping systems in relation to light quantity and red: far-red ratio. Field Crop Res. 2014;155:245–53.
Yang F, Liao D, Wu X, Gao R, Fan Y, Raza MA, et al. Effect of aboveground and belowground interactions on the intercrop yields in maize-soybean relay intercropping systems. Field Crop Res. 2017;203:16–23.
Yang J, Tan H, Cai Y. Characteristics of lactic acid bacteria isolates and their effect on silage fermentation of fruit residues. J Dairy Sci. 2016;99:5325–34.
Yang J, Tang K, Tan H, Cai Y. Characterization of lactic acid bacteria isolated from banana and application for the silage fermentation of defective banana. Microorganisms. 2022;10(1185):1–13.
Yang S, Zhao Y, Xu Y, Cui J, Li T, Hu Y, et al. Yield performance response to field configuration of maize and soybean intercropping in China: a meta-analysis. Field Crop Res. 2024;306:109235.
Yao M, Cui K, Tang W, Shi J, Zhang R, et al. Effects of different plant density on silage maize yield and partial agronomic characters in the northwest plateau of Sichuan. J Southwest Univ National (Nat Sci Ed). 2022;48(1):1–7.
Zeng T, Li X, Guan H, Yang W, Yan Y. Dynamic microbial diversity and fermentation quality of the mixed silage of maize and soybean grown in strip intercropping system. Bioresour Technol. 2020;123655
Zeng T, Wu Y, Xin Y, Chen C, Du Z, Li X, et al. Silage quality and output of different maize–soybean strip intercropping patterns. Fermentation. 2022;8(4):174.

Chapter 18
Application of Smart Agricultural Technology in Silage Production

Yasuhiro Aoki

Abstract In Japan, the number of dairy farms is declining, while the number of animals per dairy farm is significantly increasing. The aging population of the dairy farmers has led to a serious labour shortage. Due to the necessity for daily milking, many dairy farmers are forced to work long hours and cannot take time off. Additionally, feed production adds to their workload. To address these challenges, the development of smart dairy farming technology is becoming increasingly important. This chapter introduces the following new technologies used in silage production: (1) aerial image analysis for measuring weed growth and planting density in forage corn fields, (2) chemical spraying using UAV, (3) estimation of corn yield and determination of harvest timing by image analysis, (4) an automatic recording and real-time monitoring system of forage harvesting and silage preparing operation, and (5) an IoT-based total mixed ration (TMR) producing system.

Keywords Dairy farming · Silage production · Smart agricultural technology · Total mixed ration centre

Smart agriculture advanced technology further promotes labour-saving and efficiency in silage production. Expanding the scale of feed production through smart agriculture enhances the production capacity of primitive livestock and increases the added value of agricultural and livestock products. Livestock production can use various smart agriculture technologies to rapidly improve production efficiency, reduce the impact on global warming and environmental load, and increase the output of high-quality livestock products. Therefore, this chapter focuses on examples of the application of smart agriculture technologies in the field of feed production and looks forward to the future of the livestock industry which will rely on emerging science and technology, such as big data, network communications, and

Y. Aoki (✉)
Tokyo University of Agriculture and Technology, Fuchu, Japan
e-mail: aokiya@go.tuat.ac.jp

Y. Cai, K. Ataku (eds.), *Cultural History and Modern Production Technology of Silage*, https://doi.org/10.1007/978-981-96-5787-2_18

artificial intelligence, to the direction of high efficiency, high quality, automation, refinement, and intelligent development.

In agricultural production sites, there are growing expectations for the introduction of smart agricultural technologies that utilize robot technology and information and communication technology (ICT). As a background, labour-saving measures are increasingly necessary due to labour shortages, caused by a decline in the number of agricultural workers and an aging population, as well as the expansion of the scale of management per household due to the decrease in the number of management entities. Furthermore, in agricultural production, many tasks still depend on skills often referred to as 'master craftsmanship', which raises concerns about smoothly passing on technology in the future. Smart agriculture technology is attracting attention to solve these problems.

To date, a range of smart agriculture technologies have been successfully implemented, including automated driving systems for tractors and rice transplanters, automated water management of paddy fields, furrow management using mowing robots, variable fertilization in paddy fields based on remote sensing, precision pesticide spraying by drones in both paddy and ordinary fields, automatic irrigation and fertilization in facility horticulture, agricultural assist suits, and visualization of the skilled farmers' techniques and decision-making in fruit tree cultivation (Ministry of Agriculture, Fisheries and Forestry, Japan (MAFF) 2019). In the dairy farming sector, technologies such as automatic milking systems (milking robots) have already been widely adopted. Additionally, novel technologies are also expected to expand rapidly, particularly in forage production and silage preparation which have been recently introduced in Japan.

This chapter first describes the problems faced by dairy farming and the current status of smart dairy technology applications in Japan. It then introduces the examples of novel technologies used in silage production in Hokkaido, the main production area of dairy farming in Japan.

1 The Current Status and Future of Dairy Farming in Japan

To understand the need for smart agriculture technologies, it is important to examine the current situation and challenges of dairy farming in Japan. From 2001 to 2023, the number of dairy farms nationwide decreased by 19,600, about 60%, from 32,200 to 12,600 (Fig. 18.1). In recent years, about 700 farmers annually have ceased operation, and this trend is expected to continue in the near future. Despite the decrease in the total number of dairy cattle, the rate of decline was not as large as the decline in the number of farms. As a result, the number of cattle per farm has increased by 1.54 times during this period (Fig. 18.2). In Hokkaido, it increased from 87.1 to 134.2 while it increased from 37.9 to 58.5 in prefectures other than Hokkaido (MAFF, Japan 2023b). It is presumed that the number of animals raised will continue to expand for a while.

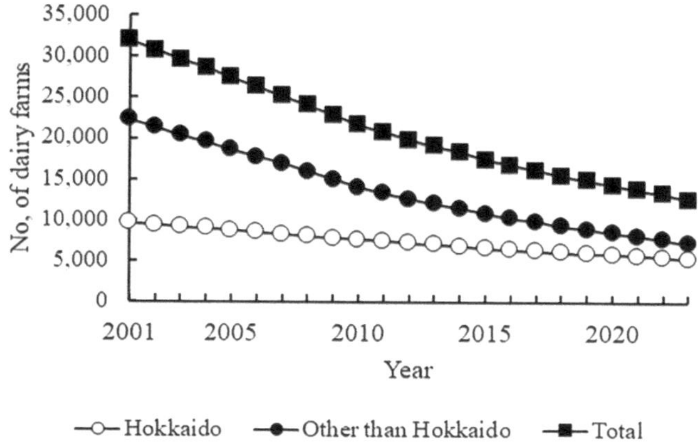

Fig. 18.1 Changes in the number of dairy farms in Japan

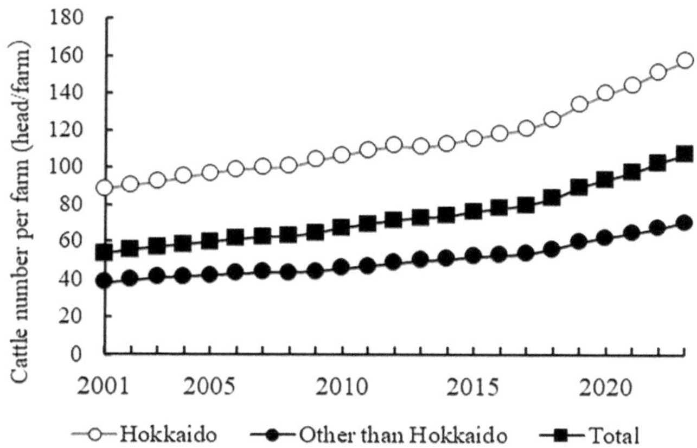

Fig. 18.2 Changes in the number of dairy cattle (cows and heifers) per farm in Japan

The proportion of dairy farmers aged 65 years and above increased from 19% in 1995 to 36.6% in 2020 (Fig. 18.3) (MAFF, Japan 2022). In agriculture as a whole, the proportion of people aged 65 years and above reached 64% in 2020, as the rate of aging in rice cultivation is particularly high. In contrast, the proportion of elderly people in dairy farming is low among agricultural sectors. However, while in 1995, the age group of 40–49 years old was the largest, the age group of 65 years and older became the largest in 2020. Although the aging rate is lower than in other sectors, the aging population is steadily progressing in the dairy industry.

Since daily milking is indispensable, dairy farmers have to work for longer periods, making difficult to take holidays. Although the average annual working hours for agriculture as a whole in owned farms were 1945 hours in 2018, it was significantly higher for dairy farming, totaling 6684 hours per year (MAFF, Japan 2023a).

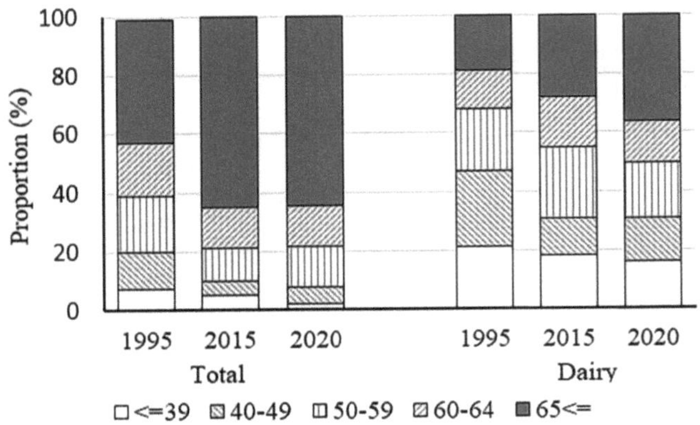

Fig. 18.3 Age structure of core persons engaged in agriculture and in dairy farming

Furthermore, the average number of holidays for a core person in dairy farming was 17.7 days annually, which is equivalent to about 1.5 days per month (Japan Dairy Council 2018).

In addition to the labour associated with daily milking, roughage production also presents an additional workload. According to a survey by the Japan Dairy Council, the average daily working hours of a core person on a dairy farm were 6.26 h. However, when feed production was factored, the total working hours increased to 8.87 hours per day, a rise of 42% (Japan Dairy Council 2018). As a measure to reduce the labour required for feed production, the services of external support organizations such as contractors that handle feed production and TMR centres (TMR plants) that supply TMR to livestock farms have been employed. However, it has become difficult for these organizations to secure human resources due to competition from other industries and the aging workforce. A survey of TMR centres revealed that more than 60% of the respondents will have difficulties in managing and operating their facilities in the future (MAFF, Japan 2016). In order to address the challenges associated with the increasing number of dairy cattle per farm, the aging population of farmers and the increase in labour burden, the development of smart dairy farming technologies has become crucial. In particular, there is an urgent need to develop technologies that will enable the organizations such as TMR centres and contractors to ensure sustainability of their operations and management in the future.

2 What Is Smart Agriculture?

Smart agriculture is a new type of agriculture that uses robot technology and ICT to achieve ultra-labour-saving and high-quality production. Japan's agriculture boasts of several key strengths including (1) mastery of craftsmanship that responds to regional characteristics such as climate and soil, (2) a wide variety of savoury and high-quality products that reflect regional distinctiveness, and (3) safe and secure agricultural products that meet consumer needs (MAFF, Japan 2019). The combination of these strengths and advanced technology is termed smart agriculture. In 2013, MAFF of Japan established a study group for the realization of smart agriculture and presented a new form of agriculture brought about by smart agriculture as (1) achieving ultra-labour-saving and large-scale production, (2) maximizing crops yield, (3) reducing the need for hard and dangerous work, (4) making agriculture more accessible and inclusive and (5) providing consumers with peace of mind and trust in agricultural products (Study Group for the Realization of Smart Agriculture 2014). Since master craftsmanship is not necessarily required, it is expected to contribute to the promotion of the sector, by encouraging people to go into farming.

2.1 *Smart Technology in Dairy Farming*

In order to address the challenges described in Chap. 2, the following technologies have been implemented at dairy production sites in Japan. The MAFF of Japan aims to enhance productivity and reduce labour by supporting the introduction of various technologies that utilize ICT and robot technology to strengthen the production base of dairy farming. Among the technologies promoted are (1) a milking robot, (2) an automatic milking unit transport system, (3) an estrus detection device, (4) a calving monitoring device, (5) an automatic feeding system for milk replacers, (6) an automatic feed pushing device and (7) a data management system. The Smart Agricultural Technology Catalog (Livestock) has been published by MAFF Japan for the purpose of widely disseminating the smart technologies currently being developed or sold to the dairy farming sites. However, smart technologies that contribute directly to forage production have lagged behind these technologies.

2.2 *Examples of Smart Agriculture Technologies for Forage Production and Silage Preparation Introduced at a TMR Centre*

The contents below are based on the research project titled ‘Demonstration of smart herd management system for highly lactating cows by using the TMR Center for high-quality self-supplied feed production’ (2019–2020) supported by Development and Improvement of Smart Agriculture Technology grants from the Project of the Bio-oriented Technology Research Advancement Institution of National Agriculture and Food Research Organization (NARO). The research was conducted by the Hokkaido Agricultural Research Centre (HARC) of NARO, the Hokkaido Agricultural Research Institute, agricultural cooperatives, and private companies.

The author, whose former affiliation was HARC of NARO, participated as a representative of this project.

The purpose of this project was to introduce various smart technologies into a series of production situations, from self-supplied feed production and TMR production at the TMR centre to TMR feeding and milk production at the dairy farms, and to verify their technical and economic effects. The TMR centre, which served as the demonstration site, managed about 1000 hectares of pasture and corn fields in the Kenebetsu area of eastern Hokkaido and supplied TMR to more than 1300 dairy cows.

In this paper, the author focuses on technologies related to silage and TMR production among the results obtained in this project.

2.2.1 Measurement of Weed Growth and Planting Density in Forage Corn Fields by Aerial Image Analysis

In order to reduce the labour required for weeding in the production of corn for feed, a technology to quantify the weed situation in the field was developed. As a result, using drone-based aerial photography information, the extent of weed overgrowth was successfully visualized together with the location information in the field. Additionally, the weed growth area (number of weed spots, 1 m2 unit) for each field was calculated. This enabled the quantitative identification of areas with high weed density and those with minimal weed (Fig. 18.4). This approach makes it possible to objectively determine if it is necessary to apply the entire herbicide to the field or whether it is more appropriate to spray it intensively in some areas.

The planting density of each field was also estimated using drone images of the post-germination of corn to accurately assess the rate of plant loss. We confirmed that it can be estimated with a high accuracy; the error from the actual measured values from the ground survey was within 2%. By listing the fields in order of

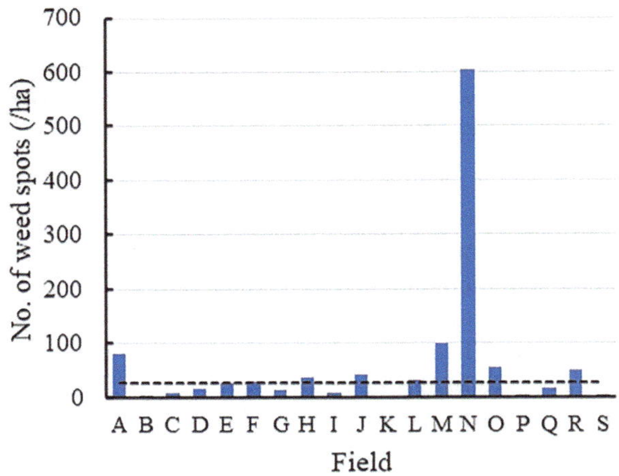

Fig. 18.4 Number of weed spot estimated based on the analysis of drone aerial photography information. No. of weed spots: the number of spots (1 × 1 m) where weeds were recognized. Dashed line: median

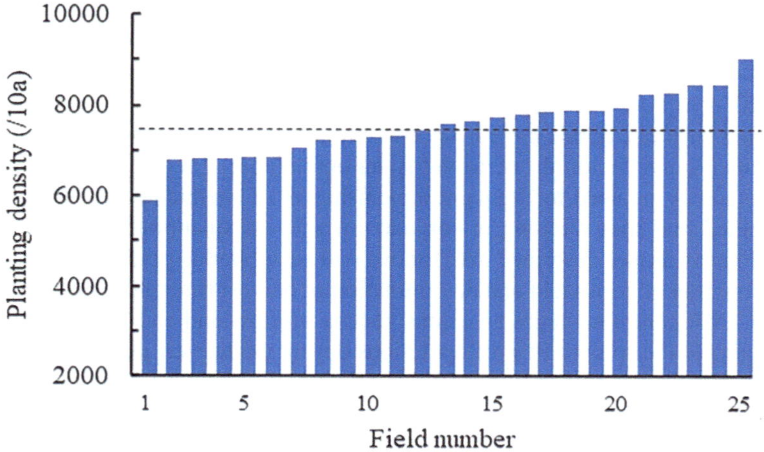

Fig. 18.5 Planting density estimated based on the analysis of drone aerial photography information. Dashed line: average value

planting density estimated based on drone aerial imagery (Fig. 18.5), it is possible to identify fields where the density is very low compared to the average value. Further investigation showed that the low planting density in those fields could be attributed to the irregular shape of the field, small field size and excessive dry soil at the time of sowing. By sharing this information with the operators, it is possible to ensure more targeted management practices, such as adhering to recommended tractor operating speed during sowing and frequently monitoring the sowing conditions. Some operators noted that such information would be useful for the training for inexperienced operators to help them better understand field conditions.

2.2.2 Chemical Spraying Work Using UAV

The damage caused by Northern leaf blight on corn can reach more than 20% of nutrient yields in significant cases (Hayashi 2014). To avoid this damage and improve silage quality, fungicide application is often. However, by August, when conditions are favourable for spraying, the height of the corn often prevents large spraying machines from accessing the fields. As a result, fungicide spraying in recent years is being carried out using unmanned helicopters to control Northern leaf blight. This project aimed to assess the effectiveness and economic feasibility of using unmanned helicopters for fungicide application. However, during the study period, the occurrence of the disease was extremely low. As a result, an experimental flight was conducted to investigate the time required for preparation and actual work for spraying, the consumption of necessary materials, cost estimation per hectare, etc. A comparison with the report on nutrient yield and economic loss associated with Northern leaf blight (Hayashi 2014) revealed that spraying fungicide by an unmanned helicopter is not economical when the damage is minor; it is better when the degree of yield loss is predicted to be about 10% or more of the normal level.

In the near future, drones are expected to be used to spray herbicides on forage crops and pastures. In this project, spot weeding in pasture was demonstrated using an automatic navigation drone to verify its operational efficiency. Although challenges such as the need for frequent replenishment of chemicals were identified, it is expected that spot weeding in conjunction with the quantitative grasp of weed conditions by drones described in (1) above will be applied to practical applications.

(2) Estimation of corn yield and determination of harvest timing by image analysis.

The TMR centre manages an extensive area of grassland and forage fields. It is important for the smooth flow of operations during the harvest period to determine the start date for the preparations of bunker silos and the harvest timing of dispersed fields. This adjustment is often based on the experience of a skilled person. However, in the current situation, where the inheritance of technology is even in jeopardy, it is desirable to proceed with work plans without relying on skills. Thus, a method was introduced to predict the yield and optimal harvest timing of forage corn based on image analysis information and verified its practicality.

For yield prediction, the above-ground volume was estimated from the integral value of the grass height using drone aerial images and analysing the relationship between the above-ground volume and the harvested corn yield (average weight of dump trucks loaded with corn × number of dump trucks). As a result, a strong positive correlation was observed between the estimated above-ground volume of each field and the actual yield (Fig. 18.6). In addition, the error in predicting harvested corn yield from above-ground volume was calculated to be about 16%. It was concluded that the corn yield can be predicted with practical reliability.

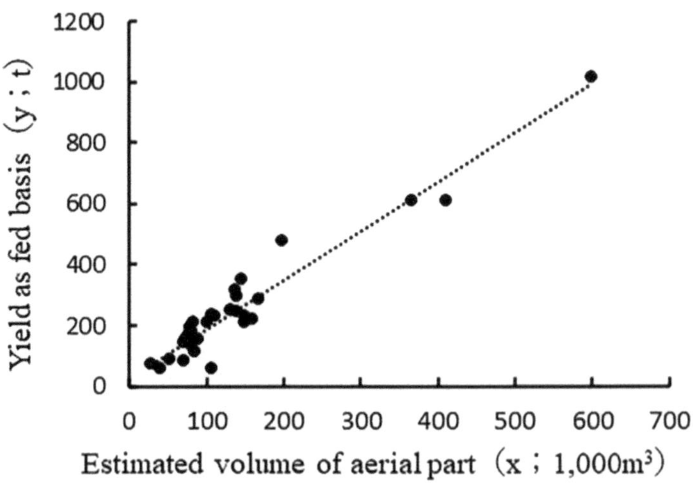

Fig. 18.6 Relationship between above-ground volume estimates and harvested yield of corn per field based on drone aerial image analysis

For the prediction of the optimal harvest timing, satellite images were analysed to visualize the difference in ripeness between fields and changes within the same field based on the normalized vegetation index (NDVI). However, since almost no data could be collected due to persistent cloudy weather throughout the year, it would be difficult to apply this technique for practice at the moment.

2.2.3 Automatic Recording and Real-Time Monitoring System of Forage Harvesting and Silage Preparing Operation

During forage harvesting and silage preparation, operations such as harvesting silage materials with a forage harvester, transporting silage materials by dump trucks and packing in bunker silos by machines (wheel loaders, etc.) proceed at the same time over an extensive area. In a large TMR centre, dozens of dump trucks run simultaneously. During the operation, there may be some fields where several dump trucks are waiting for the harvester, and there are other fields where the harvester is waiting for the dump to stop harvesting for several minutes. In order to save labour, it is desirable to eliminate such loss of operation time as much as possible and to proceed in parallel without delay. A system was introduced to automatically record work status and monitors it in real time, 'Reposaku' (Ezowin Co., Ltd., Shibetsu, Hokkaido) (Figs. 18.7 and 18.8), and also to verify its effectiveness. In the first year of the demonstration, a person in the management office of TMR centre monitored the Reposaku information and instructed the driver who stood by to move to another field using a mobile phone. In the second year, aiming at making it even faster, a tablet was equipped on each vehicle, allowing each operator to check the work status directly. By communicating with each other without instructions from the centre, the dump work time in the second year was reduced by about 10% compared to the first year (Fig. 18.9). In addition, by visualizing the compacting operation by wheel loader in the bunker silo (Fig. 18.10), the packing density increased, and the uneven compaction pressure was reduced (Fig. 18.11), making it possible to perform the silage preparation operation appropriately. With proper compaction, lactic

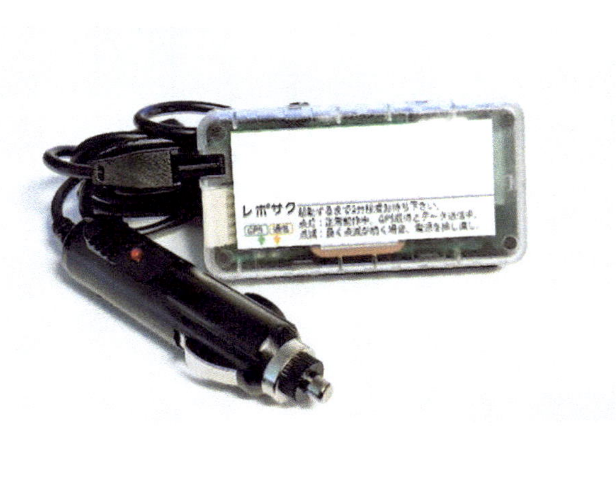

Fig. 18.7 Automatic recording and real-time monitoring system for harvest preparation work "Reposaku" (Ezowin Co., Ltd., Japan)

Fig. 18.8 Example of work flow line recording by 'Reposaku'. (The image is provided by Mr. Hiroshi Ohno, CEO of Ezowin Co., Ltd)

Fig. 18.9 Transport time of dump track after the introduction of tablets on the vehicle to obtain the information of 'Reposaku'

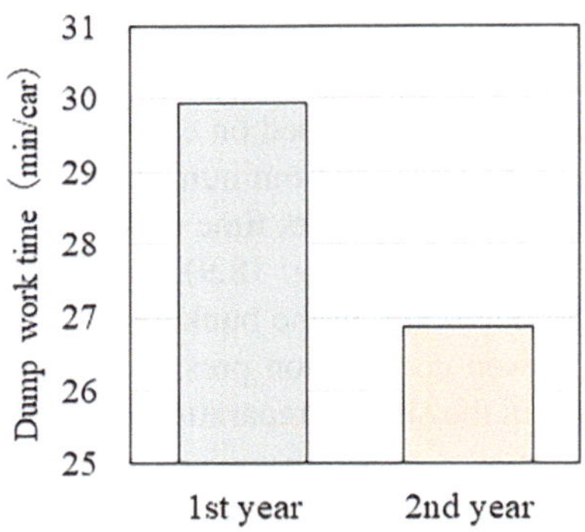

acid fermentation, which is active under anaerobic conditions, proceeds smoothly. When lactic acid fermentation predominates, the pH drops rapidly, inhibiting the activity of butyric acid, which does not like acidic conditions. With this visualization, it is possible to recognize where the pressure is insufficient in the bunker silo and that, in some places, the compaction was repeated too much. In addition to eliminating silage waste, it is also expected to help inexperienced operators to understand the necessary steps for producing high-quality silage.

It is light and compact and can be used simply by inserting the power supply part into a cigar socket. The image is provided by Mr. Hiroshi Ohno, CEO of Ezowin Co., Ltd.

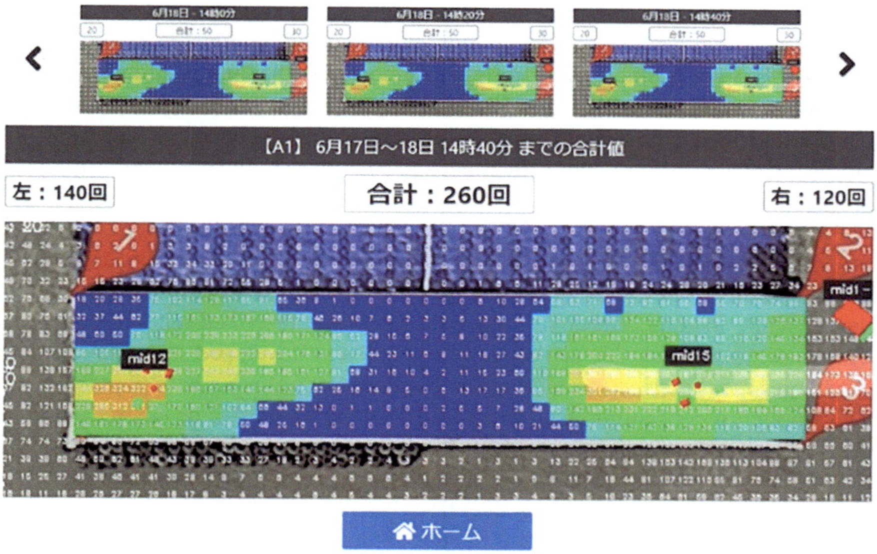

Fig. 18.10 Example of visualization of compacting operation in the bunker silo by 'Reposaku'

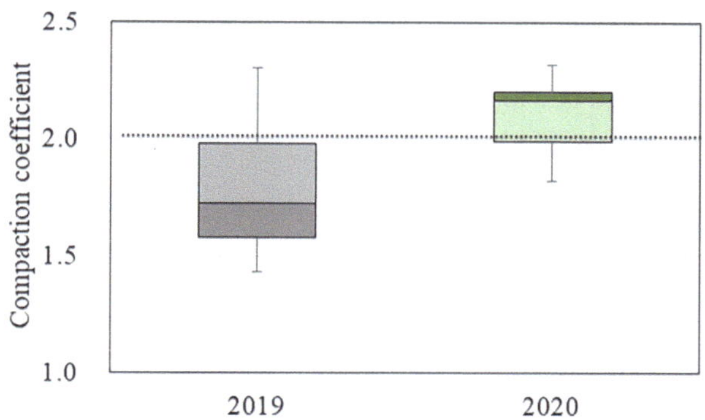

Fig. 18.11 Variation in packing density of silage after the visualization of compacting operation (Data and Fig. from Aoki)

Compaction coefficient is the ratio of silage materials put in banker silo to the volume of silage materials after compaction (Ogoshi 2005).

From 2024, 'Reposaku' will be changed to a new type with improved accuracy down to centimetres.

2.2.4 IoT-Based TMR Producing System

TMR is produced based on the silage preparation method described above with the addition of concentrate feed, vitamin and mineral supplements, etc. In large TMR centres, TMR production is often carried out using a large truck equipped with a feed mixing machine. The truck moves throughout the facility to receive materials from concentrate feed tanks, bunker silos, etc., while continuously mixing the feed. During this process, the truck driver (1) disembarks at the tank position of the concentrate feed, (2) manually opens the tank lid, (3) inputs the prescribed amount of concentrate feed while checking the weighing device attached to the truck, (4) closes the tank lid, (5) re-enters the truck and (6) moves to the next tank and repeats this process as many times as necessary. The height of the driver's seat in the truck at the TMR centre is about 1.4 m from the ground, and repeated process of getting on and off imposes a physical burden especially for elderly workers, where this strain is often more evident. Additionally, although the driver monitors the weighing device while loading, errors such as exceeding the prescribed amount can occur. Therefore, to address this challenge, a system was developed that automatically discharges the prescribed amount when the truck stops at a predetermined position. The amount of concentrate feed to be loaded at each tank can be entered online at the TMR centre's management office. As a result, there is no need to disembark at the concentrate feed tank position, and the number of times drivers walk (45% decrease) and get on and off (55% decrease) has been significantly reduced, and the operation time required to receive concentrate feed has been reduced by 12%. It was also confirmed that the weighing errors had been eliminated. Additionally, the system has also helped to minimize the need for drivers to go in and out of the cabin, especially during extreme temperature conditions in summer and winter, and hence eliminating the strain caused by the temperature difference. Furthermore, the system proved to be beneficial in terms of labour safety, as no fall-related accidents occurred due to icy road surfaces in the winter.

3 Conclusion

As described above, the introduction of smart technologies at the TMR centres has demonstrated significant benefits for silage production and the TMR manufacturing process.

In the dairy farming sector in Hokkaido, while the number of 'megafarms' (large-scale dairy farms) is increasing, family-owned farms still account for more than 90% of the total number of dairy farms (Department of Agriculture and Hokkaido Government 2023). Although the volume of milk shipped by dairy farms producing less than 1000 tons per year has decreased in recent years, it still accounts for more than 40% of total milk shipments in Hokkaido (Department of Agriculture and Hokkaido Government 2023). It is expected that the role of family-owned farms in milk shipments will continue significantly in the near future. To sustain milk production, it is crucial for family-owned farms to adopt smart farming technologies. However, current technologies involve high initial investment and operational costs, making them difficult to introduce to small-scale farmers. In this regard, if TMR centres and contractors adopt smart technologies, they could support many family farms. While cost reduction is essential, these smart technologies can be introduced more easily at the TMR centres compared to individual family farms. Further studies are required to ensure that the implementation of smart technologies in feed production and silage preparation effectively contributes to the sustainable development of family dairy farming.

References

Department of Agriculture, Hokkaido Government. Situation surrounding dairy farming and livestock farming in Hokkaido. 2023.
Hayashi T. Degree of yield decline of corn due to northern leaf blight—situation and estimation method. Res Commun Konsen Agric Exp Stn. 2014;23:5–6.
Japan Dairy Council. The current situation of dairy farming in Japan based on "National Basic Survey on Dairy Farming". 2018.
MAFF, Japan. Situation surrounding the TMR center. 2016.
MAFF, Japan. Deployment of smart agriculture. 2019.
MAFF, Japan. Census of agriculture and forestry. 2022.
MAFF, Japan. Statistical survey on farm management and economy. 2023a.
MAFF, Japan. Survey of livestock statistics. 2023b.
Ogoshi A. Compaction method of silage materials for large-scale banker silo. Res Commun Konsen Agric Exp Stn. 2005;14:1–2.
Study Group for the Realization of Smart Agriculture. Interim summary of the results of the Study Group for the Realization of Smart Agriculture. 2014.

Chapter 19
Silage Development Vision and Carbon Neutrality in Livestock Husbandry

Yimin Cai

Abstract The decarbonisation of industrial and agricultural production, including livestock farming, has become a global trend, and many countries are aiming for carbon neutrality by 2050. Consequently, future efforts in feed preparation, processing and livestock production will prioritise carbon-neutral research and development. This chapter introduces the field of feed production and livestock rearing, emphasising the use of agricultural by-products to create feed. The goal is to enhance livestock-rearing technology and contribute to a green, decarbonised society. This approach not only promotes harmony with nature but also fosters sustainable practices that allow for resource cycling and synergy between farming and livestock operations.

Keywords Carbon neutrality · Development vision · Livestock husbandry · Silage

Climate change driven by greenhouse effect is a major challenge bedeviling the human society today (Intergovernmental Panel on Climate Change, 2022; United Nations 2022). The greenhouse effect has led to an abnormal increase in the global temperatures and triggered a series of natural disasters that are seriously threatening the environment on which human beings depend (Van Aalst 2006; El Zein and Chehayeb 2015; Mora et al. 2018). There are six greenhouse gases that are being controlled by the Kyoto Protocol which include carbon dioxide (CO_2), nitrous oxide (N_2O), methane (CH_4), hydrofluorocarbons (HFCs), perfluorocarbons (PFCs) and sulphur hexafluoride (SF_6) (Reilly et al. 2002). However, based on the percentage contribution to global warming, CO_2, N_2O and CH_4 are the three gases that have the greatest capacity to contribute to the greenhouse effect and account for the largest percentage (Caillol 2011; Kweku et al. 2018). Currently, there is a global consensus

Y. Cai (✉)
Zhengzhou University, Zhengzhou, China

Japan International Research Center for Agricultural Sciences, Tsukuba, Japan
e-mail: caiyimin@zzu.edu.cn

Y. Cai, K. Ataku (eds.), *Cultural History and Modern Production Technology of Silage*, https://doi.org/10.1007/978-981-96-5787-2_19

to combat global climate change and develop green and low-carbon agricultural production technologies (Kim 2012; Huang et al. 2022).

About 20% of global greenhouse gases (GHGs) come from agricultural emissions (Bennetzen et al. 2016; Caro et al. 2014). The main sources of agricultural GHG emissions are from four-part processes comprising CH_4 emissions from paddy fields, N_2O emissions from agricultural land, CH_4 emissions from animals' enteric fermentation and emissions from animal excreta disposal (Chataut et al. 2023; Wu et al. 2024; Han et al. 2019). The livestock sector contributes 15% of the global GHG emissions accounting for 7.1 billion tonnes of CO_2 equivalent/year (IPCC 2022), of which 5.7 billion tonnes of the CO_2 equivalent/year came from ruminant production (Ervin 2013). The economic cost of GHGs emitted by ruminants to ecosystems and human health is estimated to be about $679 billion and $13 billion, respectively (Du et al. 2018). Therefore, livestock production has far-reaching negative impacts on climate change and is increasingly becoming a factor of focus in environmental decision-making.

Greenhouse gas emission of CO_2, CH_4 and N_2O is from crop residue treatment and silage preparation (Fig. 19.1). Moreover, the composition and quality of feeds have an important influence on the emission of GHGs from livestock, and GHGs are the main metabolic pathway for nutrient losses and energy wastage during the digestion and metabolism of diets (Fig. 19.2), e.g. nearly 10% of the total energy of

Fig. 19.1 Gaseous emissions of carbon dioxide resulting from crop residue incineration treatment (Photos by Cai)

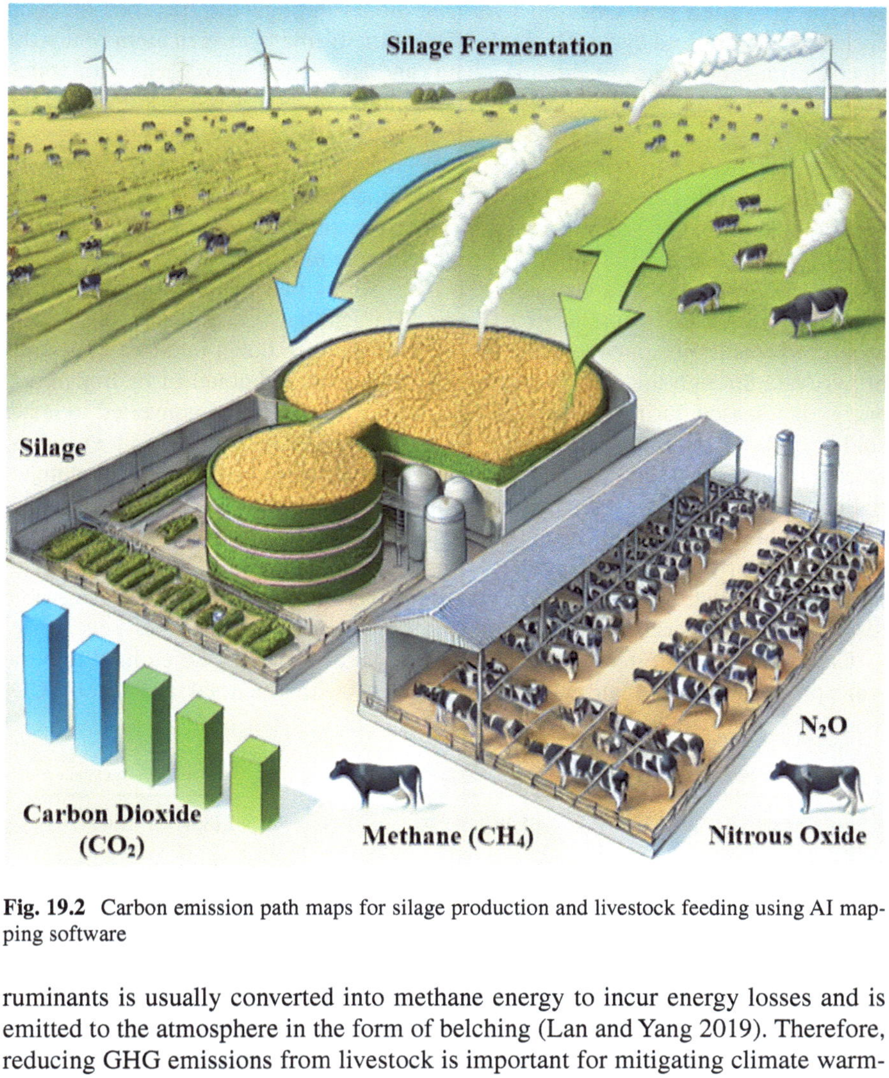

Fig. 19.2 Carbon emission path maps for silage production and livestock feeding using AI mapping software

ruminants is usually converted into methane energy to incur energy losses and is emitted to the atmosphere in the form of belching (Lan and Yang 2019). Therefore, reducing GHG emissions from livestock is important for mitigating climate warming and improving livestock productivity.

Recent research results show that silage also produce significant amounts of greenhouse gases during ensiling. Silage fermentation consists of two conditions with a distinct biochemical reaction process: aerobic and anaerobic respiration. Aerobic respiration, generally at the beginning of ensilage, comprises two biochemical reactions, one dominated by spoilage microorganisms and the other by plant cellular respiration. These two biochemical reactions use the residual air in the silage material to completely oxidise and decompose energy substances such as glucose into acetyl coenzyme A through the glycolytic metabolic pathway, catalysed by enzymes. Subsequently, the dicarbonyl acetyl group of acetyl CoA (acetyl

CoA) is converted into CO_2 and hydrogen atoms through the tricarboxylic acid cycle or the citric acid cycle (Du et al. 2024). Finally, the hydrogen atoms enter the electron transport chain (respiratory chain) and are passed to oxygen, with which water is produced. Simultaneously, a large amount of heat is released through the oxidative phosphorylation that occurs along with the electron transport process to produce adenosine triphosphate (ATP) molecules, which leads to silage fermentation losses (Nolfi-Donegan et al. 2020; Ahmad et al. 2018; Deshpande and Mohiuddin 2020). The second stage is anaerobic respiration, also known as fermentation. As shown in Fig. 19.3, lactic acid fermentation, ethanol fermentation and acetic acid fermentation are usually the main forms of silage fermentation, performed by lactic acid bacteria, yeast and acetic acid bacteria, respectively. Under anaerobic conditions, one molecule of glucose is broken down into two molecules of pyruvate through the glycolytic pathway, and a small amount of hydrogen and energy is released during this process. Pyruvic acid can then be converted to lactic acid or broken down into alcohol and CO_2 under the catalysis of different enzymes. During this process, soil methanogenic bacteria that is attached to the ensiled material could, under strict silage anaerobic environment conditions, utilise the anaerobically oxidised hydrogen for energy, using CO_2 as an electron acceptor, which leads to the production of CH_4 gas (Conrad 2020). Therefore, it is important to effectively regulate greenhouse gas emissions from forage during silage's aerobic respiration and anaerobic fermentation stages.

Carbon neutrality refers to equalising the total amount of CO_2 emitted into the atmosphere during human production activities with the total amount of CO_2

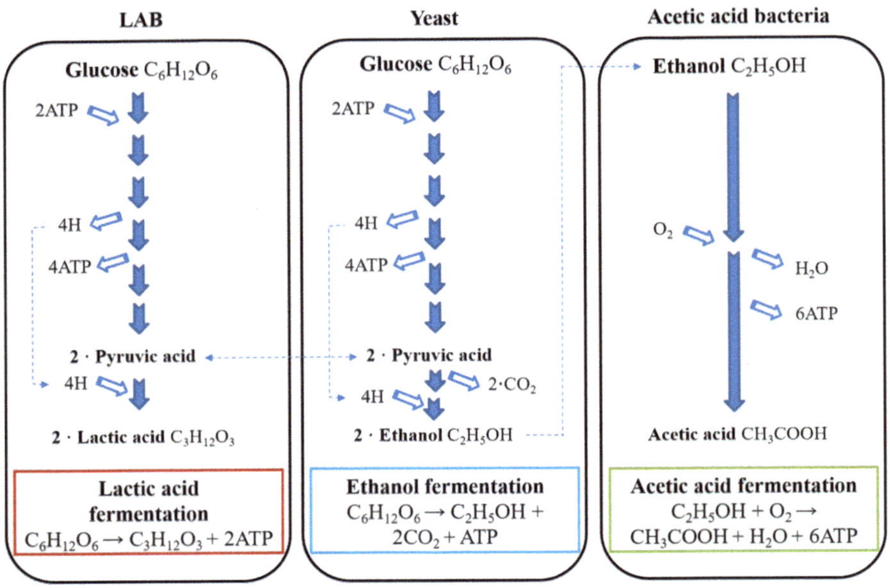

Fig. 19.3 Microorganisms related to silage and their main fermentation forms (Pictures from Cai)

absorbed from the atmosphere. Carbon neutrality in agriculture and livestock refers to a net-zero state of greenhouse gas emissions from agricultural and livestock production. Therefore, how silage is prepared to minimise GHG emissions during silage production is important for achieving the goal of carbon neutrality in the livestock sector (Fig. 19.4). Hence, improving the efficiency of feed and livestock production can play an important role in achieving carbon neutrality in the livestock sector and in promoting the sustainable development of agriculture and livestock.

Currently, research on carbon neutrality in animal husbandry focuses on greenhouse gas emission reduction from enteric fermentation of livestock and livestock excreta, and many important milestones have been achieved based on existing research results.

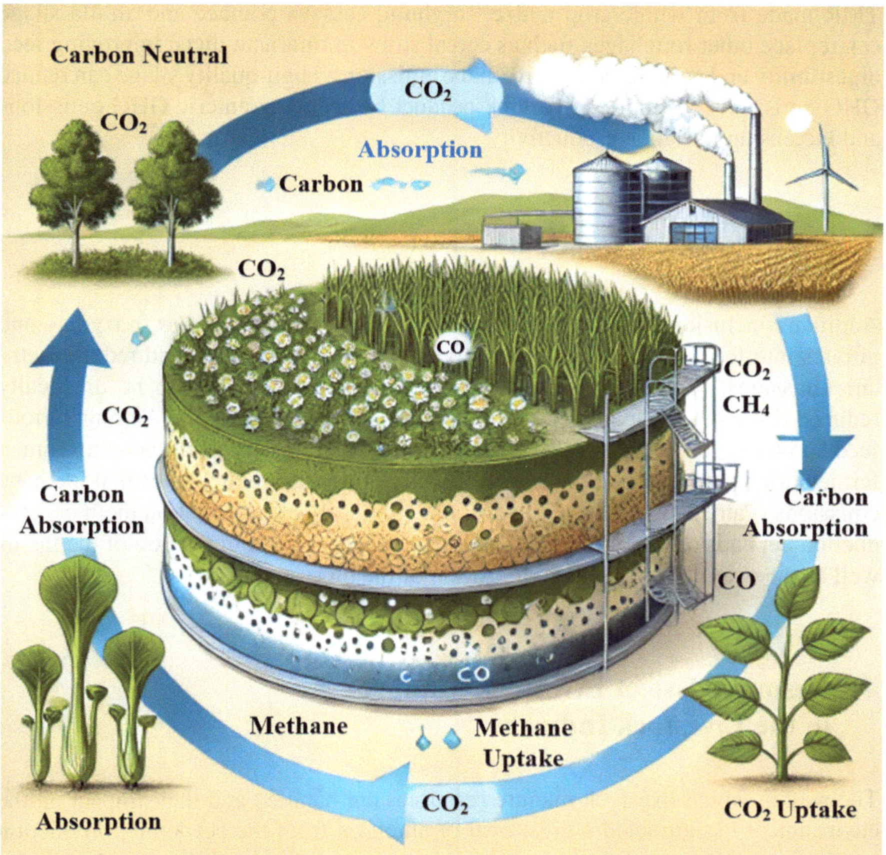

Fig. 19.4 Carbon neutral image of animal husbandry created by AI mapping software

1 Nutrient Regulation of Feed

Reducing the ratio of neutral detergent fibre to non-fibrous carbohydrates in feed and adding soluble carbohydrates to improve feed quality have been found to reduce methane emissions from enteric fermentation. High-quality feed improves palatability and intake, shortens the residence time of the feed in the rumen and reduces the proportion of methane produced. Total mixed rations (TMR) prepared according to the nutritional requirements of the animals increase daily weight gain, shorten feeding cycles and reduce the total emissions of greenhouse gases such as methane and carbon dioxide.

2 Optimisation of Feed Mix

TMR made from whole-crop maize, sorghum, cassava pomace and alfalfa silage can replace other roughages such as cereal straw in ruminant diets, improving feed digestibility and reducing rumen methane emissions. High-quality silage can reduce GHG emissions per unit of livestock product by reducing enteric GHG emissions and increasing livestock productivity.

3 Additive Regulation

Additives inclusions such as *probiotics*, plant extracts, *prebiotics*, enzymes and nitrates into feeds can inhibit the activity of *rumen methanogens* and reduce methane emissions, but the effects of their long-term addition tend to be drastically reduced due to the adaptability of the rumen microbial ecosystem. Among various feed additives, the addition of a certain amount of fat or fatty acids can alter rumen fermentation patterns and is one of the dietary options to reduce intestinal methane emissions (Patra 2013, 2014). However, the inhibitory effect of fat on methane production depends on the concentration, type and fatty acid composition of the fat, as well as the overall nutritional composition of the diet.

4 Resource Use of Livestock Manure in the Livestock Industry

The way in which livestock manure resources are handled and their impact on the environment has attracted a great deal of attention from the academic, production and utilisation sectors. Currently, livestock manure treatment technologies are relatively mature, and many studies have focused primarily on the resource attributes of manure, such as the production of fertilisers, vermicompost and farm bedding. However, the research aspects of manure treatment methods and the accompanying pollution potentials, such as the potential emission of greenhouse gases (GHGs) and other hazardous gases, have been insufficiently investigated (Zhang et al. 2023). It is well known that manure fermentation processes emit substantial amount of GHG, and the choice of manure collection and handling technologies can affect the potential GHG emissions. For example, degradation of organic matter in manure under anaerobic conditions produces CH_4 and N_2O (Yasmin et al. 2022), which exacerbate the GHG emissions during manure storage. Typically, CO_2 is emitted during compost production, and anaerobic fermentation during composting can lead to uncontrolled emissions of CH_4 and N_2O. GHG emissions can be reduced during compost production through some strategies such as lowering pH, compaction and mulching, adding appropriate conditioners or using storage tanks. Similarly, techniques such as vermiculture (biotransformation and uptake of nutrients from manure) or the addition of biochar to compost have shown potential for reducing NH_3, CH_4 and N_2O emissions (Qin et al. 2015; Puyuelo et al. 2014).

At present, carbon neutral research in animal husbandry mainly focuses on improving animal production efficiency, greenhouse gas emission reduction and livestock manure resource utilisation.

In the future, to achieve the desired goals in mitigating GHG emission from livestock, carbon neutral research in the global animal husbandry industry shall focus on the following aspects: (1) conducting innovative research on the use of resources from agricultural by-products as well as carrying out the basic and applied research related to the efficient rearing of livestock, so as to promote the sustainable development of green and low-carbon animal husbandry (Fig. 19.5); (2) comprehensive use

Fig. 19.5 Methods for efficient utilisation of crop by-product resources (Photos by Cai)

of modern information technology to monitor the whole process of greenhouse gas emissions in the animal husbandry industry and carrying out scientific assessment on the greenhouse gas emissions and emission reduction of various links and the potential of their impact on climate change; (3) creating a carbon neutral standardisation system for the agricultural and livestock industry and formulating a carbon-neutral standardisation system for agriculture and animal husbandry, formulating methods and technical standards for monitoring, emission reduction and assessment of greenhouse gases in agriculture and animal husbandry and establishing a reportable, measurable and verifiable management system.

References

Ahmad M, Wolberg A, Kahwaji CI. Biochemistry, electron transport chain. 2018.

Bennetzen EH, Smith P, Porter JR. Agricultural production and greenhouse gas emissions from world regions. The major trends over 40 years. Glob Environ Change. 2016;37:43–55.

Caillol S. Fighting global warming: the potential of photocatalysis against CO2, CH4, N2O, CFCs, tropospheric O3, BC and other major contributors to climate change. J Photochem Photobiol C: Photochem Rev. 2011;12(1):1–19.

Caro D, Davis SJ, Bastianoni S, Caldeira K. Global and regional trends in greenhouse gas emissions from livestock. Clim Chang. 2014;126(1):203–16.

Chataut G, Bhatta B, Joshi D, Subedi K, Kafle K. Greenhouse gases emission from agricultural soil: a review. J Agr Food Res. 2023;11:100533.

Conrad R. Methane production in soil environments. Anaerobic biogeochemistry and microbial life between flooding and desiccation. Microorganisms. 2020;8(6):881.

Deshpande OA, Mohiuddin SS. Biochemistry, oxidative phosphorylation. 2020.

Du Y, Ge Y, Ren Y, Fan X, Pan K, Lin L, et al. A global strategy to mitigate the environmental impact of China's ruminant consumption boom. Nat Commun. 2018;9(1):4133.

Du Z, Nakagawa A, Fang J, Ridwan R, Astuti WD, Sarwono KA, et al. Cleaner anaerobic fermentation and greenhouse gas reduction of crop straw. Microbiol Spectr. 2024;12(7):e0052024.

El Zein AL, Chehayeb NA. The effect of greenhouse gases on earth's temperature. Int J Environ Monit Anal. 2015;3(2):74.

Ervin D. Environment and food: by Colin. Sage. 2013;16:164–6.

Han M, Zhang B, Zhang Y, Guan C. Agricultural CH4 and N2O emissions of major economies: consumption-vs. production-based perspectives. J Clean Prod. 2019;210:276–86.

Huang X, Yang F, Fahad S. The impact of digital technology use on farmers' low-carbon production behavior under the background of carbon emission peak and carbon neutrality goals. Front Env Sci. 2022;10:1002181.

Intergovernmental Panel on Climate Change. Climate change: a threat to human wellbeing and health of the planet. Taking action now can secure our future. IPCC; 2022.

Kim CG. The impact of climate change on the agricultural sector: implications of the agro-industry for low carbon, green growth strategy and roadmap for the east Asian region. 2012.

Kweku DW, Bismark O, Maxwell A, Koomson D, Danso K, Buanya BA, et al. Greenhouse effect: greenhouse gases and their impact on global warming. J Sci Res Rep. 2018;17(6):1–9.

Lan W, Yang C. Ruminal methane production: associated microorganisms and the potential of applying hydrogen-utilizing bacteria for mitigation. Sci Total Environ. 2019;654:1270–83.

Mora C, Spirandelli D, Franklin EC, Lynham J, Kantar MB, Miles W, et al. Broad threat to humanity from cumulative climate hazards intensified by greenhouse gas emissions. Nat Clim Chang. 2018;8(12):1062–71.

Nolfi-Donegan D, Braganza A, Shiva S. Mitochondrial electron transport chain: oxidative phosphorylation, oxidant production, and methods of measurement. Redox Biol. 2020;37:101674.
Patra AK. A meta-analysis of the effect of dietary fat on enteric methane production, digestibility and rumen fermentation in sheep, and a comparison of these responses between cattle and sheep. Livest Sci. 2014;162:97–103.
Patra AK. The effect of dietary fats on methane emissions, and its other effects on digestibility, rumen fermentation and lactation performance in cattle: a meta-analysis. Livest Sci. 2013;155(2–3):244–54.
Puyuelo B, Gea T, Sanchez A. GHG emissions during the high-rate production of compost using standard and advanced aeration strategies. Chemosphere. 2014;109:64–70.
Qin W, Hu C, Oenema O. Soil mulching significantly enhances yields and water and nitrogen use efficiencies of maize and wheat: a meta-analysis. Sci Rep. 2015;5(1):16210.
Reilly J, Mayer M, Harnisch J. The Kyoto protocol and non-CO2 greenhouse gases and carbon sinks. Environ Model Assess. 2002;7:217–29.
United Nations. Climate change the greatest threat the world has ever faced. UN Expert Warns; 2022.
Van Aalst MK. The impacts of climate change on the risk of natural disasters. Disasters. 2006;30(1):5–18.
Wu X, Zhang Y, Han Y, Zhang Y, Zhang Y, Cheng X, et al. Advances in methane emissions from agricultural sources: part I. Accounting and mitigation. J Env Sci. 2024;140:279–91.
Yasmin N, Jamuda M, Panda AK, Samal K, Nayak JK. Emission of greenhouse gases (GHGs) during composting and vermicomposting: measurement, mitigation, and perspectives. Energy Nexus. 2022;7:100092.
Zhang L, Ali MM, Lin H, Zhang S, Jin S, Zhu ZP, et al. Livestock and poultry manure management from the perspective of carbon neutrality in China. FASE. 2023;10(3):341–62.

MIX
Papier aus verantwortungsvollen Quellen
Paper from responsible sources
FSC® C105338

If you have any concerns about our products,
you can contact us on
ProductSafety@springernature.com

In case Publisher is established outside the EU,
the EU authorized representative is:
Springer Nature Customer Service Center GmbH
Europaplatz 3, 69115 Heidelberg, Germany

Printed by Libri Plureos GmbH
in Hamburg, Germany